ALLE ZEIT WACH 1842

Spinal Cord Monitoring

Edited by J. Schramm and S. J. Jones

With 139 Figures and 63 Tables

Springer-Verlag
Berlin Heidelberg New York Tokyo

Professor Dr. Johannes Schramm
Neurochirurgische Klinik der Universität
Schwabachanlage 6, D-8520 Erlangen

Stephen J. Jones, Ph.D.
The National Hospital for Nervous Diseases
Queen Square, London WC1N 3BG, Great Britain

ISBN-13: 978-3-642-70689-9 e-ISBN-13: 978-3-642-70687-5
DOI: 10.1007/978-3-642-70687-5

Library of Congress Cataloging in Publication Data. Spinal cord monitoring. Includes bibliographies and index. 1. Spinal cord—Surgery. 2. Evoked potentials (Electrophysiology). 3. Spinal cord—Wounds and injuries. 4. Spinal cord—Abnormalities. 5. Patient monitoring. I. Schramm, J. (Johannes), 1946– . II. Jones, S.J. (Stephen J.) [DNLM: 1. Evoked Potentials. 2. Monitoring, Physiologic. 3. Spinal Cord. WL 400 S75778] RD 594.3.S6697 1985 617'.482 85-22096

Softcover reprint of the hardcover 1st edition 1985

Typesetting and printing : Petersche Druckerei GmbH & Co. Offset KG, Rothenburg ob der Tauber
Bookbinding: Konrad Triltsch, Graphischer Betrieb, Würzburg
2122/3130-543210

List of Contributors

You will find the address of each first mentioned author at the beginning of the respective contribution

Preface

The two International Symposia on Spinal Cord Monitoring, held in Tokyo in 1981 [Homma S, Tamaki T (eds) (1984) Fundamentals and clinical application of spinal cord monitoring. Saikon Publishing, Tokyo] and Erlangen in 1984, were remarkable in that they stimulated a gathering of electroencephalographic, neurophysiological, orthopaedic, neurosurgical, anaesthetic and pathological practitioners, presenting experimental, clinical and surgical experiences. The factor held in common was an understanding of the need to reduce the incidence and severity of iatrogenic neurological impairment associated with certain surgical and radiographic procedures. In the past there has been a tendency for the study of human evoked potentials (EPs) to be regarded as an end in itself, or at best a discipline which can be of limited assistance in certain problems of neurological diagnosis. Symposia such as these serve the valuable function of opening electrophysiological eyes to the problems to which EP techniques might usefully be devoted in other medical spheres. They also help those who practise electrophysiology as an adjunct to their speciality to comprehend such of the more complex properties of EPs as are relevant to their particular interests.

The title of the Symposia and of this volume arose as a result of an historical accident. The pioneers of intraoperative neuronal monitoring were mostly surgeons specialising in the correction of spinal deformities or the removal of tumours. Their first meetings, which naturally emphasized the spinal cord as a structure at risk, were not widely publicised and the proceedings are hard to come by today [Nash CL, Brodkey JS (eds) (1977) Clinical application of the spinal cord monitoring for operative treatment of spinal diseases, Cleveland. Nash CL, Brown RH (eds) (1979) Spinal Cord Monitoring Workshop, Data acquisition and analysis, St. Louis]. Gradually, the field has broadened to include monitoring of spinal and supraspinal conduction in a number of surgical, radiological and clinical situations. A similar development has taken place in monitoring the visual and auditory pathways.

In addition to the short-term assessment of pathways at risk from surgical or radiological manoeuvres, monitoring should also be understood to include the long-term evaluation of spinal cord function in certain pathologic conditions, for example myelopathy and spinal trauma. The scope of this volume additionally covers "once-off" preoperative and perioperative investigations of conduction in pathologic spinal cords. It is of great importance that the non-surgical assessment of sensory pathway function be well represented, giving an indication of the diagnostic and prognostic potential of EP recordings. Techniques now in routine application range from the noninvasive recording of cortically and subcortically generated potentials from the skin, through direct stimulation and recording of activity in the spinal cord, to the detection of descending motor tract potentials following transcutaneous cortical stimulation.

As to the course of future development, there have been calls for increased standardisation of monitoring techniques and for the establishment of definite criteria whereby conduction defects may be identified at an early and, it is hoped, reversible stage. One feels, however, that techniques have been adapted and modified in different centres with good reason, in order optimally to address the particular problems associated with the patients under investigation. Premature standardisation in a rapidly expanding field can only lead to stagnation. More effort should certainly be devoted to establishing, comprehending and if possible reducing the degree of uncertainty inherent in electrophysiological recordings made under the non-ideal conditions of the operating room, but any request to define rigid "safety" limits should, at this stage, be resisted. Intraoperative monitoring of the spinal cord is a technique in evolution, and intervention criteria are still being established empirically. There is no reason why such criteria should not eventually become well-defined, and the continued application of different techniques will be helpful in achieving this goal.

J. Schramm
S. J. Jones

Acknowledgments. The publication of this volume has been made possible due to stubstantial support from Nicolet Biomedical, West Germany. Additional funding was provided by E. Merck and Pfrimmer.

Contents

Physiology

Pharmacological Influences

Traumatic Cord Lesions

Monitoring—Cortical Evoked Potentials

Monitoring—Spinal Evoked Potentials

Atraumatic Cord Lesions

Physiology

Fundamentals of Sensory and Motor Functions of the Spinal Cord

M. Zimmermann[1]

The Second International Symposium on Spinal Cord Monitoring is another opportunity to foster clinical and basic science aimed at improving diagnostic and therapeutic measures in neurosurgery. Here I wish to give a survey on the basic functions of the spinal cord and to show that the spinal cord is not just a bundle of cables transmitting information from the peripheral nervous system to the brain, and vice versa.

Summary and Conclusions

An outline is given of the multitude of neuronal processing mechanisms in the spinal cord, with respect to somatosensory and motor information. The automatic control of joint position has been considered the modular element of the motor system, onto which act voluntary and involuntary programmes for movements and posture. The spinal sympathetic nervous system reveals a remarkably rich pattern of responses, and dysfunction may result in persisting pain states due to deleterious positive feedback. Several inhibitory mechanisms, pre- or postsynaptic, can be discerned in studies of sensory transmission in the dorsal horn. Some of these might be related to the sense of touch and others to pain related information. Inhibition of pain transmission can be produced by several procedures, such as systemic morphine, morphine microinjection into the brainstem, electrical stimulation of peripheral nerves or focal electrical stimulation in the brain. We have learned from these results of basic research to develop concepts for clinical use, as for example in relief of chronic pain.

I hope this survey illustrates that the function of the spinal cord is more than to relay sensory information in a 1:1 measure, and to produce stereotypical motor and sympathetic reflexes. The spinal cord is an integrated part of the central nervous system with a considerable level of sophistication, endowing us with functions important in daily life. I conclude with an epigram recently produced by Alan Basbaum from the University of California at San Francisco:

Oh praise the Lord
for having the spinal cord.

[1] II. Physiologisches Institut, Universität Heidelberg, Im Neuenheimer Feld 326, D-6900 Heidelberg

Spinal Cord Monitoring
Ed. by J. Schramm and S. J. Jones

The Spinal Stretch Reflex, a Motor System Module

The initial concern is to give a short outline of the motor function of the spinal cord (Fig. 1). We first consider the well known stretch reflex which is elicited by a tendon tap with the reflex hammer (Schmidt 1983; Houk 1980). The ensuing phasic contraction of a muscle is due to a rather basic monosynaptic excitatory connection of the muscle spindle Ia afferent fibre with the alpha-motoneurons of the same muscle. The muscle spindle sensory receptor is excited by the short stretch produced by the hammer and the afferent impulses excite the alpha-motoneurons which in turn result in a reflex contraction of the muscle. The stretch reflex is the functional element of motor control in the spinal cord. Apart from this phasic "tendon jerk" we know that the stretch reflex has an important tonic function which serves as a control circuit for muscle length (Houk 1980; Zimmermann 1983). This automatic control of muscle length seems to be of particular importance in the regulation of posture.

Another feature of motor control is the interaction in the spinal cord between pairs of antagonistic muscles (Fig. 1). The two stretch reflexes of the extensor and flexor muscles of a particular joint are interconnected in two ways. One is mechani-

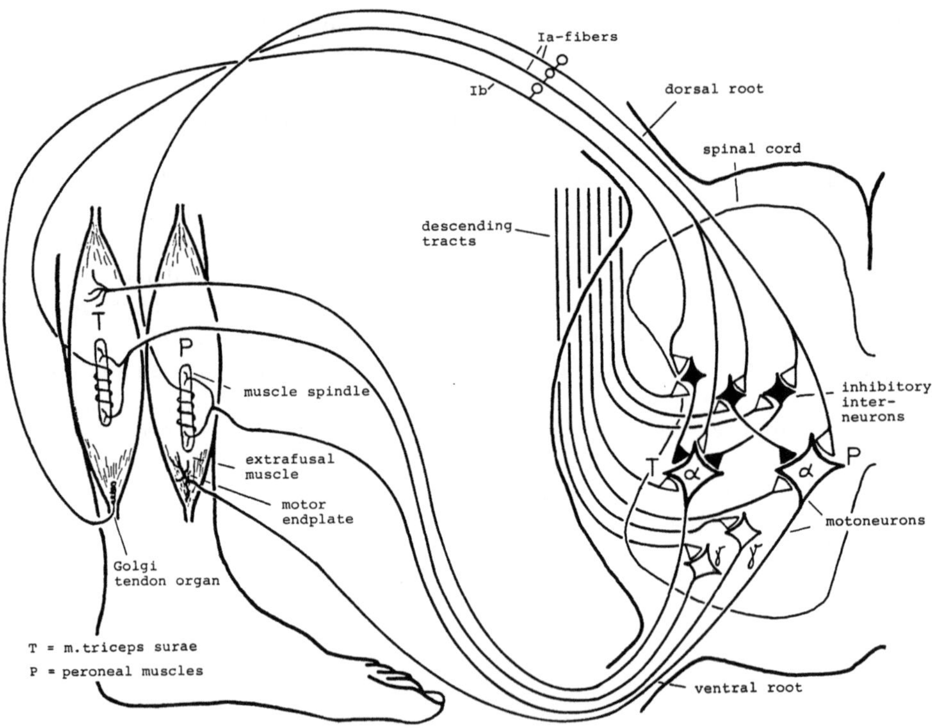

Fig. 1. The spinal stretch reflex. Schematic survey of (monosynaptic) stretch reflexes of muscles which are antagonistic on the ankle joint. The antagonistic inhibitory interactions between the two stretch reflexes are indicated by inhibitory interneurons. The autogenic inhibition is indicated for the reflex of M. triceps surae *(T)*. Descending tracts indicate where supraspinal influences modulate the stretch reflexes in order to produce movement and changes in posture

cal: the flexor and extensor muscles are coupled by their attachment to the same joint. The other interaction is neuronal: reciprocal inhibitory connections exist between the stretch reflexes of the antagonistic muscles. Whenever the Ia-fibers of the flexor muscle are excited via their gamma-motoneurons, the extensor alpha-motoneurons are inhibited, by activation of inhibitory interneurons, and vice versa. This twofold interaction of antagonistic muscles can be considered a system for automatic control of joint position (Houk 1980). Thus, joint position control, a fundamental part of the motor system, is prewired in the spinal cord. It is involved both in the automatic adjustment of posture and in the control of movements.

The third point which I would like to emphasize is the existence of supraspinal influences on the spinal motor system, which are effected by descending impulses onto the pools of alpha- and gamma-motoneurons (Fig. 1). Several pathways descending in the white matter transmit impulses from the brain to the spinal cord. For example, the pyramidal tract axons either excite or inhibit motoneurons and other spinal neurons. It is in this way that the brain commands changes in length of a muscle or changes of joint position. In terms of control theory, the supraspinal influences on the spinal control system have been considered to determine the set-point of the spinal length control system (Houk 1980; Zimmermann 1983).

Processing of Sensory Information in the Dorsal Horn

The historical picture of Fig. 2 represents Cajal's view of the dorsal horn as he saw it in his marvellous Golgi preparations of the cat's spinal cord more than 80 years ago. We see the afferent fibers entering the spinal cord via the dorsal root and coursing back (E in Fig. 2) towards the superficial laminae of the dorsal horn. These fibers branch into extensive terminal arborizations (F) which have synaptic connections with the dendrites of dorsal horn cells (A, C, D). We can discriminate several types of dorsal horn cell: Large dorsal horn cells in the deeper laminae (A) and smaller dorsal horn cells (C, D) in the superficial laminae (Brown 1981). Considerable processing of afferent information of somatosensory, viscerosensory and motosensory origin occurs in the dorsal horn. The dorsal horn can be subdivided into 6 laminae, according to cytoarchitectonic criteria (Rexed 1952). The Aβ-fibres (large myelinated fibres) which enter via the dorsal root have synaptic contacts with large neurons in laminae IV to VI, the deeper dorsal horn. In addition, most of these afferent large myelinated fibres have collaterals which ascend in the dorsal columns towards the medulla oblongata. The afferent information carried by Aβ-fibers is mainly related to low threshold mechanoreceptors, serving the sense of touch. The small myelinated or Aδ-fibres and the non-myelinated or C-fibres have their endings in the superficial laminae (I to III), i.e. the substantia gelatinosa Rolandi. Many of these fibres carry pain-related information that comes from nociceptors. Therefore, the substantia gelatinosa and superficial dorsal horn are considered a major relay for pain-related information.

The spinal cord is not just a simple relay where afferent information is transmitted in a 1:1 fashion to motoneurons or to ascending tracts; rather, it is a sophisticated central nervous processor for afferent information (Willis and Coggeshall 1978). Many of the spinal neurons have inhibitory functions, and the interaction between

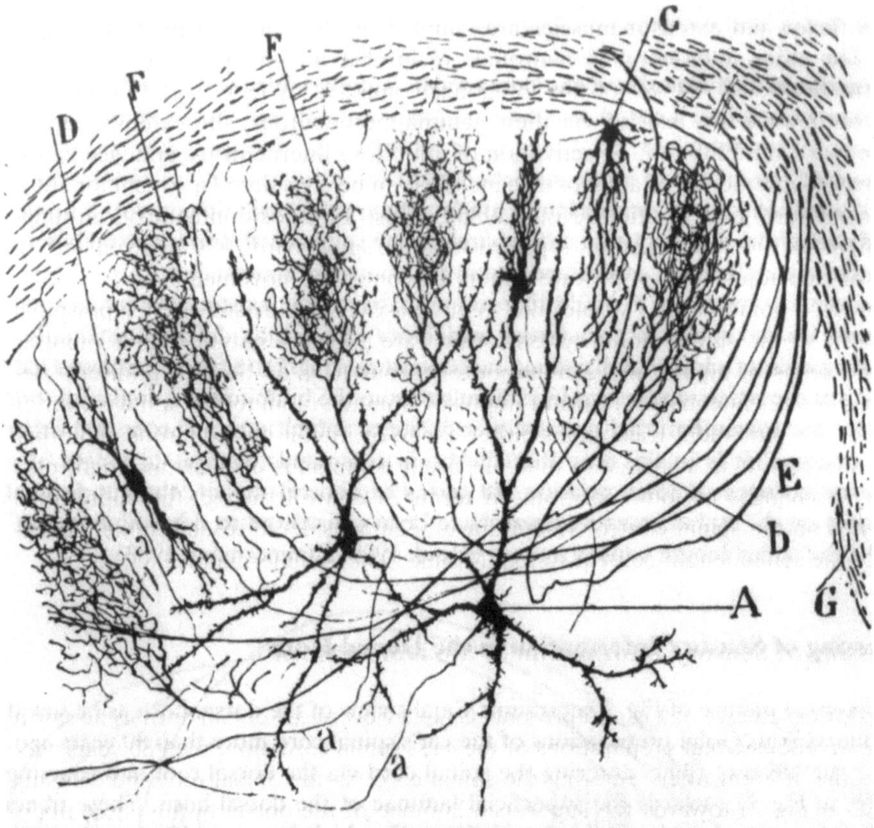

Fig. 2. Histology of the cat's dorsal horn. Large afferent fibres *(E)* enter via the dorsal root and form terminal arborizations *(F)*. Large *(A)* and small *(C, D)* dorsal horn neurons have dendrites in the region of the terminal arborizations of afferent fibres. (From Cajal 1952)

inhibitory and excitatory synapses in the spinal cord is the basis for somatovisceral information processing (see Fig. 5). Apart from the local segmental inhibitory interactions, we know that several descending tracts from brain to the dorsal horn exert inhibitory influence. For example, the pyramidal tract is not only a motor tract; many terminals of the pyramidal tract end in the dorsal horn and have a predominantly inhibitory effect on somatic information. Apart from the pyramidal tract several other descending tracts have been identified, coming mainly from the brainstem, which are concerned in the control of afferent information in the spinal dorsal horn.

Of the many ascending pathways, best known among neurosurgeons is the anterolateral or ventrolateral tract, considered a pain pathway. However, this is not a true and complete picture. From animal experiments it is clear that the anterolateral tract conveys a great deal of information on tactile stimuli, and also on cold and warm stimuli, in addition to pain information (Willis and Coggeshall 1978). Neurophysiological analysis of the ascending axons in the anterolateral tract shows that most can be excited from low threshold cutaneous mechanoreceptors. A proportion of the neurons have, in addition to the input from low threshold mechanoreceptors, input from nociceptors. This confluence of nociceptive and non-nociceptive afferent information is characteristic for many neurons of the dorsal horn. There are only a few which have an exclusive input from nociceptors. It is still not clear what the contributions are of these two kinds of neurons carrying nociceptive information (Zimmermann 1984). From these findings it appears that the transmission of pain information is but one of the functions of the anterolateral tract.

Compound Potentials of the Cord Dorsum

It is now proposed to analyze the potentials which can be recorded from the cord dorsum. In animal experiments we can discern three components (Fig. 3), and some clinical work reported at this symposium and previously (Homma and Tamaki 1984) shows that basically the same potentials can be recorded from the human spinal cord, although there is considerable variability. After an electrical nerve stimulus we first see a bi- or triphasic component which is the afferent compound action potential of the dorsal root fibres. Therefore, this incoming or afferent volley is not a spinal potential but a peripheral nerve potential. Usually, components due to fast conducting $A\gamma$- and $A\beta$-fibres can be seen, whereas the contributions of $A\delta$- and C-fibres are masked by potentials subsequently generated in spinal neurons. Next we see a negatively-going wave which is called the N-wave. According to the view of physiologists the N-wave represent postsynaptic potentials (Fig. 3B) generated in the dorsal horn neurons (Willis 1984). The cord dorsum electrode records the superimposed extracellular postsynaptic currents from both excitatory and inhibitory synapses. The earliest part of the rising N-wave represents postsynaptic currents of neurons mono-

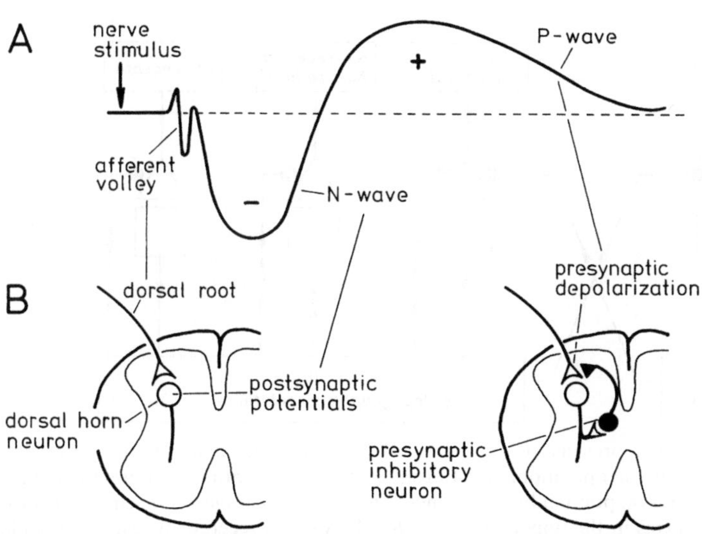

Fig. 3A, B. Survey of cord dorsum potentials. **A** Following stimulation of e.g. a cutaneous nerve a sequence of potentials can be recorded within about 100 ms from the dorsal surface of the spinal cord: the afferent volley, the N-wave (negative) and the P-wave (positive). **B** The elements of dorsal horn are shown in association with the three components of potentials shown in **A**

synaptically connected with the afferent fibres, whereas the later phases contain contributions of monosynaptically connected and higher order spinal neurons.

The next component of the cord dorsum potential is a positively-going wave called the P-wave, having a duration of about 100 ms. The P-wave reflects depolarization of the afferent fibre terminals, which is considered a sign of a presynaptic inhibitory process. The primary afferent terminals are depolarized by means of axo-axonic synapses, which impinge from inhibitory spinal interneurons onto the presynaptic terminals of afferent fibres (Fig. 3B). This primary afferent depolarization (PAD) is an inhibitory potential, resulting in a decrease of the release of excitatory transmitter from the primary afferent terminal to the dorsal horn neuron (Schmidt 1971). This presynaptic inhibition is an important mechanism for the control of afferent information in the spinal cord, in both the dorsal and the ventral horns. The P-wave consists of superimposed extracellular currents flowing during the depolarization of large myelinated fibres.

Presynaptic inhibition can be considered a negative feedback system for control of afferent information. It is organized in a rather complicated manner and in the last 20 years its functional organization in relation to the motor and the somatosensory systems has been studied in animal experiments (Eccles 1964; Schmidt 1971).

Presynaptic Inhibition of Cutaneous Afferents

An outstanding finding was that a large part of presynaptic inhibition was produced in cutaneous afferent fibres by activation of other cutaneous fibres; we have studied what the meaning might be of this strong presynaptic inhibitory interaction between cutaneous afferents which terminate in the dorsal horn.

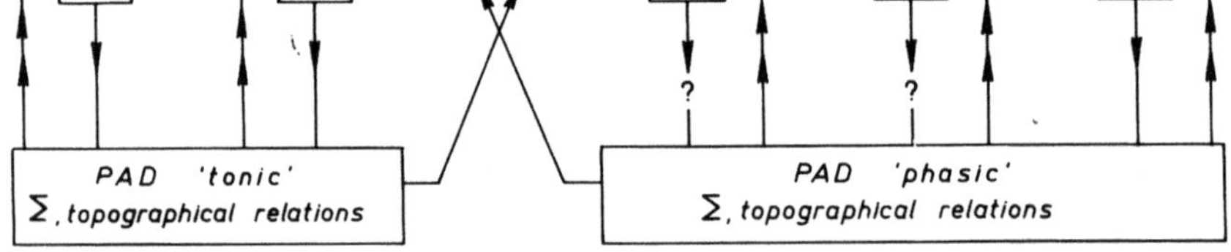

Fig. 4. Modality-specific spinal presynaptic inhibition of cutaneous mechanoreceptor afferents. Schematic diagram illustrating the operational relationships of cutaneous primary afferent depolarization *(PAD)*, reflecting presynaptic inhibitory connections within a cutaneous mechanoreceptor afferent population. The boxes in the upper row stand for the various receptor types. The middle row complex indicates the first synapses in the spinal dorsal horn onto which the receptor afferents project, the closed circles showing the presynaptic inhibitory control. In the lower row, the two specific PAD-generating systems are drawn. The number of arrows in the feedback lines denote the strength of presynaptic inhibitory control. (From Jänig et al. 1968)

What we have found is summarized in the diagram of Fig. 4. We could discriminate five types of cutaneous afferent fibres within the Aβ-group, according to the cutaneous mechanoreceptors connected (Jänig et al. 1968): the touch corpuscle of the hairy skin, the slowly adapting or SA-receptor of the glabrous skin, the rapidly adapting or RA-receptor of the glabrous skin, hair follicle receptors, and the PC-receptor, i.e. the Pacinian corpuscle receptor of the subcutis. The touch corpuscle receptor and the SA-receptor are slowly adapting; they respond to sustained skin indentations. The hair follicle receptor, the RA-receptor and the PC-receptor are rapidly adapting mechanoreceptors, responding best to moving stimuli.

The slowly and rapidly adapting receptors comprise two systems of presynaptic inhibition in the spinal dorsal horn, as indicated in Fig. 4 by the axo-axonic inhibitory synapses onto the afferent fibre terminals which have excitatory synapses onto the dorsal horn neurons to relay information from either type of receptor. We could show that presynaptic inhibitory interactions mutually occur within the group of rapidly adapting receptors, and the group of slowly adapting receptors. The cross interactions between these two groups were small or absent. The most potent input producing presynaptic inhibition in the group of rapidly adapting afferents is provided by Pacinian corpuscle (PC-) receptors. Thus, the PC-receptors which are the most sensitive mechanoreceptors of the skin (they respond best to slight vibrations of about 200 Hz) produce a powerful presynaptic inhibition on their own afferent terminals and also on afferent terminals of other rapidly adapting mechanoreceptors.

Obviously, this complex system of presynaptic inhibition provides a feedback mechanism to control the flow of afferent information into the spinal cord. It appears that the sensitivity of the sensory channel from rapidly adapting mechanoreceptors of the skin is adjusted appropriately to account for different intensities of sensory stimulation. A similar negative feedback system exists for the group of slowly adapting cutaneous mechanoreceptors. Thus, the analysis of the functional organization of presynaptic inhibition in the spinal cord shows that sophisticated interactions among the afferent fibres from touch receptors can be discerned behind the simple phenomenon of the P-wave.

Processing of Pain-related Sensory Information in the Spinal Cord

Recent neurophysiological and neuropharmacological investigations have contributed much to our knowledge on how information related to painful stimuli is processed in the spinal cord (Zieglgänsberger 1984; Zimmermann 1984), as illustrated in Fig. 5. Excitation of nociceptors is conveyed to the dorsal horn neurons by nonmyelinated or C-fibres and thin myelinated or Aδ-fibres. They have excitatory synaptic contacts to dorsal horn neurons. The dorsal horn neurons share several functions. One is to relay the filtered afferent information onto ascending tracts, e.g. the anterolateral tract (we know at least three additional ascending tracts which carry pain information to the supraspinal centers). Other functions of the dorsal horn neurons are to distribute sensory information coming in from the skin and inner organs to the sympathetic neurons in the lateral column and to the motoneurons in the ventral horn. This involvement of the cutaneous and visceral afferent fibres in sympathetic and motor control has important implications in mechanisms of chronic pain.

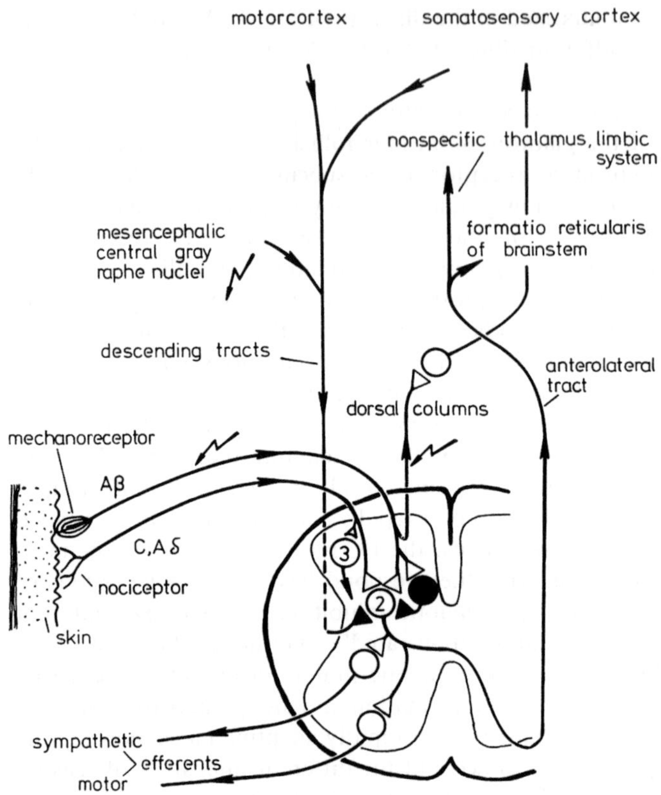

Fig. 5. Functional diagram of dorsal horn neurons. Afferent fibers synapse onto dorsal horn neurons, which in turn conduct their impulses to segmental motor and sympathetic afferents, and to ascending tracts. Class 2 dorsal horn neurons have converging inputs from Aβ-fibers (from low threshold mechanoreceptors) and Aδ- and/or C-fibres (from nociceptors). Class 3 neurons have exclusive nociceptive input. Inhibitory effects onto the dorsal horn neurons are indicated by filled synapses and neurons. *Arrows* indicate sites where electrical stimulation can produce inhibition of afferent transmission in the dorsal horn. (Modified from Zimmermann 1979)

Apart from these excitatory transmissions of afferent information in the dorsal horn we know that major inhibitory processes occur in the spinal dorsal horn; these are symbolized in Fig. 5 by filled neurons and synapses. Some of these inhibitions originate from local segmental inhibitory neurons, including also presynaptic inhibitory neurons as shown in Fig. 4. A most prominent inhibitory influence in the dorsal horn originates in supraspinal centers. Many descending axons are inhibitory to dorsal horn neurons (Fig. 5). There is much evidence that the spinal processing of afferent information is controlled from the brain by descending inhibition. The emphasis of recent research was on the transmission of afferent information from nociceptors. However, descending inhibition in the dorsal horn probably is also important for information from sensitive cutaneous mechanoreceptors, which subserve the sense of touch.

The variety of excitatory and inhibitory interactions on dorsal horn neurons is evident also from pharmacological and biochemical studies of the neurotransmitters

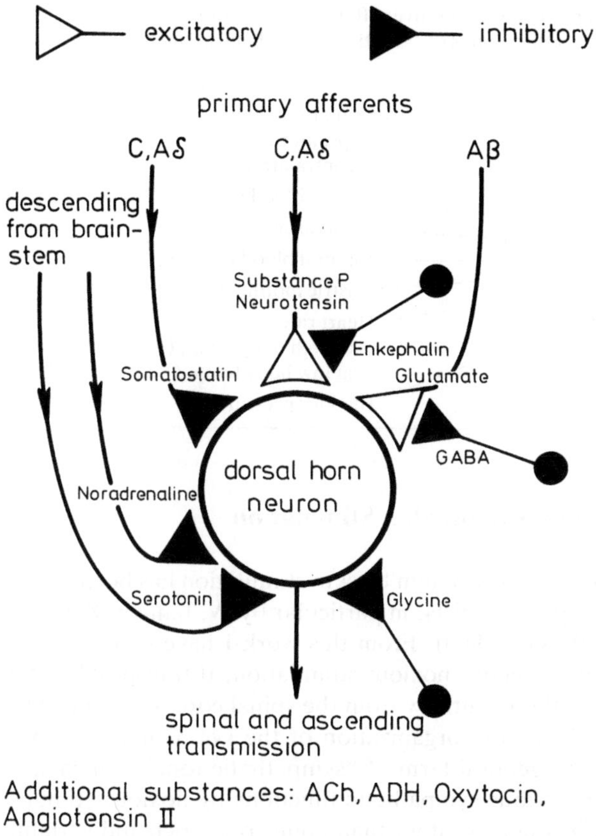

Fig. 6. Pharmacology of the dorsal horn neurons. Schematic survey of putative transmitter substances at excitatory and inhibitory synapses on dorsal horn neurons. The diagram summarizes findings from pharmacological, biochemical and histochemical research. (From Zimmermann 1979)

and neuromodulators involved (Fig. 6). Whereas neutral aminoacids, such as glutamate or aspartate, are considered to be excitatory synaptic transmitters related to afferent Aβ-fibres, the neuropeptide substance P is thought to be involved in synaptic excitations of dorsal horn neurons by nociceptive afferents (C-fibres and Aδ-fibres). Inhibitory transmitters which have been identified so far in the spinal cord (Curtis and Johnston 1974) are gamma-aminobutyric acid (GABA), glycine, serotonin, noradrenaline, enkephalin, and, more recently, dynorphin and other endogenous opiates (Willis 1982; Zieglgänsberger 1984). Many other substances have been found histochemically in the dorsal horn and therefore have been hypothesized to have neurotransmitter or neuromodulator functions.

There are many ways to interfere therapeutically with synaptic transmission in order to inhibit pain information in the spinal cord or brain: the synthesis or the axoplasmic transport of the excitatory transmitter or its precursors can be reduced; the release of the transmitter from (or its reuptake into) the presynaptic terminal can be manipulated, or the postsynaptic action or enzymatic degradation of the neurotransmitter can be antagonized. All these mechanisms could be utilized to modulate synaptic transmission, e.g. of pain-related signals. Thus far, of these possibilities only one has come to clinical use in the treatment of pain, that is the activation of spinal inhibition by exogenous opiates.

Table 1. Patterns in anesthetized cats and rats of vegetative reflexes upon noxious skin stimulation. (+) Increase, (−) decrease. Other kinds of skin stimuli (mechanical, thermal) result in different reflex patterns

Noxious stimulation	Skin blood flow ipsilateral (+) contralateral (−)
	Muscle blood flow (−)
	Sweat excretion (+)
	Visceral blood flow (−)
	Gastric motility (−)
	Heart rate at high body temp. (+) at low body temp. (−)
	Bladder pressure (+)

Patterns of Sympathetic Reactions upon Skin Stimulation

The activation of the sympathetic nervous system by skin stimulation has been intensively studied recently by several physiologists, in particular by W. Jänig in Kiel and by A. Sato in Tokyo (Jänig 1979; Sato 1980). From this work I have extracted the sympathetic effects produced by cutaneous noxious stimulation. It transpired that a highly organized pattern of sympathetic outflow from the spinal cord is produced by noxious stimulation (Table 1). Thus, the organization of the sympathetic reflexes cannot simply be described by the general term of "sympathetic tone", which suggests a uniform reflex effect on various sympathetic functions by noxious stimulation. This is a misleading term; it conceals rather than points out a rich and variable pattern of sympathetic reactions that can be found in animal and human experiments (Table 1).

There are cases of pain known to the clinician as "sympathetic reflex dystrophy" (Bonica 1953; Lankford 1980). It is now widely accepted that these states of chronic pain are enhanced and maintained by a pathophysiological action of the sympathetic system onto the nociceptors. It is in this way that some kind of vicious circle or positive feedback may be produced which results in the perpetuation of the pain. In these cases a repeated sympathetic block with a local anesthetic might give prolonged or even permanent relief of pain. Postherpetic neuralgia, some forms of amputation pain and causalgia obviously depend on such a type of dysfunction of the sympathetic system; they can often be alleviated by sympathetic local anesthesia. Similar phenomena of positive feedback have been observed in the skeletomotor system, where excessive reflex tone of muscle might enhance and maintain a pain state.

Spinal Inhibition of Nociceptive Information

In previous sections I have pointed out that inhibitory processes in the spinal cord are of crucial importance for motor and sensory functions. Here, I wish to emphasize inhibition of pain-related information in the dorsal horn (see Figs. 5 and 6). This

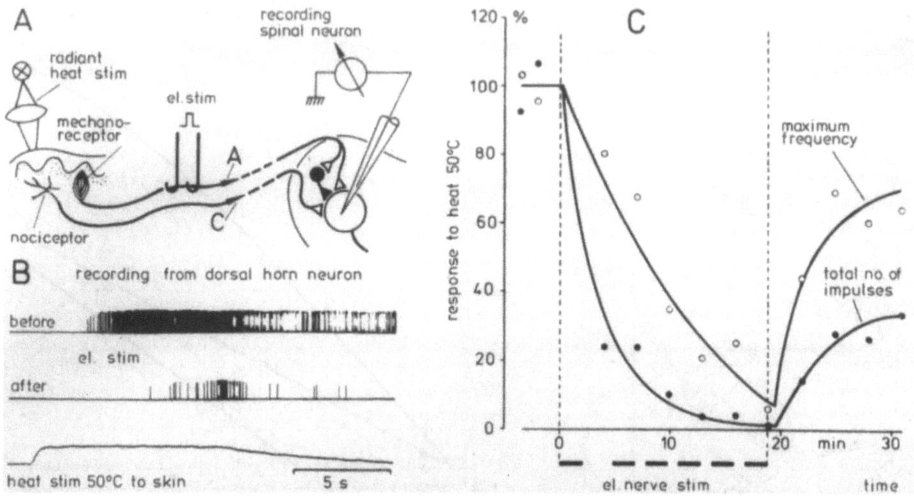

Fig. 7A–C. Suppression of nociceptive dorsal horn neuronal responses by electrical nerve stimulation. **A** Survey of experiment: Discharges are recorded from single dorsal horn neurons upon cutaneous nociceptor stimulation by radiant heat; these responses are inhibited by repetitive electrical stimulation of A-fibres of a cutaneous nerve. **B** Oscillographic records of responses of dorsal horn neuron upon noxious skin heating (50°C, 10 s), before and after repetitive electrical nerve stimulation. **C** Time course of maximum frequency of discharge (○) and total number of impulses (●) per heat-evoked response of the dorsal horn neuron, during and after the electrical A-fibre stimulation; repeated 2 min periods of electrical nerve stimulation are indicated below the abscissa. Discharge measures were relative to control values recorded before the period of nerve stimulation. (From Dickhaus et al. 1978)

question was studied in experiments on anesthetized animals (cats) in which, for example, cutaneous nociceptors were activated by experimental painful stimuli such as noxious skin heating. Many dorsal horn neurons are excited by noxious skin heating (Fig. 7), indicating that cutaneous nociceptive afferent Aδ- and/or C-fibres have excitatory synapses with these neurons. Typically, a discharge is evoked by noxious skin heating to 50°C during 10 s, over a receptive field 5 mm in diameter. These nociceptive discharges of dorsal horn neurons can be reduced by a variety of manipulations, physiological or pharmacological. For example, many analgesic drugs with an action on the central nervous system produce a reduction of heat-evoked responses. In Fig. 7 the afferent fibres of cutaneous nerves were stimulated electrically in a repetitive manner. As a result, the responses to noxious skin heating were depressed for a prolonged period. Build-up of and recovery from inhibition had time courses of 10–20 min (Dickhaus et al. 1978). This inhibition might be at least a partial mechanism of the pain relief obtained in patients by transcutaneous electrical nerve stimulation (TENS).

Another method of inducing inhibition of nociceptive dorsal horn discharges is to stimulate in the periaqueductal gray (PAG) and at other sites in the brainstem (Carstens et al. 1980; Willis 1982). Stimulation for example within the PAG elicits a powerful inhibition in dorsal horn neurons of information related to a painful skin stimulus (Fig. 8). This inhibition indicates the presence of descending inhibitory pathways either originating in or passing through the stimulated brain site. From

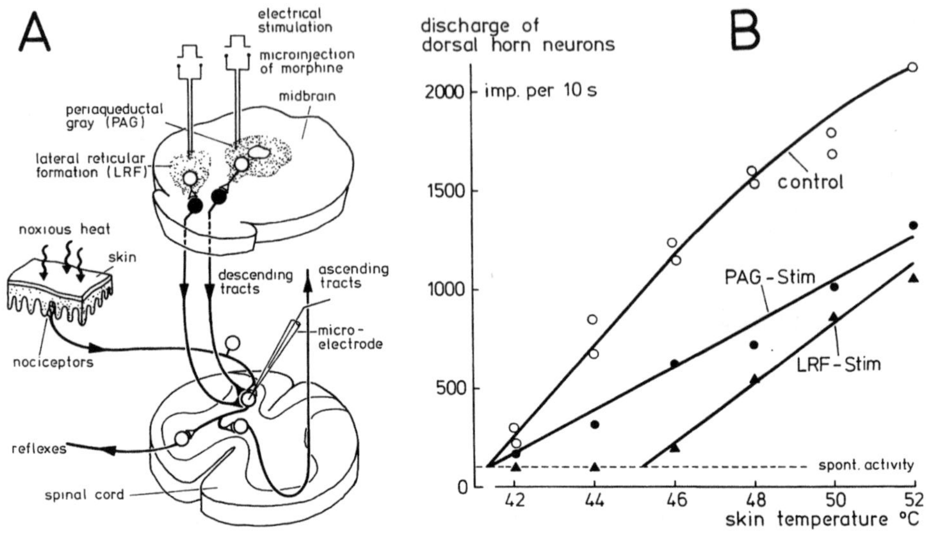

Fig. 8A, B. Descending inhibition of nociceptive transmission in the spinal dorsal horn. **A** Diagram of experiment. Responses of dorsal horn neurons are recorded to noxious skin heating. These can be inhibited by repetitive electrical stimulation in the midbrain periaqueductal gray *(PAG)* or lateral reticular formation *(LRF)*, as well as by microinjection of morphine in PAG. **B** Two types of descending inhibitory influences on a dorsal horn neuron. Lines plot the encoding of intensity of noxious skin heating in a single dorsal horn neuron, as affected by stimulation in the PAG and LRF. (**B** from Carstens et al. 1980)

spinal lesion studies it has been concluded that inhibitory influences elicited by stimulation in the PAG are conducted by descending axons in the dorsolateral funiculus of the spinal white matter. As to the neurotransmitter mediating this descending inhibition, many experiments show that serotonin is involved. In other experimental paradigms, endogenous opiates (enkephalin, dynorphin) have been found to contribute. On the other hand, microinjection of morphine into the PAG (e.g. 10 micrograms) can also activate descending inhibitory pathways, and this may be one of the mechanisms of morphine action in the central nervous system (Gebhart et al. 1984).

Acknowledgements. I gratefully acknowledge the editing of the manuscript by Frau Hannelore Ehlers, and the graphic work by Frau Almuth Manisali.

References

Bonica JJ (1953) The management of pain. Lea & Febiger, Philadelphia

Brown AG (1981) Organization in the spinal cord. Springer, Berlin Heidelberg New York

Cajal SR (1952) Histologie du système nerveux de l'homme et des vertèbres, tome I. Instituto Ramon y Cajal, Madrid (reprint)

Carstens E, Klumpp D, Zimmermann M (1980) Differential inhibitory effects of medial and lateral midbrain stimulation on spinal neuronal discharges to noxious skin heating in the cat. J Neurophysiol 43:332–342

Curtis RD, Johnston GAR (1974) Amino acid transmitters in the mammalian central nervous system. Ergebn Physiol 69:97–188

Dickhaus H, Pauser G, Zimmermann M (1978) Hemmung im Rückenmark, ein neurophysiologischer Wirkungsmechanismus bei der Hypalgesie durch Stimulationsakupunktur. Wien Klin Wochenschr 90:59–64

Eccles JC (1964) Presynaptic inhibition in the spinal cord. Progr Brain Res 12:65–91

Gebhart GF, Sandkühler J, Thalhammer JG, Zimmermann M (1984) Inhibition in spinal cord of nociceptive information by electrical stimulation and morphine microinjection at identical sites in midbrain of the cat. J Neurophysiol 51:75–89

Homma S, Tamaki T (eds) (1984) Fundamentals and clinical application of spinal cord monitoring. Saikon Publishing, Tokyo

Houk J (1980) Principles of system theory as applied to physiology. In: Mountcastle VB (ed) Medical physiology, 14th edn, vol 1. Mosby, St. Louis, Toronto, pp 225–267

Jänig W (1979) Reciprocal reaction patterns of sympathetic subsystems with respect to various afferent inputs. In: Brooks CMC, Koizumi K, Sato A (eds) Integrative functions of the autonomic nervous system. University of Tokyo Press, Tokyo, Elsevier/North-Holland Biochemical Press, Amsterdam, pp 263–274

Jänig W, Schmidt RF, Zimmermann M (1968) Two specific feedback pathways to the central afferent terminals of phasic and tonic mechanoreceptors. Exp Brain Res 6:116–129

Lankford LL (1980) Reflex sympathetic dystrophy. In: Omer GE, Spinner M (eds) Management of peripheral nerve problems. Saunders, Philadelphia, pp 216–244

Rexed B (1952) The cytoarchitectonic organization of the spinal cord of the cat. J Comp Neurol 96:415–495

Sato A (1980) Physiological studies of the somato-autonomic reflexes. In: Haldeman S (ed) Modern developments in the principles and practice of chiropractice. Appleton-Century-Crofts/Prentice Hall, Engelwood Cliffs, pp 93–105

Schmidt RF (1971) Presynaptic inhibition in the vertebrate central nervous system. Ergebn Physiol 63:20–101

Schmidt RF (1983) Motor systems. In: Schmidt RF, Thews G (eds) Human physiology. Springer, Berlin Heidelberg New York, pp 81–110

Willis WD (1982) Control of nociceptive transmission in the spinal cord. In: Autrum H, Ottoson D, Perl ER, Schmidt RF (eds) Progress in sensory physiology, vol 3. Springer, Berlin Heidelberg New York

Willis WD (1984) Evoked spinal cord potentials in the cat and monkey: Use in the analysis of spinal cord function. In: Homma S, Tamaki T (eds) Fundamentals and clinical application of spinal cord monitoring. Saikon Publishing, Tokyo, pp 3–19

Willis WD, Coggeshall RE (1978) Sensory mechanisms of the spinal cord. Plenum Press, New York London

Zieglgänsberger W (1984) Opioid actions on mammalian spinal neurons. Intern Rev Neurobiol 25: 243–275

Zimmermann M (1979) Peripheral and central nervous mechanisms of nociception, pain, and pain therapy: facts and hypotheses. In: Bonica JJ, Liebeskind JC, Albe-Fessard DG (eds) Advances in pain research and therapy, vol 3. Raven Press, New York, pp 3–32

Zimmermann M (1983) Cybernetic aspects of the nervous system and sense organs. In: Schmidt RF, Thews G (eds) Human physiology. Springer, Berlin Heidelberg New York, pp 315–328

Zimmermann M (1984) Neurobiological concepts of pain, its assessment and therapy. In: Bromm B (ed) Pain measurement in man. Elsevier, Amsterdam New York Oxford, pp 15–35

Spinal Cord Monitoring — A Review of Current Techniques and Knowledge

K. SHIMOJI[1], Y. MARUYAMA, H. SHIMIZU, H. FUJIOKA, and K. TAGA

Summary

The epidural recording of human SCP could provide extensive information on spinal cord function during anesthesia and surgery when several technical procedures are properly handled.

Three kinds of human SCPs can be recorded from the epidural space: the segmental and conductive SCPs, and the SCP produced by the descending volley. The segmental SCP consists of an initially positive spike (P_1) followed by sharp negative (N_1) and slow positive (P_2) potentials, which are very similar in waveform to those in animals. The conductive or ascending SCP is composed usually of three spike-like potentials, sometimes followed by a negative slow wave. The SCP produced in the lumbosacral enlargement by the descending volley manifests itself as initial spikes followed by sharp negative-positive waves which are alike in waveform to the segmentally evoked SCP.

Anesthetics differentially affect each component of the segmental SCP. Hypoxia and acidosis suppress the SCP studied in the rat. The suppression is most pronounced in the P_2 wave and "the heterosegmental potential." The N_1 amplitude is also decreased by hypoxia, with prolongation of its duration.

The simultaneous recording of other electrical activities along the sensory pathways, such as the SEP from the scalp, may provide a more accurate evaluation of spinal cord function during surgery.

This paper surveys several basic and technical problems in recording human spinal cord potential (SCP) from the epidural space (ES) and the effects of various procedures on the SCPs in order to evaluate accurately the spinal cord function during anesthesia and surgery.

1. The Method of Recording the SCP from the Epidural Space in Man

Figure 1 shows the schematic representation of introducing the recording electrodes into the spinal ES, which is taken from our preliminary report (Shimoji et al. 1971). Procedures to introduce the electrode into the posterior ES are based on the technique of continuous epidural block, used in the operating room and in pain clinic.

The subjects are placed in a lateral position and are well flexed in order to open the inter-spaces of the vertebral column. After making a skin wheal and injecting

[1]Department of Anesthesiology, Niigata University School of Medicine, Asahi-machi, Niigata 951, Japan

Spinal Cord Monitoring
Ed. by J. Schramm and S. J. Jones

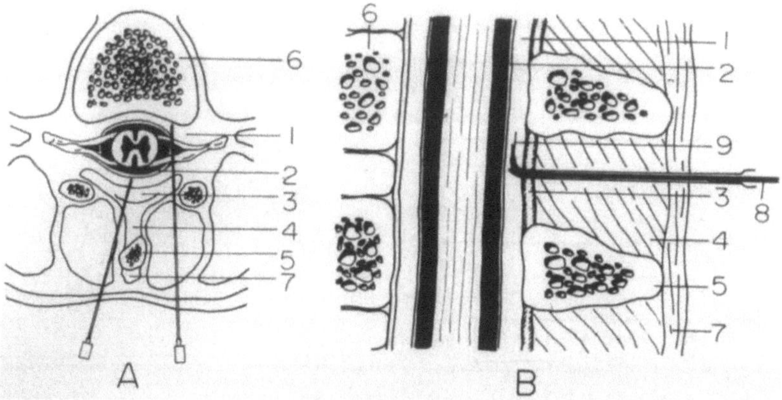

Fig. 1A, B. Schematic representation of insertion of the Tuohy needle into the posterior or anterior spinal epidural space (**A**), for introduction of the catheter electrode (**B**). **A** Horizontal plane; **B** sagittal plane. *1,* Epidural space; *2,* dura mata; *3,* ligamentum flavum; *4,* interspinous ligament; *5,* spinal process; *6,* body of vertebra; *7,* supraspinous ligament; *8,* polyethylene tube; *9,* epidural catheter electrode

local anesthetics over the selected inter-space aseptically, a 16-gauge Tuohy needle is inserted into the ES by paramedian approach, which makes it possible to reach any segment of the ES without interfering with spinal processes. The anterior ES can also be accessed by a more lateral approach as shown in Fig. 1. Placement of the tip of the needle is easily confirmed by the "loss of resistance" against the plunger of the syringe (Moore 1965). After ascertaining that neither blood nor spinal fluid is aspirated, a catheter electrode is inserted through the Tuohy needle to 2–5 cm beyond the bevel rostrally.

a) Assessment of the Position of the Epidural Electrodes

The approximate position of the electrodes in the ES can be assessed by applying weak pulses through these recording electrodes (Shimoji et al. 1977). If an electrode is situated in the mid-sagittal plane, the stimulation causes a bilateral muscle twitching in the segment; if the electrode is located laterally in the ES, the stimulation produces a segmental twitching of the ipsilateral muscles. These procedures do not cause the subjects any discomfort when an appropriate stimulus intensity is applied. Final confirmation of the electrode position is made by X-ray photograph.

b) Elimination of ECG Artifacts

Figure 2A represents a specimen polygraphic record of the SCPs recorded from the posterior ES at the T_{12} vertebral level (the second trace) and bipolarly at the T_{12}–L_1 level simultaneously with an electroencephalogram (left F-P lead) (top trace), the evoked electromyogram of the calf muscle (the fourth trace) and electrocardiogram (ECG) (bottom trace). The stimuli are applied to the right posterior tibial nerve at the popliteal fossa. As demonstrated in this figure, the polygraphic records of the SCPs are contaminated by large ECG artifacts which may interfere with the evoked SCPs. To eliminate these ECG artifacts, the evoked potentials are obtained at inter-

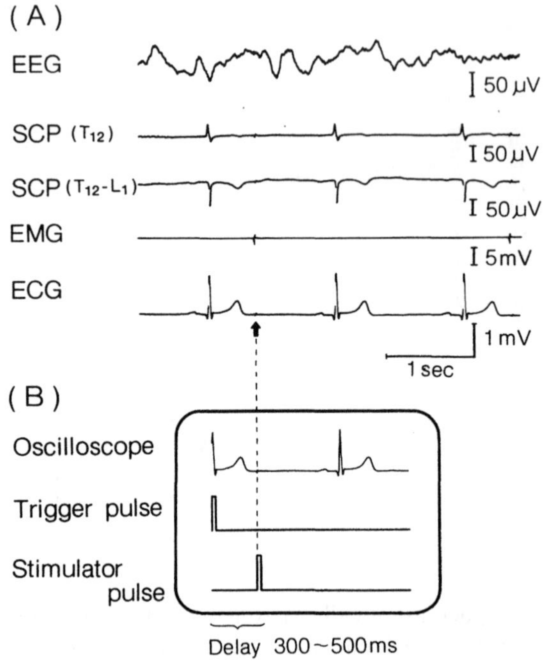

Fig. 2A, B. Experimental arrangements for elimination of ECG artifacts. See the text for explanation

vals between T and P waves of ECG (Shimoji et al. 1976; Maruyama et al. 1982; Shimizu et al. 1982). This is accomplished by the following arrangement as demonstrated schematically in Fig. 2B. The ECG is displayed on the screen of an oscilloscope with the sweep triggered by every other QRS complex. A pulse is generated at the start of each sweep and is fed into the stimulator as a trigger. Thus the stimulus is delivered with a variable delay after every other QRS complex of the ECG. The SCPs, which are monitored on other channels of the oscilloscope and continuously recorded on an ink-writing polygraph, are fed to a computer for averaging. The computer is triggered by the stimulating pulse. The time constant (TC) must be set at a longer duration for the accurate recording of the slow components of the SCP. The TC adopted in our recording is usually 1.5–2.0 s.

2. The Waveforms of the SCPs Recorded from the ES in Man

a) Three Waveforms of the SCP

Figure 3 shows the fundamental patterns of the SCPs recorded from the posterior ES at the cervical and lumbar enlargements in a subject. Ulnar nerve stimulation of a moderate intensity (5 times the threshold), at which the subject does not complain of any discomfort, evokes the characteristic sequence of potential in the posterior ES at the cervical enlargement; the initially positive spike (P_1) is followed by a sharp negative (N_1) and a slow positive (P_2) wave, as shown on the top, left. Posterior tibial nerve stimulation at the popliteal space also produces the SCP with a similar waveform characteristic in the posterior ES at the lumbar enlargement. We call these

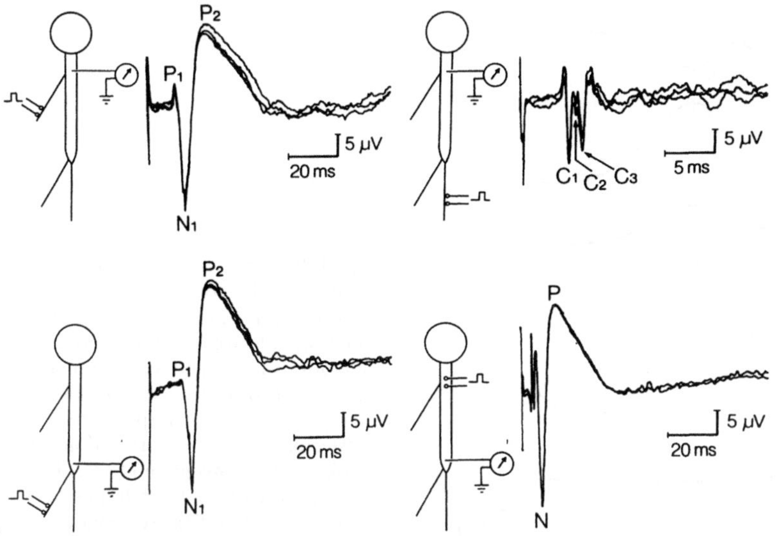

Fig. 3. Three fundamental patterns of the SCPs recorded from the ES in a normal subject. Ulnar *(top left)* and tibial nerve *(bottom left)* stimulation produces the initially positive spike, followed by a sharp negative and a slow positive wave in the corresponding spinal segments. Cauda equina stimulation elicits polyphasic spike-like potentials in the remote rostral segment *(top right)*. Cervical epidural stimulation produces the spikes and a subsequent sharp negative and a slow positive wave in the ES of the lumbar enlargement *(bottom right)*. Note that the SCP produced by the descending volley is very similar to that evoked segmentally. All evoked responses in this and subsequent figures are averaged ($N = 25$–50)

potentials segmentally evoked SCPs (Maruyama et al. 1982). The waveform characteristics of the segmental SCP recorded from the ES, coincide with those recorded from the intrathecal space by Magladery et al. (1951) and Ertekin et al. (1978). When the SCP is recorded from the rostral portion of the spinal cord far beyond the segmental nerve, the small action potential-like spikes, usually composed of three components, can be recorded (top, right). This potential, the conductive SCP, has been studied intensively by several investigators for monitoring the spinal cord function during spine or spinal cord surgery (Kurokawa et al. 1980; Tamaki et al. 1981; Maruyama et al. 1984; Schramm et al. 1984). Another characteristic type of the SCP is produced in the lumbar enlargement in response to the descending volley delivered to the rostral spinal segment from the posterior ES (bottom, right). It consists of a spike or spikes followed by a sharp negative and a slow positive wave (Shimizu et al. 1979a). The waveforms of these slow negative and positive waves of the SCP, produced in the lumbosacral enlargement by descending volleys, are very similar to those of the segmental SCP (Shimizu et al. 1979b, 1982). Those three types of SCPs are recordable from the posterior ES in man in response to the electrical stimulation of the peripheral nerve or the spinal cord.

b) Recording from the Anterior ES

The segmental SCP recorded from the posterior ES consists of an initially positive spike followed by negative and positive waves in reference recording. Figure 4

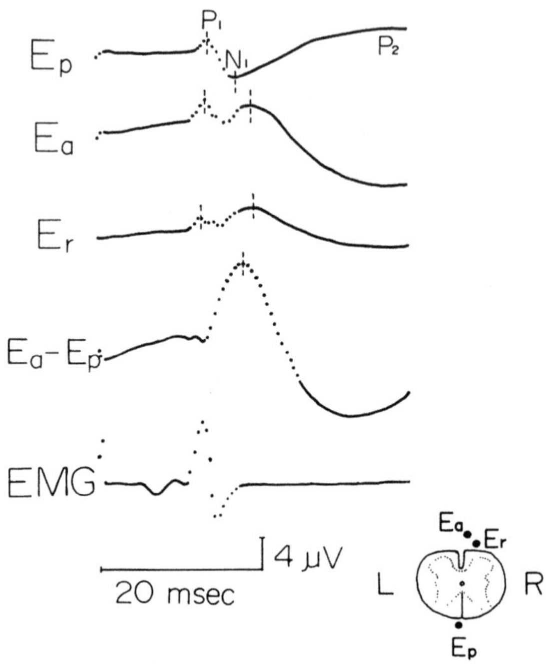

Fig. 4. Simultaneous recordings of the SCPs from the posterior *(Ep)* and anterior *(Ea, Er)* ES with a common reference electrode in the inter-spinous ligament and with the bipolar lead *(Ea-Ep)* between the anterior and posterior ES, in response to posterior tibial nerve stimulation. The evoked electromyogram *(EMG)* is also recorded from the calf muscle, showing the H-reflex. Schematic representation of the positions of three electrodes in the ES is shown in the *bottom, right.* Note the reversed polarity of the N_1 and P_2 waves in the SCP recorded from the anterior ES

demonstrates the simultaneous recordings from anterior and posterior ES at the level of lumbar enlargement in response to tibial nerve stimulation. The SCPs recorded from the anterior ES are composed of initially positive spikes and subsequent slow positive and negative deflections, making a mirror image of the N_1 and P_2 waves of the SCP recorded from the anterior ES (Shimoji et al. 1977, 1984). In the bipolar recording between the anterior and posterior ES, both potentials related to the N_1 and P_2 are, therefore, greater in amplitude than those in reference recordings, as expected.

c) Comparison with Skin Surface Recordings

Figure 5 represents the simultaneous recordings of the SCPs from the posterior ES at the level of C_7 and the somatosensory evoked potentials from the cervical skin surface at the same cervical level with various reference electrode placements, in response to ulnar nerve stimulation at the wrist. The initial spike-like potential can be recorded in all the skin surface recordings with variable polarities depending on the site of the reference electrode. It is also demonstrated that the peak latency of the initial spike in all the skin surface recordings coincides with that of the P_1 wave in the epidural recording (Shimoji et al. 1978). Subsequent small negative dips in the surface recordings share the same peak latency with the N_1 in the epidural recording. However, the slow wave corresponding to the P_2 wave has not clearly been discerned in the surface recordings. The initial spike-like potential has been recorded in all the surface recordings, while the subsequent negative wave has only been recorded in about one third of the subjects tested.

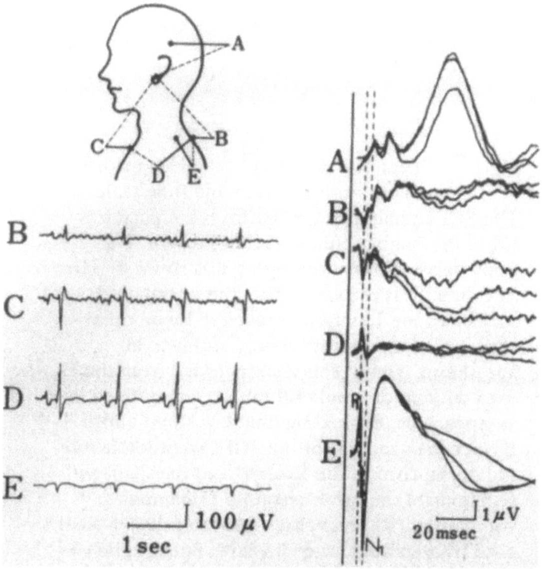

Fig. 5. Comparison of the SCP recorded from the posterior ES *(E)* with the somatosensory evoked potentials from the cervical skin *(B, C, D)* and the scalp *(A)*. Electrode placements are shown schematically in the *top left*. *Solid lines* represent active electrodes; *hashed lines* denote the reference electrodes. The polygraphic record is also shown on the *left*

3. Comparison of Human SCP with that in Animals

Are there any species differences in the waveforms of the SCPs between man and other mammalian species? The segmental SCP recorded from the posterior ES at the level of lumbar enlargement in response to tibial nerve stimulation in man (top trace) is illustrated with those recorded directly from the dorsal cord surface of the L_5 segment evoked by hindpaw stimulation in a rat (middle trace) and a rabbit (bottom trace) (Fig. 6A). It is demonstrated that the fundamental patterns and durations of the slow negative and positive waves of the SCPs are almost the same among the three species.

In the rat, however, the forepaw or hindpaw stimulation not only evokes the segmental SCP in the corresponding segment, but also produces a slow positive wave in the other segmental areas. For instance, the forepaw stimulation not only produces the segmental potential in the C_5 segment, but also evokes a slow positive wave with a long latency at the L_5 level (Fig. 6B). Similarly, the hindpaw stimulation also evokes the slow positive wave in the cervical enlargement. These "heterosegmental slow positive waves" are abolished by spinal cord transection, and it is suggested that they may reflect the PAD produced in the heterosegmental area through a feed-back loop via a supraspinal structure (Andersen et al. 1962; Engberg et al. 1968; Martin et al. 1979; Tang et al. 1969). The physiological significance of these slow potentials remains to be investigated. We have not been able to record similar potentials in man. Species differences, therefore, might exist in the feed-back control over the heterosegmental area of the spinal cord.

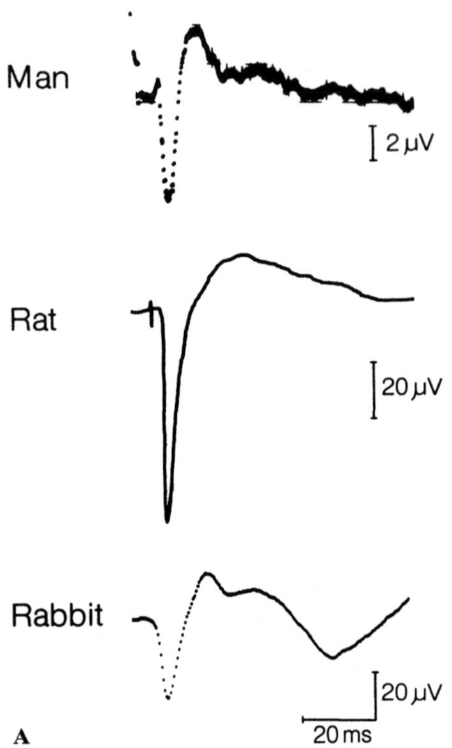

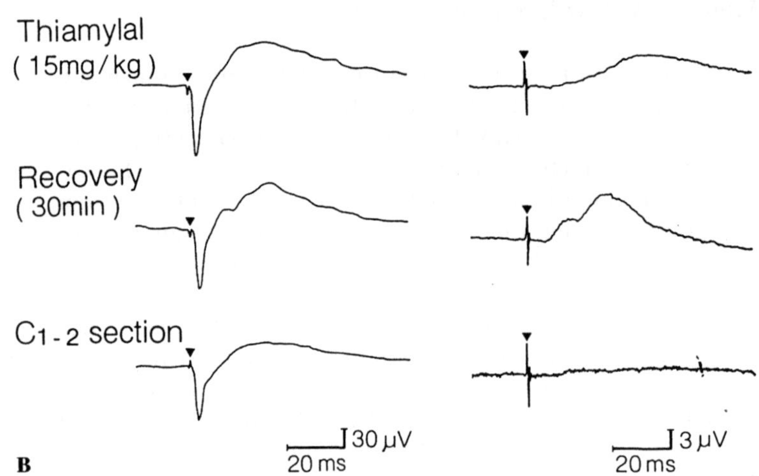

Fig. 6. A Specimen records of the SCPs recorded in man, rat and rabbit on the same time scale. The SCP in man is recorded from the posterior ES of the lumbar enlargement in response to tibial nerve stimulation during wakefulness. The SCP in a rat is recorded from the dorsal cord surface of the L_5 segment, evoked by hindpaw stimulation, under nitrous-oxide-thiamylal anesthesia. The SCP is also recorded from the L_7 cord dorsum of a wakeful rabbit, gently fixed in an apparatus, evoked by hindpaw stimulation. **B** Specimen records of the SCPs recorded from the dorsal cord of the L_5 *(left)* and the C_5 *(right)* segments of the rat in response to hindpaw stimulation (25 times threshold intensity). Spinal cord transection at the C_{1-2} level under light anesthesia (nitrous oxide-thiamylal) abolishes the "heterosegmental slow potential" *(bottom traces)*

4. Effects of Anesthetics, Hypoxia, Acidosis

a) Effects of Anesthetics

Figure 7 represents the differential effects of morphine and thiamylal on the segmental SCPs in man. Intravenous injection of morphine (1 mg/kg) suppresses the

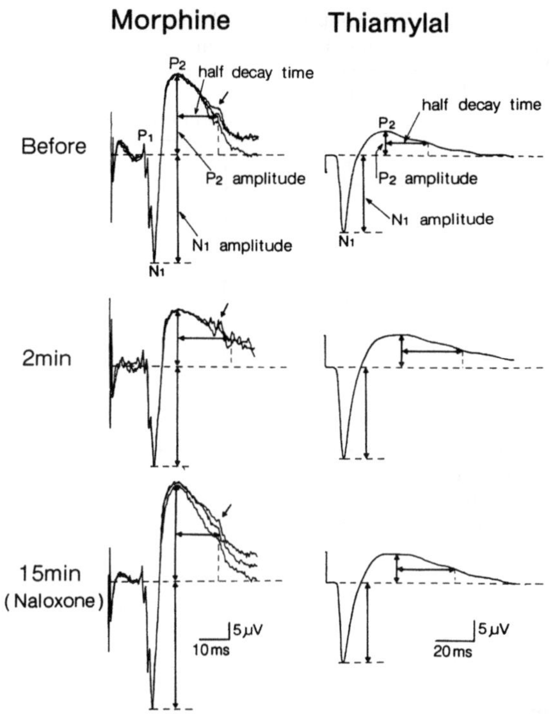

Fig. 7. The effects of morphine and thiamyal on the segmental SCP recorded from the posterior ES in man, studied in different subjects. The SCPs are evoked by tibial nerve stimulation and produced in the lumbar enlargements. The duration of the P_2 wave is estimated by the half decay time

amplitudes of all components (P_1, N_1, P_2) of the SCP recorded from the ES at the T_{12} vertebral level in response to tibial nerve stimulation. Naloxone (0.1 mg/kg) injected 15 min after morphine reverses the effect of morphine and even exceeds the control values. On the other hand, narcotic doses of thiamylal (5 mg/kg) enhance the amplitudes of N_1 and P_2 waves, and also prolong the half decay time of the P_2 wave. Thus, the data suggest that morphine suppresses the input to the cord, the activities of interneurons in the dorsal horn and also the primary afferent depolarization (PAD) in an intact human spinal cord, while thiamylal facilitates both the activity of interneurons and PAD. Table 1 summarizes the effect of moderate doses of anesthetics on each component of the segmental SCP in man (Kaieda et al. 1981; Kano et al. 1974; Maruyama et al. 1980; Shimoji et al. 1972, 1974, 1975, 1984). It is evident that each component of the SCP is differentially affected by the anesthetics. Generally speaking, the amplitude of the N_1 wave is augmented by anesthetic drugs with strong hypnotic effects, such as thiamylal, ketamine, halothane, enflurane and isoflurane, and suppressed by the strong analgesic drugs, morphine, fentanyl and nitrous oxide. The amplitude of the P_2 wave is suppressed by all anesthetics except thiamylal and diazepam. Thiamylal enhances the amplitude of the P_2 wave, while diazepam has no significant effect on it.

b) Effect of Hypoxia

Figure 8 shows the sequential changes produced by hypoxia in the waveforms of the SCPs recorded directly from the dorsal cord at the levels of L_5 (left) and C_5 (right) in

Table 1. Effects of moderate doses of anesthetics on each component of the segmental spinal cord potential, recorded from the posterior epidural space in man

	P_1 amplitude	N_1 amplitude	N_1 duration	P_2 amplitude	P_2 duration
Thiamylal	→[a]	↑	↑	↑	↑
Diazepam	↓	→	→	→	↑
Ketamine	→	↑	→	↓	↓
Morphine	↓	↓	→	↓	↑
Fentanyl	↓	↓	→	↓	↑
N_2O	→	↓~→	→	↓	↑
Halothane	→	↑	↑	↓	↑
Enflurane	→	↑	↑	↓	↑
Isoflurane	→	↑	↑	↓	↑

P_1 reflects the action potential along the spinal root
N_1 is thought to represent the activity of interneurons in the dorsal horn
P_2 is believed to record the primary afferent depolarization (PAD)
[a] →: no change ↑: increased ↓: decreased

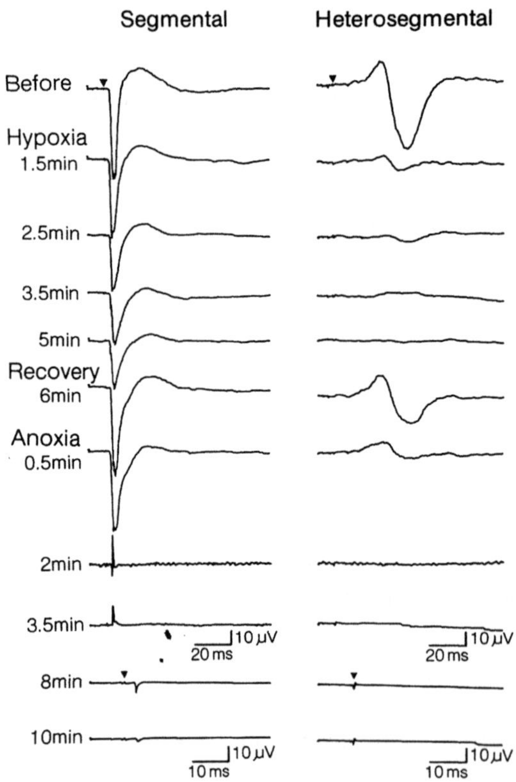

Fig. 8. The effect of hypoxia on the segmental (L_5) *(left)* and heterosegmental (C_5) *(right)* SCPs in the rat. The electrical stimuli (25 times threshold) were delivered to the hindpaw. Hypoxic (3% oxygen in nitrous oxide) and anoxic (100% nitrous oxide) gases were administered to the rat anesthetized with pentobarbital sodium (50 mg/kg, i.p.). Note the different time-scale of the bottom two traces of each record

response to hindpaw stimulation in the rat under pentobarbital (50 mg/kg, i.p.)-nitrous oxide (75%) anesthesia. It is shown that the P_2 wave is the most susceptible to hypoxia and the slowest to recover. The N_1 wave is also decreased in amplitude and prolonged in duration by hypoxia. The P_1 wave is most resistant to hypoxic insult and sometimes even potentiated transiently at the terminal stage of hypoxia. Furthermore, the slow positive-negative deflections produced in the remote segment ("heterosegmental SCP") which have not been recordable in the human spinal cord, are also very vulnerable to hypoxia as shown on the right in Fig. 8.

c) Effect of Acidosis

Acidosis also affects the SCP. Figure 9 reveals the depressant effect of metabolic acidosis on the SCPs in the rat. The rat underwent laminectomies at cervical, thoracic and lumbar regions for recording the SCPs under nitrous oxide-pentobarbital anesthesia. A certain amount of blood was lost during the surgery, but blood transfusion was not carried out due to the good general condition with a slight metabolic acidosis (base excess −7.0 mEq/L). About 3 h after the surgery, mean arterial pressure decreased from 80 to 60 mm Hg, with amplitude reduction in the SCPs. Blood gas analysis showed a profound metabolic acidosis (base excess −21.0 mEq/L). Half correction was made for acidosis by sodium bicarbonate (0.84 mEq/L). As shown in the third traces of the records, the amplitudes of the N_1 and P_2 waves are increased by this correction with an insignificant effect on the heterosegmental SCP. Another half correction of acidosis, preceded by the injection of calcium chloride, produces further recovery of the segmental SCP and also affords substantial amplitude increase in the heterosegmental SCP. During this period mean arterial pressure increased slightly (from 60 to 70 mm Hg).

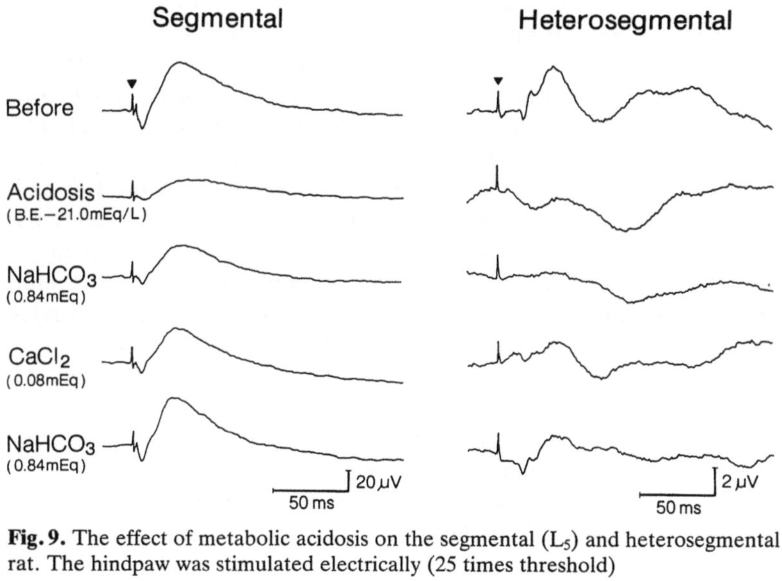

Fig. 9. The effect of metabolic acidosis on the segmental (L_5) and heterosegmental (C_5) SCPs in the rat. The hindpaw was stimulated electrically (25 times threshold)

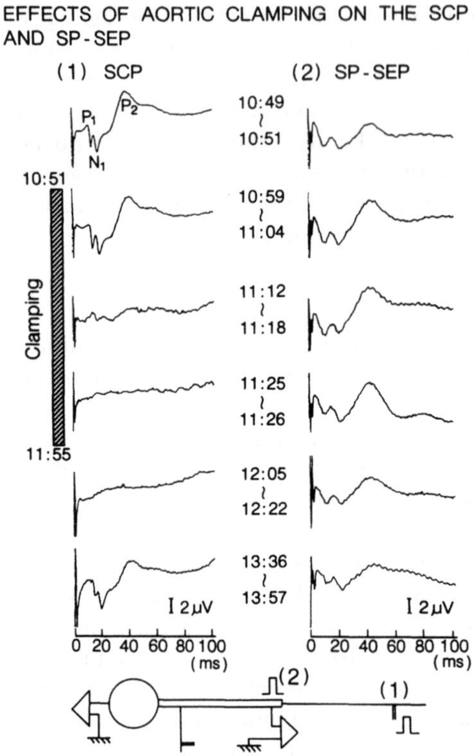

Fig. 10. Simultaneous recordings of the SCP and the somatosensory evoked potential (SP-SEP) in a patient during surgery. The SCP and SP-SEP were evoked by tibial nerve and spinal cord stimulation, respectively

5. Significance of the Simultaneous Recordings of Other Electrical Activities

An example of spinal cord monitoring by the simultaneous recording of the segmental SCP and SEP in the Y-graft operation on abdominal aneurysm is presented in Fig. 10. The segmental SCP was recorded from the posterior ES at the L_1 vertebral level in response to tibial nerve stimulation, while the SEP, evoked by spinal cord stimulation from the ES of the L_1 vertebral level, was recorded from the contralateral parietal region. After clamping of the abdominal artery rostral to the branching of Adamkiewicz's artery, the segmental SCP gradually disappeared, while the SEP remained unchanged. This finding indicates that the disappearance of the segmental SCP by arterial clamping is not due to the functional loss of the lumbar enlargement but to the impairment of afferent nerves. Thus, observation of the SCP with simultaneous recording of other electrical activities along the sensory tract may provide more valid information for monitoring spinal cord function during surgery.

During spine surgery, the epidural electrode which is situated near the manipulation site could be dislodged by the force exerted on the spine, and this may result misjudgment of cord function. The simultaneous recording of other electrical activities along the sensory tract may eliminate this wrong diagnosis. Figure 11 shows an example of simultaneous recordings of conductive SCP and SEP during Luque's procedure on congenital scoliosis. Stimulation of the cauda equina from the ES produces the characteristic polyphasic spikes in the posterior ES at C_7 and the SEP

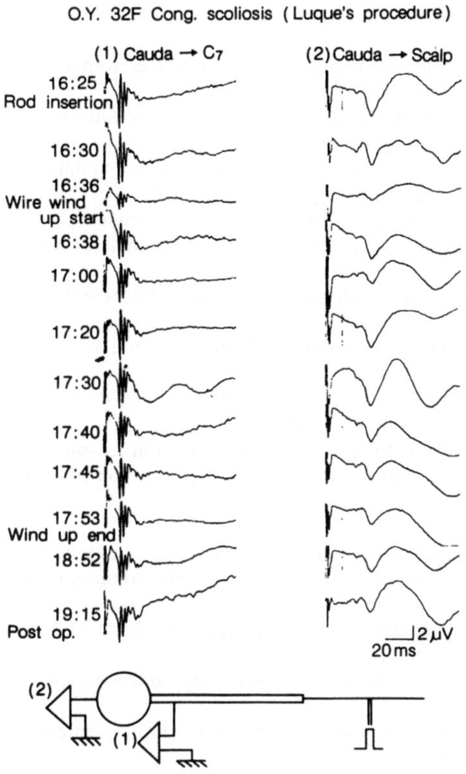

Fig. 11. Simultaneous recordings of the conductive SCP (Cauda equina → C_7) and SEP (Cauda equina → Scalp) in a patient undergoing spinal surgery. The cauda equina (L_{3-4}) was stimulated from the posterior ES

at the vertex. When the wire wind-up started, the amplitude of the SCP was suppressed and simultaneous decrease in the early components of the SEP occurred. This reveals that the reduction of the amplitude of the conductive SCP was caused by functional change in the spinal cord due to this wind-up procedure, and not by electrode displacement. Immediate reduction of the windup force led to recovery of the SCP. In addition, the amplitude of the conductive SCP is varied substantially during the course of surgery by an unknown factor; the SEP, even its early component, however, is more variable during the procedure in spite of the steady state of anesthesia. Surgical stimulation of the spinal cord might itself be one of the major causes of these variations in electrical activity.

Acknowledgments. We thank Mr. Yukio Sato for his assistance in the experiments and Mr. Patrick J. Runkel for his kind comments on the manuscript. This work was supported in part by grants from the Ministry of Education and the Ministry of Public Welfare, Japan.

References

Andersen P et al. (1962) Presynaptic inhibitory action of cerebral cortex on the spinal cord. Nature 194:740–741

Engberg et al. (1968) Reticulospinal inhibition of interneurones. J Physiol 194:225–236

Ertekin C (1978) Evoked electrospinogram in spinal cord and peripheral nerve disorders. Acta Neurol Scand 57:329–344
Kaieda R et al. (1981) Effects of diazepam on evoked electrospinogram and evoked electromyogram in man. Anesth Analg 60:197–200
Kano T, Shimoji K (1974) The effects of ketamine and neuroleptanalgesia on the evoked electrospinogram and electromyogram in man. Anesthesiology 40:241–246
Kurokawa T (1980) Functional spinal cord monitoring in spinal surgery through evoked spinal cord action potential measurement (in Japanese). Clinical Electroenceph 22:464–470
Magladery JW et al. (1951) Electrophysiological studies of nerve reflex activity in normal man. IV. The two-neurone reflex and identification of certain action potentials from spinal roots and cord. Bull Johns Hopkins Hosp 88:499–519
Martin RF et al. (1979) Primary afferent depolarization of identified cutaneous fibers following stimulation in medial brain stem. J Neurophysiol 42:779–790
Maruyama Y et al. (1980) Effects of morphine on human spinal cord and peripheral nervous activities. Pain 8:63–73
Maruyama Y et al. (1982) Human spinal cord potentials evoked by different sources of stimulation and conduction velocities along the cord. J Neurophysiol 48:1098–1107
Maruyama Y et al. (1984) Spinal cord function monitoring by spinal cord potentials during spine and spinal surgery. In: Homma S, Tamaki T (eds) Fundamentals and clinical application of spinal cord monitoring. Saikon, Tokyo, p 191
Moore DC (1965) Regional block. Thomas, Springfield, Ill, p 427
Schramm J et al. (1984) Relevance of spinal cord evoked injury potential for spinal cord monitoring. In: Homma S, Tamaki T (eds) Fundamentals and clinical application of spinal cord monitoring. Saikon Publishing, Tokyo, pp 113–124
Shimizu H et al. (1979a) Interaction between human evoked electrospinograms elicited by segmental and descending volleys. Experientia 35:1199–1200
Shimizu H et al. (1979b) Slow cord dorsum potentials elicited by descending volleys in man. J Neurol Neurosurg Psychiatry 42:242–246
Shimizu H et al. (1982) Human spinal cord potentials produced in lumbosacral enlargement by descending volleys. J Neurophysiol 48:1108–1120
Shimoji K et al. (1971) Epidural recording of spinal electrogram in man. Electroencephalogr Clin Neurophysiol 30:236–239
Shimoji K et al. (1972) Evoked spinal electrograms recorded from epidural space in man. J Appl Physiol 33:468–471
Shimoji K et al. (1974) The effects of thiamylal sodium on electrical activities of the central and peripheral nervous systems in man. Anesthesiology 40:234–240
Shimoji K et al. (1975) Presynaptic inhibition in man during anesthesia and sleep. Anesthesiology 43:388–391
Shimoji K et al. (1976) Interactions of human cord dorsum potential. J Appl Physiol 40:79–84
Shimoji K et al. (1977) Wave-form characteristics and spatial distribution of evoked spinal electrogram in man. J Neurosurg 46:304–313
Tamaki T et al. (1981) The prevention of iatrogenic spinal cord injury utilizing the evoked spinal cord potentials. Internat Orthop 4:313–317
Tang AH (1969) Dorsal root potentials in the chloralose-anesthetized cat. Exp Neurol 25:393–400

Comparison of Spinal Cord Evoked Potentials Elicited by Spinal Cord and Peripheral Nerve Stimulation

H. TAKANO[1], T. TAMAKI, T. NOGUCHI, and K. TAKAKUWA

Summary

Two techniques of epidural spinal cord monitoring, SCEP, produced by spinal cord stimulation and spinal SEP following peripheral nerve stimulation, were evaluated in animal studies of the effect of compression and distraction. Although the two responses were affected in parallel, SCEP was the more sensitive and is therefore preferred for clinical application.

Introduction

To monitor spinal cord functions intraoperatively, several combinations of stimulation and recording techniques are currently utilized. There are two methods which use epidural recording electrodes. One is for potentials elicited by spinal cord stimulation (spinal cord evoked potential: SCEP) and the other is for potentials generated by peripheral nerve stimulation (spinal SEP). We have been using SCEP for intraoperative monitoring since 1972 (Tamaki et al. 1972, 1981, 1984). Recently the spinal SEP, which was recorded by Shimoji in 1971 with a clinically applicable method (Shimoji et al. 1971), has been used extensively for intraoperative spinal cord monitoring by Jones et al. (Jones et al. 1982, 1983). To evaluate the usefulness of these potentials, we recorded both simultaneously in our surgical cases.

During intraoperative spinal cord monitoring of Harrington Instrumentation on an idiopathic scoliosis patient, we noticed 30% attenuation of amplitude of SCEP after distraction of the spine, while no recognisable change of the spinal SEP was observed. Although this patient did not exhibit any neurological deficits after the operation, it did appear that the sensitivity to spinal cord insult might not be the same between the two methods.

Therefore we performed experimental studies to evaluate the feasibility of both potentials for intraoperative spinal cord monitoring, comparing the manner in which both potentials were affected by spinal cord lesions.

Methods

17 adult mongrel cats were anesthetized with nitrous oxide and oxygen at 2 and 1 liter/min, respectively, and 1.5% Fluothane through an intubation tube, and immobilized by introducing pancuronium bromide.

[1]Department of Orthopaedic Surgery, Faculty of Medicine, Toyama Medical and Pharmaceutical University, 2630 Sugitani, Toyama 930-01, Japan

Spinal Cord Monitoring
Ed. by J. Schramm and S. J. Jones

For the recording of SCEP, stimulation was performed at high thoracic level and evoked potentials were recorded at L_{5-6}. The electrodes used for this purpose were bipolar tube-type electrodes inserted in the epidural space.

To elicit spinal SEP, the sciatic nerves were exposed bilaterally and stimulated just distal to the hip joints, being wrapped around by specially designed bipolar stimulating electrodes. Spinal SEPs were recorded at high thoracic level by the epidural bipolar electrodes, which was thought to be a practical method of recording low amplitude potentials in the operating theatre. This level was the same as that used for spinal cord stimulation to elicit SCEPs.

Constant current pulses of 0.3 ms duration, delivered at a frequency of 30 pulses per second, were used in both methods. Intensities were supramaximal, about 5 times the threshold. Bipolar recording using a differential amplifier with a bandwidth of 32 to 3 kHz was performed. Generally 50 to 100 consecutive responses were averaged.

With regard to the techniques used to jeopardize the spinal cord, we employed compression from the dorsal side and distraction. For the spinal cord compression study in 10 cats, we used a tonometer at the level of T_{9-10} to produce a quantitative compression from the epidural space. The compression force was increased gradually over 60 to 90 min until the amplitude of SCEP was reduced by more than 50% of the control level.

For the distraction study in 7 cats, the distraction instrument designed by Nordwall and colleagues (1979) was placed at the T_9 and T_{10} vertebral bodies. The ribs were cut and the facet joints excised, and the disc and the anterior longitudinal ligament were divided. Distraction was applied to an extent of 5 mm initially, and increased in increments of 2 mm every 10 min.

Results

SCEP's consisted of an initial spike wave and following polyphasic waves. The initial spike waves of control recordings were fairly constant in amplitude and latency.

Spinal SEPs consisted of 3 or 4 negative deflections. In the present study almost all recordings could be divided into 2 components, these being a fast component consisting of 1 or 2 negative deflections with peak latencies of 5–7 ms, and a following late component. The amplitude of the fast component of the spinal SEP fluctuated by 10 to 20% of the control level.

In this experiment, we measured the amplitude of the spike wave of the SCEP and the fast component of the spinal SEP as an indicator of spinal cord function.

Dorsally applied compression to the thoracic spinal cord produced gradual reduction in amplitude and increase in peak latency of both potentials. In the initial stage of dorsally applied compression little change occurred in either potential, but once both potentials started to reduce in amplitude, responses were altered significantly by a slight degree of additional compression. Therefore compression was carefully applied until the SCEP started to show alteration in the waveform, whereupon small increments of compression (0.5–1 g) were added. SCEPs and spinal SEPs tended to diminish proportionately against the application of compression from the dorsum, although amplitudes decreased more markedly in SCEPs than in spinal SEPs.

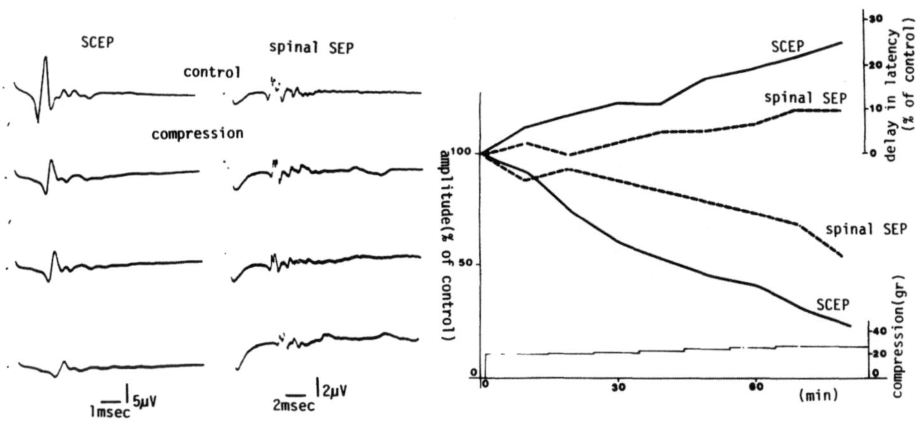

Fig. 1. Typical responses of SCEP and spinal SEP during compression. When SCEP amplitude was attenuated down to 50%, spinal SEP remained at approximately 80% of control amplitude. SCEP was more markedly affected by applied compression than spinal SEP, in amplitude and latency

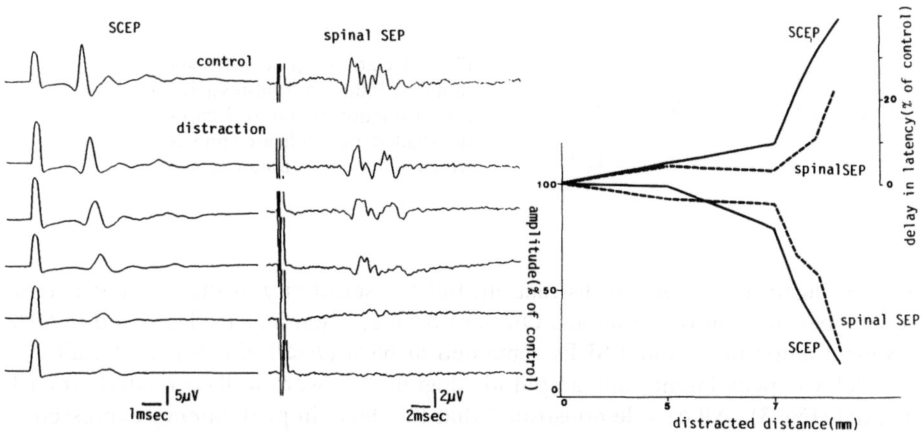

Fig. 2. Potential change during spinal cord distraction. Both potentials showed parallel responses against applied distraction. When distraction was increased to 7 mm, both potentials fell in amplitude and increased in latency rapidly. The sensitivity of SCEP was relatively higher in this animal

Similar phenomena were observed in 9 out of 10 animals. Latency also increased more markedly in SCEPs (Fig. 1), but there was good correlation between latency delay and amplitude degradation of each potential. On average, when SCEPs fell to 50% of control amplitude, spinal SEPs remained at 65% of the control level ($P<0.02$) (Fig. 3).

The effects of spinal cord distraction were studied in 7 cats. In the initial phase of distraction both potentials changed only slightly, but once the waveforms started to change they decreased in amplitude and increased in latency rapidly. The late component of the spinal SEP tended to diminish quickly after the SCEP spike wave was reduced to 50% of control amplitude. SCEP and spinal SEP showed parallel

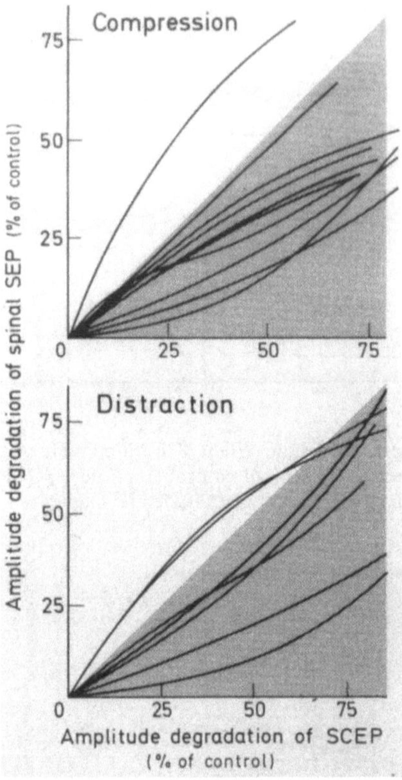

Fig. 3. Relationship of SCEP and spinal SEP during compression *(top)* and distraction *(bottom)*. Lines in the shaded area indicate animals in which SCEPs decreased to a greater degree

responses against the applied distraction, but the sensitivity of the SCEP was relatively higher in 5 out of 7 animals. On the average, when SCEPs decreased to 50% of control amplitude, spinal SEPs remained at 65% ($P<0.05$) (Fig. 2). In all animals, delay in peak latency and amplitude degradation were well correlated for each potential (Fig. 3). All cats demonstrated that the delay in peak latency expressed as a percentage of the control value was also more marked for the SCEP.

Discussion

Bennett (1984) reported the close correspondence between ascending and descending cord potentials with compressive and ischemic insult. In the present study the SCEP, which has been regarded as containing potentials mediated by both ascending and descending tracts, was compared with the ascending cord potentials. As an overall conclusion, both potentials showed a tendency to diminish proportionately, although the SCEP was likely to be more quickly attenuated. As yet it is not possible to determine the neurophysiological mechanism of this phenomen. The spike wave of the SCEP is conducted through large fibers present in all quarters of the spinal cord (Harada 1983; Toyoda et al. 1984). Therefore, a lesion applied to any part of the spinal cord may effect the initial spike wave of the SCEP. On the other hand, the spinal SEP is thought to be conducted mainly in the dorsal and dorsolateral columns (Cohen et al. 1981; Jones et al. 1982; Satomi et al. 1984). The conduction velocity of the early spike wave of the descending SCEP was calculated as 73 ± 9.4 m/s by Tsuyama et al. (1978), and 86.4 ± 3.6 m/s by Shimoji et al. (1984). With regard to the spinal SEP, Jones et al. (1982) estimated that the fastest evoked spinal cord activity, perhaps originating in the lateral tracts, was conducted at 65–80 m/s. Consequently there may not be any marked difference in the diameter of nerve fibers which compose the spike wave of the SCEP and the fastest component of the spinal SEP, although the number of large fibers responsible for the potential is apparently greater in the SCEP. There have been debates on the role of mechanical factors versus ischemia in the pathophysiology of spinal cord compression and distraction (Gelfan et al. 1956; Schramm et al. 1979; Dolan et al. 1980; Larson et al. 1980; Cusick et al. 1982). Large diameter fibers are more sensitive to mechanical factors than fine fibers, and large fibers which contribute to the spike wave of the SCEP are likely to exist in all quarters of the spinal cord. If the mechanical factor were the prevalent cause of both compressive and distractive insult, the results observed in the present study could be explained. However, in order to obtain more reliable proof of this assumption, further work is required.

As previously mentioned, both evoked potentials diminished proportionally in this study. Therefore, each potential can be used as an alternative method to the other, but care must be taken as the criterion to be taken as an alarm signal is different. In other words, the sensitivity to early insult to the spinal cord is different for the two potentials. From the standpoint of clinical application, we regard the SCEP as more appropriate for use as an indicator of spinal cord function, on account of being simpler in waveform, larger in amplitude, more stable in amplitude and latency and higher in sensitivity to spinal cord insult.

References

Bennett MH (1984) Correlations of ascending and descending conduction block following ischemic and compressive insult to the spinal cord of cat. In: Homma S, Tamaki T (eds) Fundamentals and clinical application of spinal cord monitoring. Saikon Press, Tokyo, p 125

Cohen AR, Young W, Ransohoff J (1981) Intraspinal localization of the somatosensory evoked potential. Neurosurgery 9:157–162

Cusick JF, Myklebust J, Zyvolski M, Sances A Jr, Houterman C, Larson SJ (1982) Effects of vertebral column distraction in the monkey. J Neurosurg 57:651–659

Dolan EJ, Transfeldt EE, Tator CH, Simmons EH, Hughes KF (1980) The effect of spinal distraction on regional spinal cord blood flow in cats. J Neurosurg 53:756–764

Gelfan S, Tarlov IM (1956) Physiology of spinal cord, nerve root and peripheral nerve compression. Am J Physiol 185:217–229

Harada Y (1983) Study on experimental spinal cord injuries by distraction using spinal cord evoked potentials (in Japanese). J Jpn Orthop Ass 57:685–701

Jones SJ, Edgar MA, Ransford AO (1982) Sensory nerve conduction in the human spinal cord: epidural recordings made during scoliosis surgery. J Neurol Neurosurg Psychiat 45:446–451

Jones SJ, Edgar MA, Ransford AO, Thomas NP (1983) A system for electrophysiological monitoring of the spinal cord during operations for scoliosis. J Bone Joint Surg 65-B:134–139

Larson SJ, Walsh PR, Sances A Jr, Cusick JF, Hemmy DL, Mahler H (1980) Evoked potentials in experimental myelopathy. Spine 5:299–302

Nordwall A, Axelgaard J, Harada Y, Valencia P, McNeal DR, Brown JC (1979) Spinal cord monitoring using evoked potentials recorded from feline vertebral bone. Spine 6:486–494

Satomi K, Nishimoto GI (1984) Effects of selective spinal cord transections on evoked spinal potentials in cats. In: Homma S, Tamaki T (eds) Fundamentals and clinical application of spinal cord monitoring. Saikon Press, Tokyo, p 87
Shimoji K, Higashi H, Kano T (1971) Epidural recording of spinal electrogram in man. Electroenceph Clin Neurophysiol 30:236–239
Shimoji K, Shimizu H, Maruyama Y, Fujioka H (1984) Human spinal cord potentials produced by segmental ascending and descending volleys. In: Homma S, Tamaki T (eds) Fundamentals and clinical application of spinal cord monitoring. Saikon Press, Tokyo, p 46
Schramm J, Hashizume K, Fukushima T, Takahashi H (1979) Experimental spinal cord injury produced by slow graded compression. J Neurosurg 50:48–57
Tamaki T, Yamashita T, Kobayashi H, Hirayama H (1972) Spinal cord monitoring (in Japanese). Jpn J Electroenceph Electromy 1:196
Tamaki T, Tsuji H, Inoue S, Kobayashi H (1981) The prevention of iatrogenic spinal cord injury utilizing the evoked spinal cord potential. Int Orthop 4:313–317
Tamaki T, Noguchi T, Takano H, Tsuji H, Nakagawa T, Imai K, Inoue S (1984) Spinal cord monitoring as a clinical utilization of the spinal evoked potential. Clin Orthop 184:58–64
Toyoda A, Kanda K (1984) Origins of spinal cord potentials evoked by stimulation of the cat spinal cord. In: Homma S, Tamaki T (eds) Fundamentals and clinical application of spinal cord monitoring. Saikon Press, Tokyo, p 99
Tsuyama N, Tsuzuki N, Kurokawa T, Imai T (1978) Clinical application of spinal cord action potential measurement. Int Orthop 2:39–46

An Analysis of Spinal Cord Potentials Evoked by Median Nerve Stimulation

A. Kaneda[1], I. Yamaura, M. Kamikozuru, K. Shinomiya, H. Sato, and M. Yokoyama

Summary

The early components (NI and NII) of the cervical cord potentials which were evoked by median nerve stimulation and recorded from the dorsal surface of the spinal cord were analysed. On the posterior surface rostral to the entry zone of the median nerve, the NI component is derived from the activity of primary afferent fibres ascending in the dorsal columns on the stimulated side. At the entry zone the posterior root potentials appeared to be added to this. The NII component originated in the synaptic or post-synaptic potentials of the dorsal horn on the stimulated side.

Introduction

Cervical spinal cord potentials evoked by median nerve stimulation are reported to be useful in diagnosis of cervical disorders (myelopathy and radiculopathy) [6, 9]. The potentials evoked by median or ulnar nerve stimulation and recorded from the dorsal surface of the spinal cord reveal an early negative wave with two peaks, followed by a positive wave. While the positive wave has been reported to originate from primary afferent depolarization, there are various opinions concerning the origin of the early negative waves [1–5, 7, 8]. This study describes the early negative waves evoked by median nerve stimulation and recorded on the dorsal surface of the cervical spinal cord in the cat.

Materials and Methods

Twenty cats were used in this study, anesthetized with ketamine and intubated. The median nerve was exposed in the forelimb and stimulated by means of a series of impulses set at 3 impulses per second, 0.2 ms in duration and 0.2 mA in amplitude. A laminectomy was performed from C3 to T1. The dura and arachnoid were resected over the dorsal half of the cord. A polyethylene tube with 4 monopolar electrodes was placed on the midline of the cervical cord. Each electrode was adjacent to the entry zone of the posterior roots from C4 to C7. The reference electrode was placed in the back musculature. After control potentials were recorded in response to stimulation as described previously, the following study was performed:

1. Stimulation at fast frequency,
2. Transection of posterior roots (rostrally and caudally),

[1]Tokyo Medical and Dental University, 5-45, Yushima, 1-Chome, Bunkyo-Ku, Tokyo, Japan

Spinal Cord Monitoring
Ed. by J. Schramm and S. J. Jones

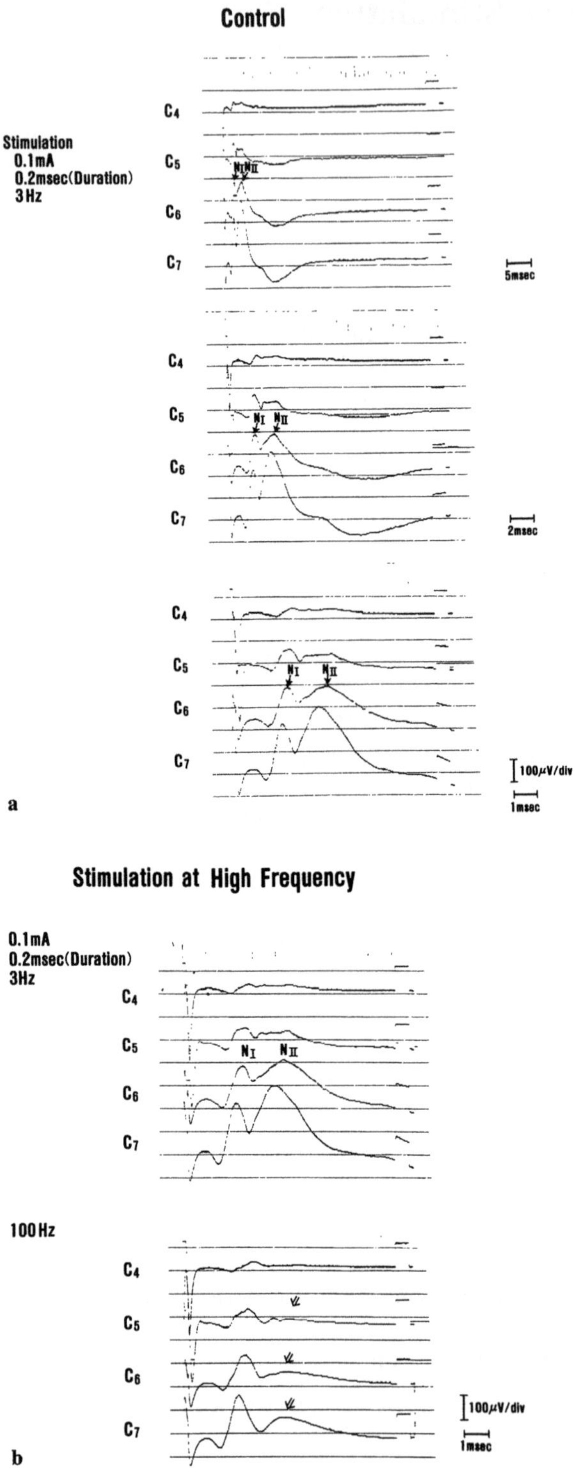

Fig. 1. a Cervical spinal cord potentials evoked by median nerve stimulation. The analysis time was 50 ms, 20 ms and 10 ms in the upper, middle and lower tracings, respectively. **b** The NII waves were remarkably diminished by stimulation at 100 impulses per second

3. Transection of the spinal cord,
4. Alcohol injection of the spinal cord.

The DISA digital 1500 system was used for the stimulation and recordings.

Results and Discussion

The analysis times used were 50ms, 20ms and 10ms. The early component of cervical spinal cord potentials evoked by median nerve stimulation was recorded at each cervical segment as a negative wave having two peaks (NI and NII). The inter-peak latency between NI and NII was approximately 2.0ms. Both NI and NII waves increased in latency as the recording electrode was moved rostrally. The conduction velocity of the NI wave was approximately 50m per second (Fig. 1a). When the median nerve was stimulated at 100 impulses per second, the NI component decreased in amplitude by 20% but NII was more markedly attenuated (Fig. 1b). The NI wave was therefore thought to reflect mainly non-synaptic potentials, while NII was thought to originate in synaptic or post-synaptic potentials.

The posterior roots were divided rostrally at the point of entry to the spinal cord in rostral-caudal sequence. The potentials did not change following division of the posterior roots of C4, C5 and C6. Both NI and NII decreased in amplitude by 30% after division of the C7 posterior root. After division of the C8 posterior root, NI was diminished by 80% and NII was also greatly reduced. Finally, following division of the T1 posterior root, both NI and NII disappeared (Fig. 2).

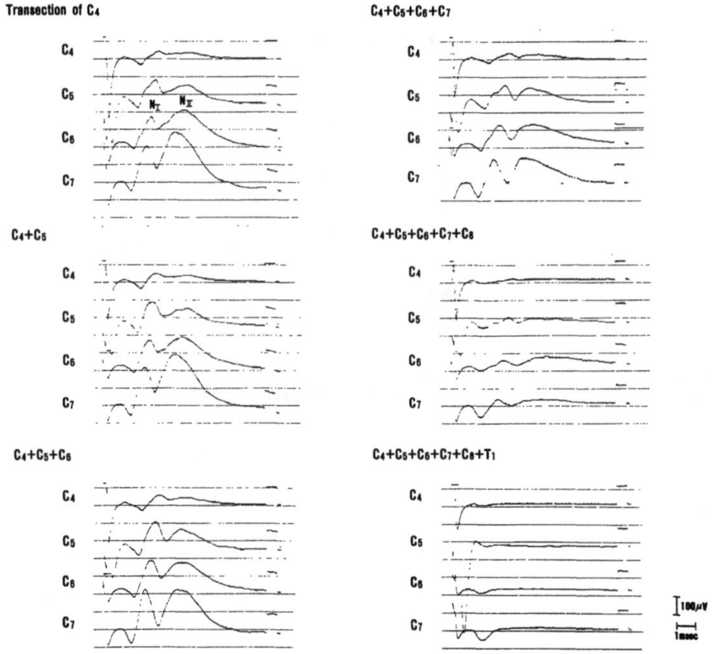

Fig. 2. Transection of posterior roots (rostral-caudal)

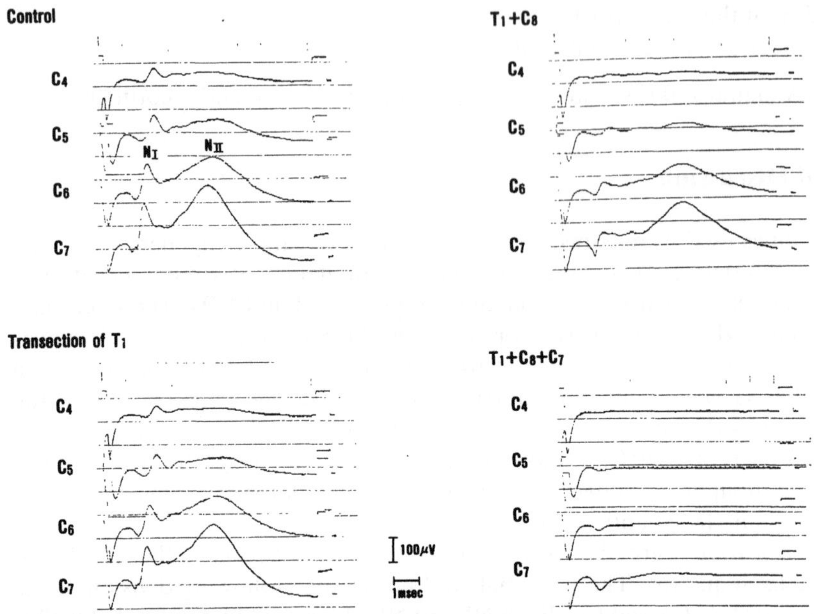

Fig. 3. Transection of posterior roots (caudal-rostral). According to these results, cervical spinal cord potentials evoked by median nerve stimulation appear to originate in the posterior roots of C7, C8 and T1 (especially C8)

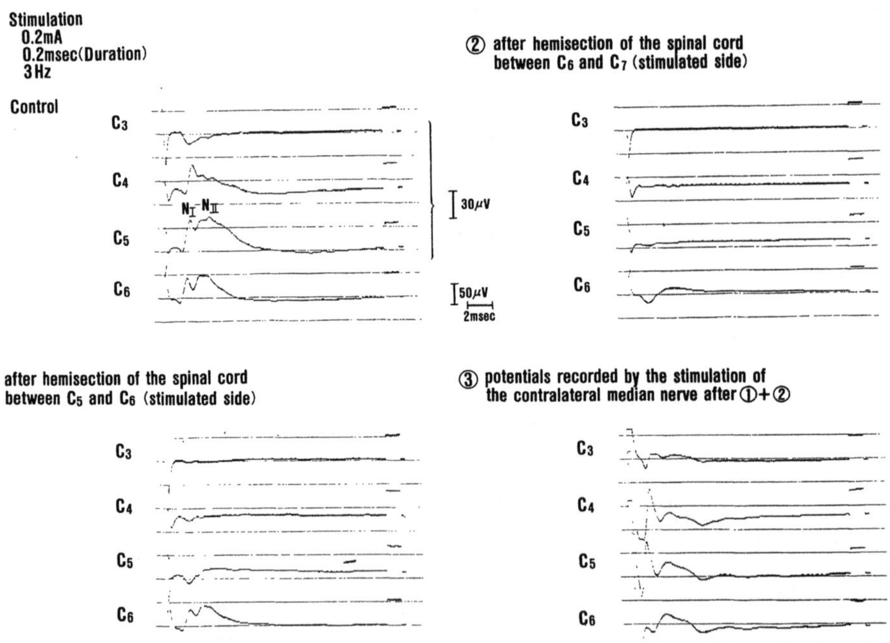

Fig. 4. Hemisection of the spinal cord on the stimulated side

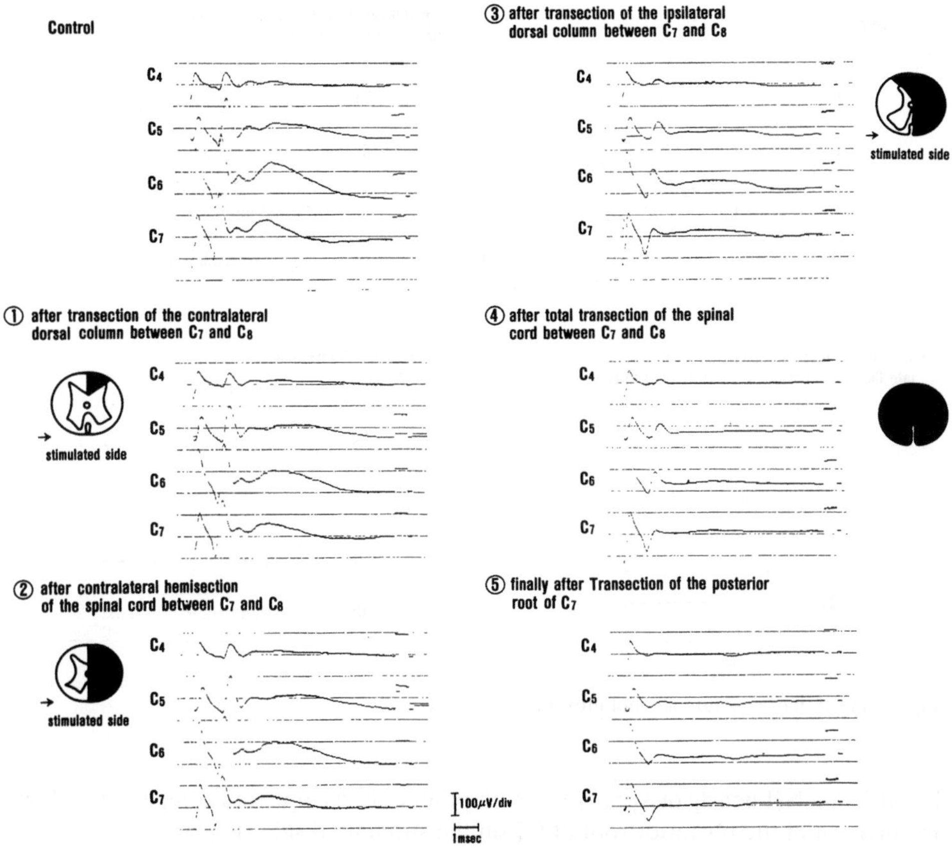

Fig. 5. Hemisection of the spinal cord on the side contralateral to stimulation. These results suggest that NI and NII reflect posterior root potentials and spinal cord potentials respectively, both on the stimulated side

The posterior roots were also divided in caudal-rostral sequence. On division of the T1 posterior root, NI decreased in amplitude by 10% but the NII component did not change. After division of the posterior root of C8, NI and NII were diminished by 80% and 30%, respectively. Finally, NI and NII disappeared entirely after division of the C7 posterior root (Fig. 3). According to these results, cervical spinal cord potentials evoked by median nerve stimulation in the cat appear to originate in the posterior roots of C7, C8 and T1 (especially C8). After hemisection of the spinal cord between C5 and C6 on the stimulated side, the spinal cord potentials disappeared rostral to C5. At C5 NI was of inverted (positive) polarity. After hemisection of the spinal cord between C6 and C7 on the stimulated side, NII disappeared and NI became positive in polarity. When the median nerve was stimulated on the contralateral side, evoked spinal cord potentials were recorded at every cervical segment (Fig. 4). After hemisection of the spinal cord between C7 and C8 on the side contralateral to the stimulus, both NI and NII decreased in amplitude by 20%. Following this, when the dorsal columns were transected on the stimulated side NI decreased

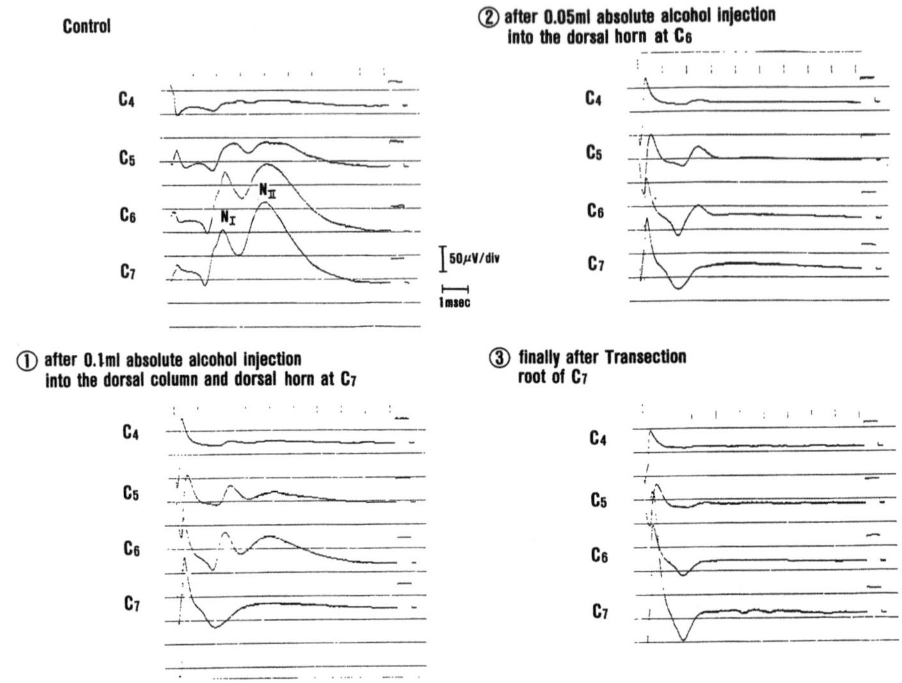

Fig. 6. Injection of absolute alcohol into the spinal cord

by 60% and NII nearly disappeared. Both potentials disappeared completely following division of the posterior root of C7 on the stimulated side (Fig. 5).

After myelotomy at the midline of the spinal cord from C6 to C7, no change in the evoked spinal cord potentials was observed. These results suggest that NI and NII reflect posterior root and the spinal cord potentials respectively, both generated on the side ipsilateral to stimulation.

After injection of the dorsal columns and dorsal horns at C7 with 0.1 ml absolute alcohol, NI recorded at C7 became positive in polarity and NII was remarkably reduced in amplitude. Both NI and NII were decreased in amplitude by 50% at C4, C5 and C6. A transverse section of the spinal cord at the injected level showed that parts of the dorsal columns and dorsal horns were poorly stained. After a further 0.05 ml of absolute alcohol was injected into the dorsal horns at C6, NI decreased by 15% and 50% of amplitude at C5 and C6, respectively. NII disappeared at both levels. A transverse section of the spinal cord at the injected level showed that part of the dorsal horns was poorly stained. Finally, after division of the C7 posterior root the evoked potentials disappeared at each cervical segment (Fig. 6). The results suggest that NI reflects the potentials of primary afferent fibres ascending in the dorsal columns on the stimulated side, while NII originates in the synaptic or postsynaptic potentials of the dorsal horn on the stimulated side.

When the recording electrodes were attached to the spinal cord semicircularly on the stimulated side between C6 and C7, the more dorsally the electrodes were located, the more both NI and NII increased in amplitude. This suggests that the

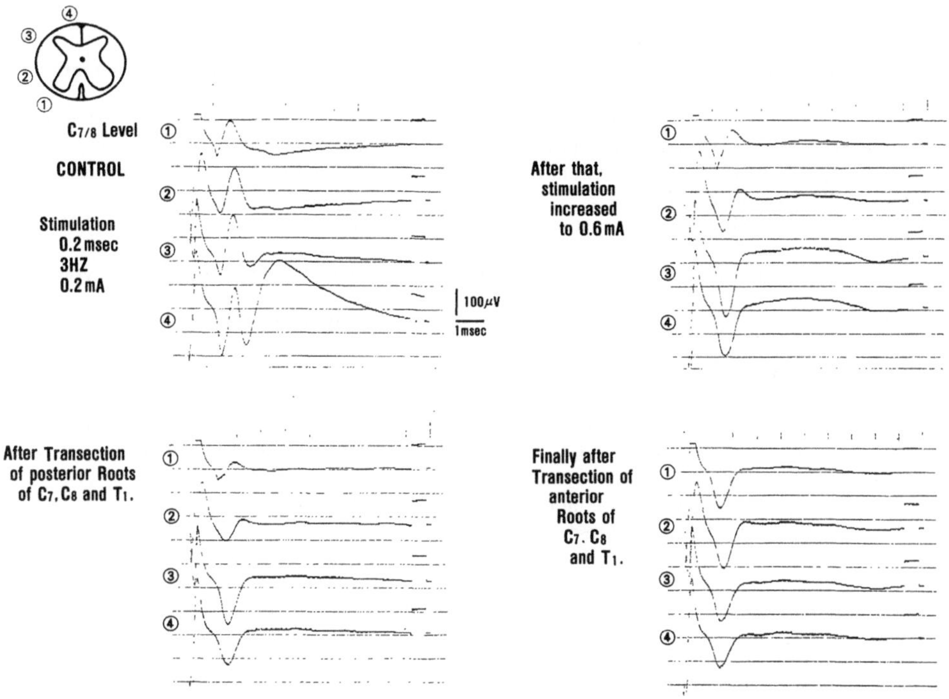

Fig. 7. Spinal cord potentials recorded by anteriorly located electrodes, suggesting that a portion of N1 originates in activity of the ventral roots

evoked spinal cord potentials at this level reflect the activity of the dorsal columns and of the dorsal horns at the rostral end of the entry zone of the median nerve. When the recording electrodes were attached to the spinal cord semicircularly on the stimulated side between C7 and C8, NI did not increase in amplitude at dorsal location as much as was seen at C5 and C6. This suggests that the NI wave originates not only in the spinal cord, but also in posterior roots of C7 and C8. Conversely, the NII wave increased in amplitude as the recording electrodes were located more dorsally. This supports the belief that NII originates in the dorsal horns of the spinal cord. When the electrodes were attached circularly to the spinal cord between C7 and C8, the potentials were hardly recorded at the lateral and ventral electrodes on the side contralateral to the stimulus. This supports the evidence that spinal cord potentials evoked by median nerve stimulation reflect the function of the stimulated side of the spinal cord.

When the recording electrodes were attached semicircularly to the spinal cord on the stimulated side, the evoked potentials recorded by the dorsally located electrodes disappeared after transection of the posterior roots of C7, C8 and T1. The NI wave recorded at ventrally located electrodes did not disappear but increased in amplitude after the stimulus was augmented (Fig. 7). This suggests that if the recording electrodes were placed ventrally, the potentials originating in the ventral roots would be recorded as NI waves.

References

1. Austin GM, McCouch GP (1955) Presynaptic component of intermediary cord potential. J Neurophysiol 18:441–451
2. Bernhard CG (1953) The spinal cord potentials in leads from the cord dorsum in relation to peripheral source of afferent stimulation. Acta Physiol Scand [Suppl] 106:1–29
3. Gasser HS, Graham HT (1933) Potentials produced in the spinal cord by stimulation of dorsal roots. Am J Physiol 103:303–320
4. Gelfan S, Tarlov IM (1955) Differential vulnerability of spinal cord structures to anoxia. J Neurophysiol 18:170–188
5. Lloyd DPC, McIntyre AK (1949) On the origins of dorsal root potentials. J Gen Physiol 32: 409–443
6. Saiki K (1979) Spinal evoked potential (SEP) obtained by stimulation on the median nerve—experimental and clinical studies. J Jpn Orthop Soc 13:1893–1913
7. Shimoji K (1972) Evoked spinal electrograms recorded from epidural space in man. J Appl Physiol 33:68–471
8. Willis WD Jr (1984) Evoked spinal cord potentials in the cat and monkey: use in the analysis of spinal cord function. In: Fundamentals and clinical application of spinal cord monitoring. Saikon Publishing, Tokyo, pp 3–19
9. Yoshida M (1973) The evoked spinal electrogram (SEG) in patients with spinal cord lesions. J Clin Neurol 13:745–752 (in Japanese)

Descending Lumbosacral Cord Potentials (DLCP) Evoked by Stimulation of the Median Nerve

Y. SARICA[1] and C. ERTEKIN

Summary

In 22 normal human subjects, descending lumbosacral cord potentials (DLCP) were recorded intrathecally after stimulation of the median nerve at the elbow. The onset of DLCP is very short in latency (mean 12.1 ms) with a prominent sharp early positive peak (mean latency 13.7 ms) followed by a sharp negative peak (mean 17.6 ms). The amplitude of the first part of DLCP varied between 0.6 and 6.7 microvolts (mean 2.3 microvolts).

The response was recorded most easily when the tip of the intrathecal electrode was positioned posterolaterally. The threshold of the response was above or around the excitation threshold of the motor nerve fibers and it could not be produced by pure cutaneous nerve stimulation. DLCP was resistant to peripheral nerve shocks at subtetanic frequencies. The mean peripheral conduction velocity of fibres responsible for the response was about 60 m/s. Occasional late and more slowly conducted deflections appeared in many cases.

It was concluded that the DLCP must originate in the descending and very fast-conducting propriospinal pathways. These are located within the anterolateral funiculus, which has an oligosynaptic anatomical organisation. The DLCP appears to be the first direct evidence in man of interlimb reflex action between the arm and leg, which is important in the coordination of movement and posture.

Introduction

The evoked electrospinogram recorded from the lumbosacral spinal cord in man is usually obtained by stimulation of the leg nerves and is composed of electrical activity in the dorsal root fibers, presynaptic activity of the dorsal columns and the segmental postsynaptic activity of the dorsal horn (Ertekin 1976, 1978). When the electrodes are placed over segments above the lumbosacral cord, i.e. the upper thoracic or cervical region, ascending evoked spinal cord potentials with longer latency can be recorded intrathecally (Ertekin 1973, 1976a, 1978), epidurally (Shimoji et al. 1972) or superficially (Cracco 1973; Cracco et al. 1975; and others). Although details concerning the origin of this activity are debatable and the results are somewhat variable depending on the methods applied, there is no doubt that it comes from the ascending long tracts.

[1]Department of Neurology, Medical School Hospital, Aegean University, Bornova-Izmir, Turkey. *Present address:* Sleep Disorders Center, Baylor College of Medicine, Houston, TX, USA

Spinal Cord Monitoring
Ed. by J. Schramm and S. J. Jones

It is also possible to record a sequence of potentials at the lumbosacral cord level, after weak electrical stimulation of the cervical enlargement in man by either intrathecal (Ertekin 1976b) or epidural (Shimizu et al. 1979) techniques. Although various afferent and efferent tracts might be activated antidromically and orthodromically, the physiological basis and anatomical structures involved in these potentials are not precisely known as yet.

"Descending lumbosacral cord potentials" (DLCP) elicited by stimulation of arm nerves have not hitherto been studied in man. Experimentally, it has long been known that there are connections between the cervical and lumbosacral enlargements which permit sensory integration of forelimb and hindlimb activity (Gernandt and Shimamura 1961; Kots 1977; Miller et al. 1973). The effects elicited by cervical spinal afferent nerve stimulation upon lumbosacral motorneurone activities are due to the impulses transmitted along the descending propriospinal projections (Lloyd 1942; Schomburg et al. 1978). There is also indirect evidence of descending propriospinal reflex activity coming from the upper limb and influencing the lumbosacral motorneurone pools in humans, as demonstrated by means of different EMG methods (Gassel and Ott 1973; Kots 1977; Piesieur-Strehlow and Meinck 1980).

In this paper, a DLCP is described and its possible physiological significance is discussed in the light of propriospinal descending interlimb reflex activity. A preliminary account is to be published elsewhere (Sarica and Ertekin 1985).

Material and Methods

Twenty-two patients (21 males and 1 female) were investigated. They were all inpatients in the wards of our neurology department. The ethical principles of the Declaration of Helsinki (1964) concerning human experimentation were followed throughout the study. The ages of subjects ranged from 18 to 55 years. None of the patients had clinical signs or symptoms related to diffuse peripheral nerve or spinal cord disease. Their spinal cord potentials were recorded at the level of lumbosacral segments following stimulation of the posterior tibial nerve and the median nerve at the elbow on both sides. In some cases the ulnar nerve at the elbow, the (median) digital nerve at the second digit and the median nerve at the wrist were also stimulated. Computer analysis times were usually 50 and 200 ms, but 500 ms was also used in some cases. Sixty-four to 128 responses were averaged for segmental lumbosacral spinal cord potentials while 128 to 256 responses were needed for DLCP. In some subjects the EMG activity of erector spine muscles was simultaneously recorded by a long concentric needle electrode inserted at the same level as the intrathecal electrode.

The position of the tip of the intrathecal recording electrode was crucially important for the DLCP; when the electrode was positioned laterally clear-cut averaged responses could be obtained. Therefore great care was taken to assure the laterality of the intrathecal electrode in the majority of cases.

The position of the intrathecal electrode tip was judged by the following criteria (Ertekin 1976a): 1) During insertion of the intrathecal electrode, rectangular electrical pulses (0.1 ms, 1–2 mA) were delivered at a rate of 1/s. When the electrode tip entered into the lateral epidural space, unilateral sensory-motor clinical phenomena were immediately observed in the lower limb ipsilateral to the electrode, whereas bilateral clinical phenomena were observed when the electrode was positioned at the midline. 2) Laterality of the clinical segmental responses during intrathecal stimulation was always associated with a DR potential of high amplitude upon stimulation of the posterior tibial nerve at the ipsilateral side. The segmental evoked responses were of the CD type bilaterally in the case of midline electrode position.

Results

When the position of the intrathecal recording electrode tip was lateral to the midline at lower thoracic levels, the majority of normal subjects revealed clear-cut DLCPs which were evoked by stimulation of the median or ulnar nerves at the elbow as shown in Fig. 1 (upper trace on the left). However, similar evoked potentials were less frequently obtained with the electrode tip in a midline position. The DLCP always appeared upon stimulation of the median nerve ipsilateral to the electrode, when the segmental responses were either of the DR type or of the higher amplitude CD type. Stimulation of the contralateral median nerve produced either no potential or smaller potentials with comparable latency values.

The shape of the DLCP was somewhat variable from case to case, but was constant in a given subject. The DLCP was almost always recorded with positivity predominating. The deep positive peak of the potential had a mean latency of 13.7 ms, varying between 12.0 and 16.5 ms in normal subjects. The mean onset time was approximately 12.1 ms following median nerve stimulation at the elbow (range 9.5–14.8 ms). Following the positive deflection there was a negative peak which appeared at a mean latency of 17.6 ms (range 13.8–22.0 ms). The first positive and

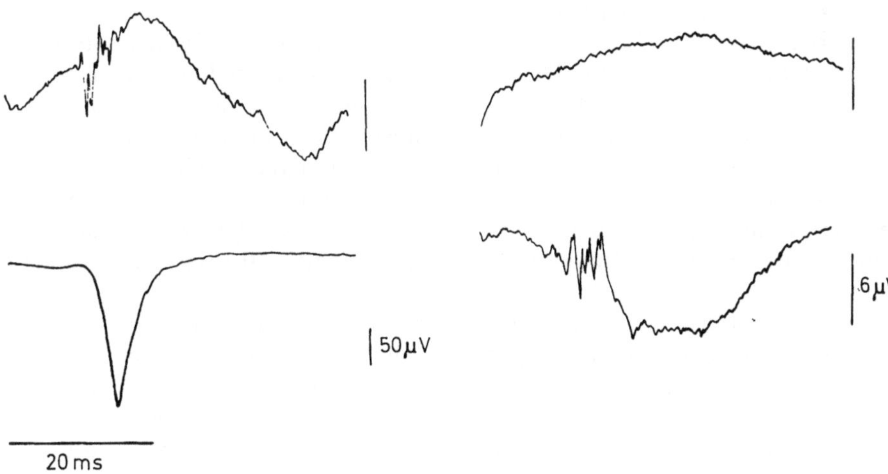

Fig. 1. Descending *(upper traces)* and segmental *(lower traces)* evoked spinal cord responses recorded at the level of the lumbosacral cord. Stimulation sites are the median nerve at the elbow and the posterior tibial nerve on the left and right sides. Note the appearance of the descending response to stimulation of the median nerve ipsilateral to the position of the intrathecal recording electrode (50 μV calibration applies only to the lower left trace)

Table 1. Latency and amplitude values of DLCP

	Latencies (ms)			
	Onset	First positivity	Negativity	Second positivity
n	22	22	22	22
Mean	12.1	13.7	17.6	27.6
SD	1.5	1.3	2.4	6.9
Range	9.5–14.8	12.0–16.5	13.8–22.0	17.0–43.0

	Amplitudes (microvolts)		
	Onset to first positivity	First positivity to negativity	Negativity to second positivity
n	22	22	22
Mean	1.5	2.3	2.5
SD	0.8	1.6	1.6
Range	0.6–3.6	0.6–6.7	0.6–8.4

n, number of cases; SD, standard deviation

the following negative peaks were very stable both interindividually and in a given subject, whereas the following second positivity and other late deflections were very labile and variable among subjects and between different recordings from the same subject. Therefore the evoked potential recorded from the lumbosacral cord after stimulation of the median nerve was a small sharp response with a very short latency. Statistical results are given in Table 1.

The peripheral origin of the DLCP was studied in some cases. Figure 2 shows the effect of stimulus strength on the DLCP. In all cases investigated, a stimulus intensity above the motor threshold of the thenar muscles invariably elicited the response. The DLCP was never observed below the threshold of motor nerve fibers. The amplitude of the evoked potential was also increased in parallel with the increase in stimulus strength, but no change in latency was seen. Stimulation of a pure sensory nerve such as interdigital branches of the median nerve did not produce any response, even if the stimulus intensity was two or three times the sensory perception threshold.

The DLCPs obtained by stimulation of the median nerve at the level of the elbow and the wrist were compared in 9 subjects. The shape of the DLCP was not changed substantially on stimulation of the median nerve at the wrist, but its latency was increased (Fig. 3). The mean conduction velocity of afferent fibers which are probably responsible for the evoked DLCP was 60.8 m/s and varied from 51 to 80 m/s.

In 6 cases, pulse trains of various frequencies were also applied to the median nerve in order to observe their effects on the potential. The results should be evaluated cautiously, because stimulation at tetanic frequencies was found unpleasant by some patients and the position of the tip of the intrathecal recording electrode may have been moved from its original position. Nevertheless, the results obtained from two

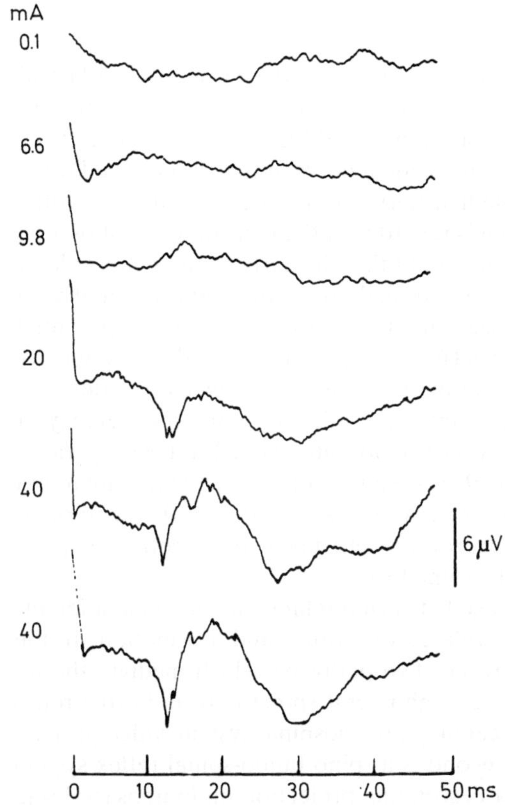

Fig. 2. Descending spinal cord evoked potential recorded at the lumbosacral level with different intensities of stimulation of the median nerve at the elbow. The numbers on the left denote the stimulus strength in mA. No thenar twitch was observed at the two lowest intensities

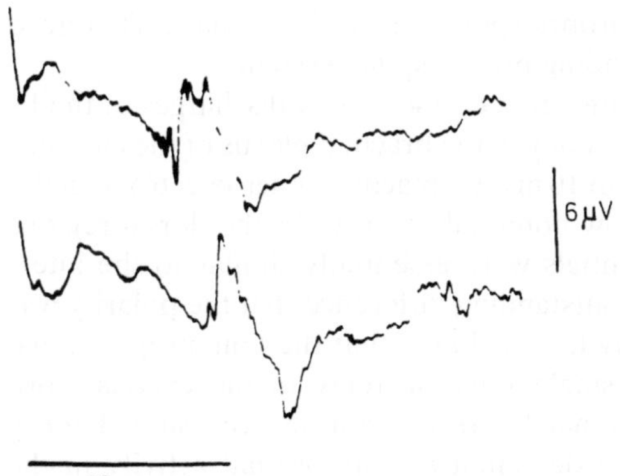

Fig. 3. Descending spinal cord evoked responses at the level of the lumbosacral cord after stimulation of the median nerve at the elbow *(upper trace)* and wrist *(lower trace)*. Note the conduction time differences between the two stimulation sites, with shorter latency from the elbow

cases were convincing. When the stimulus frequency was over 5–10 per second, the DLCP was diminished in amplitude, and the response disappeared beyond the frequencies of 30–50 per second.

Late waves following the short latency components of the DLCP were encountered in some subjects, but will not be described in this paper.

Discussion

When a needle intrathecal recording electrode is introduced into the posterolateral subarachnoid space at the level of the lumbosacral spinal cord in normal human subjects, a clear-cut, small, positive-negative sharp wave (DLCP) with a mean latency of 13.7 ms is obtained upon stimulation of the median nerve at the elbow. The DLCP is less prominent or absent when the median nerve contralateral to the recording electrode is stimulated. Since the potential is recorded at the level of the lumbosacral cord and its latency is very short, there is no doubt that the early components at least are evoked by activation of fibers descending directly from the cervical spinal cord to the lumbosacral segments. This response cannot be a volume-conducted potential originating in surrounding excitable tissues (for example, the evoked EMG activity of the erector spinae muscles), since the paravertebral muscles which are reflexively activated by stimulation of nerves in the upper extremity have an onset latency of about 50–80 ms in the case of "microreflexes or spino-bulbo-spinal reflexes" (Cracco and Bickford 1968; Ertekin and Ertekin 1975; Meier-Ewert et al. 1972; Shimamura et al. 1964). Furthermore no time-locked response was recorded in the deep paravertebral muscles when simultaneous recordings were made from both the spinal cord and the erector spinae muscles at the same level.

In animal studies, a sensory input linkage between forelimb and hindlimb afferents has been demonstrated and found to be related to posture and locomotion. In the case of forelimb nerve stimulation, there are two pathways which mediate the influence of forelimb sensory inputs on the lumbosacral spinal cord motorneurones and interneurones. One of these is a descending propriospinal system which projects exclusively to the ipsilateral cord. The second is a spino-bulbo-spinal reflex system involving a relay in the brainstem and a descending projection on lumbosacral segments (Shimamura and Akert 1965; Shimamura and Livingstone 1963). Since the DLCP recorded from the lumbosacral cord has a very short latency, it can only be compared with the directly descending propriospinal system in animals. Therefore we will focus our discussion on the descending propriospinal system.

Lloyd (1942) showed that the descending effects assessed by discharges of the L_7 motorneurones in response to an afferent volley in the brachial plexus of the cat were quite constant and rarely lasted longer than 10 ms. By placing a microelectrode in the L_7 segment of the cord he recorded a slow potential, positive in the dorsal regions and negative in the ventral. These potentials were essentially similar to the intermediary dorsal horn potentials, with the outstanding difference that the polarity was reversed in the dorso-ventral axis. Finally he concluded that the long propriospinal fibers were involved and there was a single synaptic relay in the cervical cord between primary afferent and propriospinal fibers transmitting the short-latency interlimb reflex activity. The role of the descending propriospinal activity in the reflex action of the lumbar motorneurones was later proved by further studies (Gernandt and Megirian 1961; Gernandt and Shimamura 1961; Lloyd and McIntyre 1948; Shimamura and Akert 1965; Shimamura and Livingstone 1963; and others). The DLCPs recorded from human subjects in the present study have very similar features to potentials elicited by descending volleys coming from propriospinal pathways in animals. Whatever its anatomical basis, this response must be propagated within very fast-conducting route(s), both in the periphery and along the spinal cord.

Electrical stimulation of the cervical cord has produced evoked responses of very short latency (4–7 ms) at the lumbosacral cord segments, as demonstrated by an intrathecal approach in man (Ertekin 1976b). Also, the fastest afferent nerve volley was found to reach the human cervical cord at about 6.5 ms (Ertekin 1976a). Thus the average latency of the DLCP (13.7 ms) is in accord with the transmission of afferent impulses in a fast pathway in both the periphery and the spinal cord (6.5–6–12.5 ms), having only one or two relays between the stimulation and recording sites selected in this study. This result is also compatible with the descending propriospinal system for the interlimb reflex, which is oligosynaptic (Lloyd 1942; Miller et al. 1973; Schomburg et al. 1978). Indeed the DLCP was quite resistent to repetitive peripheral stimuli and disappeared only at high frequencies; an observation compatible with oligosynaptic linkage.

The peripheral origin of DLCP was investigated and found to be related to stimulation of muscle and joint afferents, since neither excitation of pure sensory nerves nor stimulation below the motor threshold was found to produce the response. The conduction velocity along the forearm calculated for the descending response was found to be 60.8 m/s on average. This indicates that DLCP is elicited by muscle afferents (probably also joint afferents) with fast conduction velocity. This result is in accord with experimental findings that the medium and large size muscle afferents are mainly responsible for the propriospinal interlimb responses (Lloyd 1942; Lloyd and McIntyre 1948; Schomburg et al. 1978).

In conclusion, the DLCP recorded at the level of the lumbosacral cord segments exists in man, and is likely to be due to the activity of propriospinal descending pathways ipsilateral to the stimulated arm. It seems functionally important in the very fast transmission of sensory information from arm to the lumbosacral cord and is probably related to interlimb reflex integration involving posture and locomotion which had previously been indirectly demonstrated in man (Gassel and Ott 1973; Kearney and Chan 1981; Kots 1977; Piesieur-Strehlow and Meinck 1980).

References

Cracco RQ (1973) Spinal evoked response: peripheral nerve stimulation in man. Electroenceph Clin Neurophysiol 35:379–386

Cracco RQ, Bickford RG (1968) Somatomotor and somatosensory evoked responses. Arch Neurol 18:52–68

Cracco JB, Cracco RQ, Graziani L (1975) Spinal evoked response in infants and children. Neurology (Minneap) 25:31–36

Declaration of Helsinki (1964) Recommendations guiding doctors in clinical research. Wld Med 11:281

Ertekin C (1973) Human evoked electrospinogram. In: Desmedt JE (ed) New developments in electromyography and clinical neurophysiology, vol 2. Karger, Basel, pp 344–351

Ertekin C (1976a) Studies on the human evoked electrospinogram. I—The origin of the segmental evoked potentials. Acta Neurol Scand 53:3–20

Ertekin C (1976b) Studies on the human evoked electrospinogram. II—The conduction velocity along the dorsal funiculus. Acta Neurol Scand 53:21–38

Ertekin C (1978) Comparison of the human evoked electrospinogram recorded from the intrathecal, epidural and cutaneous levels. Electroenceph Clin Neurophysiol 44:683–690

Ertekin N, Ertekin C (1975) Long loop myostatic and spino-bulbo-spinal reflexes. Aegean Med J 3:79–88

Gassel MM, Ott KH (1973) Patterns of reflex excitability change after widespread cutaneous stimulation in man. J Neurol Neurosurg Psychiat 36: 282–287

Gernandt BE, Megirian D (1961) Ascending propriospinal mechanisms. J Neurophysiology 24: 364–376

Gernandt BE, Shimamura M (1961) Mechanism of interlimb reflexes in the cat. J Neurophysiology 24: 665–676

Kearney RE, Chan CWY (1981) Interlimb reflexes evoked in human arm muscles by ankle displacement. Electroenceph Clin Neurophysiol 52: 65–71

Kots YM (1977) The organisation of voluntary movements. Neurophysiological mechanisms. Plenum Press, New York, pp 189–212

Lloyd DPC (1942) Mediation of descending long spinal reflex activity. J Neurophysiology 5: 435–458

Lloyd DPC, McIntyre AK (1948) Analysis of forelimb-hindlimb reflex activity in acutely decapitated cats. J Neurophysiology 11: 455–470

Meier-Ewert K, Muemme U, Dahm J (1972) New evidence favouring long loop reflexes in man. Arch Psychiat Nervenkr 215: 121–128

Piesieur-Strehlow B, Meinck HM (1980) Response patterns of human lumbosacral motoneuron pools to distant somatosensory stimuli. Electroenceph Clin Neurophysiol 48: 673–682

Sarica Y, Ertekin C (to be published) Descending lumbosacral cord potentials evoked by stimulation of the median nerve. Brain Res

Schomburg ED, Meinck HM, Haustein J, Roesler J (1978) Functional organisation of the spinal reflex pathways from forelimb afferents to hindlimb motoneurones in the cat. Brain Res 139: 21–33

Shimamura M, Akert K (1965) Peripheral nervous relation of propriospinal and spino-bulbo-spinal reflex systems. Jap J Physiol 15: 638–647

Shimamura M, Livingstone RS (1963) Longitudinal conduction systems serving spinal and brainstem coordination. J Neurophysiology 26: 258–272

Shimamura M, Mori S, Matsushima S, Fujimori S (1964) On the spino-bulbo-spinal reflex in dogs, monkeys and man. Jap J Physiol 14: 411–421

Shimoji K, Kano T, Higashi H, Marioka T, Henschel EO (1972) Evoked spinal electrograms recorded from epidural space in man. J Appl Physiol 33: 468–471

Shimizu H, Shimoji K, Maruyama Y, Sato Y, Harayama H, Tsubaki T (1979) Slow cord dorsum potentials elicited by descending volleys in man. J Neurol Neurosurg Psychiat 42: 242–246

Extraspinal Stimulation and Recording: A Method for Operative Monitoring

M. H. BENNETT[1] and O. N. AKIN

Introduction

Although the monitoring of cortical Somatosensory Evoked Potentials (SEPs) during operations on the spine and spinal cord of humans has proven to be a useful surgical adjunct [3, 4, 7, 8, 26], the technique's well known limitations and the occasional reports of false findings [1, 16, 21] have led to active investigations and operative monitoring applications of the recording of spinal cord evoked potentials (SCEPs). The level of activity in this area is well documented by the previous [10] and present international symposium on the subject.

While the recording of ascending SCEP to peripheral nerve stimulation by epidural [20] bone [15] or skin [6] placement of recording electrodes offers certain advantages over SEP monitoring, it is only with the techniques employing spinal cord stimulation that measures of activity conducted within the ventral half of the spinal cord become available [12]. Because of the differences in source and distribution of the blood supply to dorsal and ventral halves of the spinal cord [25] and the differential effects produced by interrupting that supply [2], monitoring of activity mediated ventrally is of profound clinical significance.

Stimulation of the spinal cord and SCEP monitoring by means of epidural electrodes inserted with a catheter has been most frequently applied. It has proven to be a safe and effective technique used in hundreds of cases [11, 19, 22, 24]. Less often, direct placement of electrodes on the pial surface [13] or in the disc space and yellow ligament have been used [17]. In the expectation that a less invasive technique would carry even lower risk of possible iatrogenic injury and still provide the necessary information for monitoring, a technique of spinal cord stimulation and SCEP monitoring with electrodes located close to, but outside the spine was investigated.

We now report preliminary findings on descending and ascending SCEPs obtained with extraspinal stimulating and recording electrodes.

Methods

Cats and monkeys were anesthetized with sodium pentobarbital (35 mg/kg), injected with atropine, immobilized with gallamine triethiodide and artifically respired. A femoral artery catheter was used for blood pressure measurement and withdrawal of blood for analysis. Necessary steps were taken to maintain PO_2, PCO_2, pH, and base

[1] University of Pittsburgh, School of Medicine, Pittsburgh, PA 15213, USA

Spinal Cord Monitoring
Ed. by J. Schramm and S. J. Jones

concentrations within the normal ranges. Body temperature was maintained with a heating pad.

The sciatic nerve was stimulated by either placement of platinum needle electrodes (Grass) within the epineurium of the nerve or by hypodermic needles percutaneously applied at the sciatic notch. The spinal cord was stimulated by extraspinal hypodermic needles or insulated, 3 mm exposed tip, stainless steel wires or by direct epidural placement of stimulating electrodes. Stimuli consisted of 50 microsecond square wave pulses delivered at 10 Hz with intensity adjusted to produce a maximal amplitude of the first peak complex. Recordings were performed with epidural or extraspinal wire electrodes or stereotaxically directed tungsten electrodes (50 micron exposed tip). Standard amplification with a 3 Hz to 3 kHz bandwidth, signal averaging and recording techniques were employed. Lesions of the spinal cord were made with radio-frequency currents or direct surgical techniques. Paraffin embedded, hematoxylin and eosin stained sections of the spinal cord were used to assess lesions and determine the location of intraspinal recording electrodes.

Spinal cord evoked potentials in humans in response to peroneal nerve stimulation were recorded during operations in which a balanced anesthetic technique with muscle blockade was used. Recordings were performed with extraspinal wire electrodes directed 3–5 mm parasagitally just caudal to the facet or toward the intervertebral foramen at the T_{10-11} vertebral junction. Stimuli consisted of 0.1 to 0.5 ms square wave pulses delivered at a 3 to 10 Hz rate. Standard signal averaging techniques were employed.

Variations in the above stimulating and recording parameters or arrangements are presented where appropriate.

Results

The ascending spinal cord evoked potentials (SCEPs) recorded epidurally at C_2 in response to sciatic nerve stimulation and the descending SCEPs recorded epidurally at L_1 in response to C_2 spinal cord stimulation with extraspinally located electrodes in a cat are shown in the top two traces in Fig. 1.

The descending response consisted of an initial positive-negative peak complex followed by one, two or more smaller positive-negative complexes. The ascending response to sciatic nerve stimulation consisted of an initial positive-negative complex and an extended series of peaks. Based on the latency of the first positive peaks, the conduction velocities of these responses ranged from 57 to 71 m/s for the ascending responses and from 95 to 117 m/s for the descending responses. Stimulation of both the cervical spinal cord and either sciatic nerve, timed to produce collison of responses of common elements, demonstrated independence of the ascending and descending responses. Hemisection of the cord from the midline laterally reduced the amplitude of the descending response by close to 100%, 50% or had virtually no effect upon it. These results depended entirely upon the location of the spinal cord stimulation electrodes producing primarily unilateral or bilateral activation of the descending elements. With both stimulating electrodes near the intervertebral foramen of each side bilateral stimulation was obtained. With the anode placed subcutaneously several centimeters laterally unilateral stimulation was produced. In

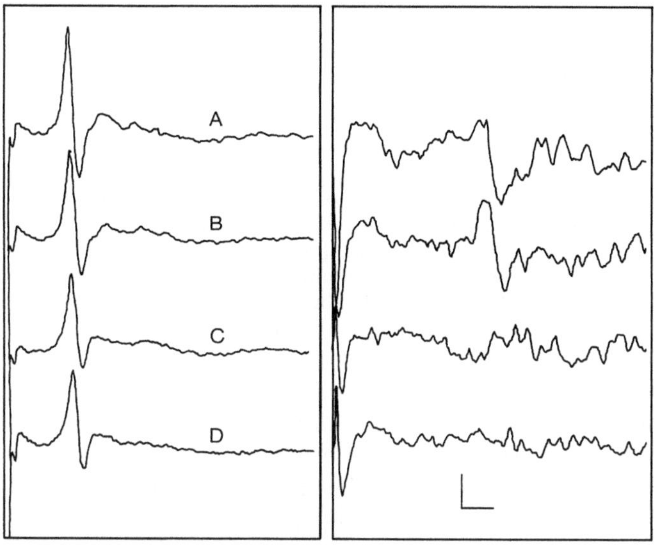

Fig. 1. Descending SCEPs recorded at L_1 in response to C_2 extraspinal stimulation *(left)* and ascending SCEPs recorded at C_2 to sciatic nerve stimulation *(right)*. *A* Control, *B* following attempted dorsal hemisection, *C* dorsal hemisection extended to central canal, *D* fifteen minutes following second cut. Calibration: Vertical; 6.4 μV *(left)* and 1.0 μV *(right)*, positivity upward. Horizontal; one millisecond

addition to the observations following hemisection, it was observed that the descending response was lost or significantly reduced with stimulus polarity reversal when using the unilateral electrode placement. Such an effect did not occur with electrodes across the spine near each foramen. Transection of the dorsal half of the spinal cord, histologically verified to extend to the level of the central canal, abolished the ascending responses, whereas 70 to 80% of the pre-section descending response remained. Similarly, radio-frequency lesions of the dorsal columns which abolished the ascending responses had no effect upon the descending responses. The effects of the dorsal hemisection on the SCEPs are shown in Fig. 1. The first shallower cut abolished much of the later peaks of the ascending response without affecting the initial positive-negative spike and produced a 15% reduction of the descending response. The second cut, which extended the lesion to the central canal produced the differential loss of response described above. In a variation of the basic preparation, a hemisection at the midthoracic level was followed by insertion of an exploring electrode into the intact half of the cord at the same level. Various components of the descending response rose to maxima as the electrode was advanced and several peaks not seen at the cord surface were observed. Maxima occurred most frequently deep within the ventral and ventrolateral funiculi of the cord and once in the area of the cortico- and rubrospinal tracts. In eight experiments, discrete radio-frequency lesions were made at these maxima points and the effect upon the descending response recorded at lumbar levels was observed. When the lesions produced a change, it amounted to a 10–50% reduction of the descending response remaining after the prior contralateral hemisection. Many such lesions were without effect.

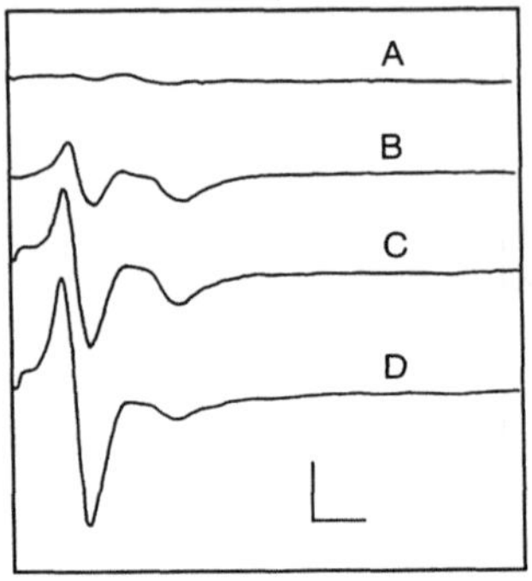

Fig. 2. Ascending SCEPs recorded at T_{4-5} in response to extraspinal stimulation about L_{2-3} vertebral junction. *A* At the midline in front of vertebral process, *B* same but close to the ligamentum flavum, *C* near intervertebral foramen, *D* at the intervertebral foramen. Calibration: Vertical; 64 μV, positivity upward. Horizontal; one millisecond

Effective stimulation of the cord and the recording of SCEPs was a function of electrode location. The vertebral junction produced more effective stimulation than locations more caudal or rostral. In Fig. 2 are shown the effects of various extraspinal stimulus locations about the L_{2-3} vertebral junction upon the ascending SCEP recorded at T_{4-5}. As can be seen, the largest response was obtained when the stimulating electrode was in the vicinity of the intervertebral foramen. The same results were seen for the descending response with cervical stimulation. The positioning of the SCEP recording electrodes was not as sharply defined with approximately equally large responses being obtained at the foramen and in the area near the facet. In all these latter tests the recording reference or stimulating return electrode was subcutaneous 3–4 cm lateral to the active electrode.

Other characteristics of the descending response should be noted. A test pulse following an equally intense conditioning stimulus by up to 15 ms augmented the amplitude and reduced the latency of the peaks following the initial positive-negative peak complex. This effect was most pronounced for the third positive peak (P_3) in which a sharp peak could be elicited on the rising phase of P_3. Also it was noted that the same waveform of descending SCEP elicited by extraspinal cervical stimulation was elicited by direct thoracic cord stimulation. These responses are shown in Fig. 3 in which the cervical stimulation is delayed with respect to the thoracic cord stimulation. At interstimulus intervals in which the ascending thoracically elicited response would collide with the descending cervically elicited response, the latter response was totally occluded.

Monkeys were used to model the electrode arrangement proposed for operative monitoring in which wires, insulated except for the tip, were placed percutaneously. High amplitude responses were readily obtained. The descending response elicited at C_2 and recorded at L_2 was calculated to have a conduction velocity of 100 m/s for the first positive peak. The response consisted of an early P-N-P sequence of peaks followed by a large late negative peak. The ascending response to sciatic stimulation recorded at T_{5-6} consisted of two initial positive-negative peak sequences followed by smaller multiple peaks. The conduction velocity of the first positive peak was calculated as 79 m/s and the second at 65 m/s.

Finally, partial application of these techniques to the clinical setting was provided by two cases of operations for mobilization of the peroneal nerve compressed by the pyramus muscle. A balanced anesthetic technique was used with N_2O, a narcotic

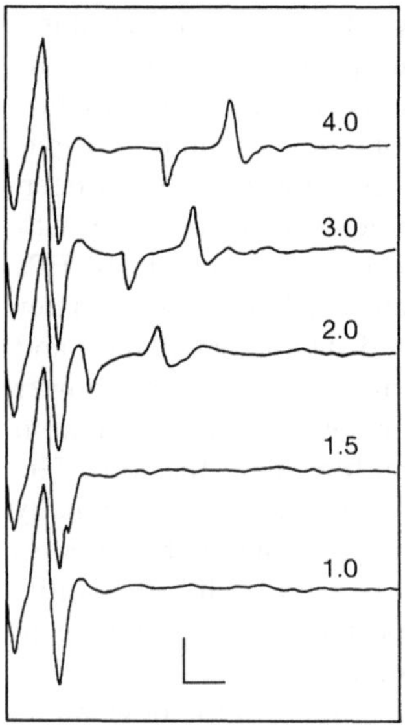

Fig. 3. Descending SCEPs recorded at L_4 in response to T_5 epidural spinal cord stimulation followed by C_2 extraspinal stimulation. Milliseconds delay of C_2 stimulation indicated on each trace. Calibration: Vertical; 80 μV, positivity upward. Horizontal; one millisecond

(Fentanyl) and a muscle blocking agent (Pavulon) since cortical SEPs were also recorded. Electrodes were placed percutaneously below the head of the fibula for stimulating the common peroneal nerve at a rate of 3–10 Hz. The active recording wire electrodes were percutaneously directed toward the T_{10-11} vertebral junction. One lead was directed toward the posterior surface of the spine 5 mm lateral to the spine ipsilateral to the side of stimulation. For comparative purposes the second lead was directed toward the ipsilateral intervertebral foramen. High amplitude responses were obtained in both patients and consisted of an initially positive triphasic wave. The largest peak to peak amplitude was approximately 2.2 μV in one patient and 0.7 μV in the other. There was no significant difference in waveform or amplitude of the SCEPs obtained at the two extraspinal recording electrode positions in each patient. The conduction velocities of these responses calculated for the prominent negative peak averaged 50 m/s. No changes were seen in the ascending SCEPs or cortical SEPs throughout these procedures and both patients maintained normal peroneal nerve function postoperatively.

Discussion

The primary findings of this work consist of the demonstration that descending SCEPs in cats and monkeys can be readily elicited by electrodes placed near, but exterior to the spine and that the primary peaks of this response are conducted by axons located mainly in the ventral half of the spinal cord. The descending SCEP elicited by epidural stimulation of the cord dorsum had a similar waveform and completely occluded the descending response obtained by extraspinal stimulation. This indicates that the descending neuronal elements activated by extraspinal stimulation were also activated by direct cord stimulation. The findings of independence of the descending response and the ascending SCEP by interaction and lesioning experiments indicates that a component of the descending response attributable to the dorsal columns was not identified by these methods. This result is at variance with results of other works in cats where strong evidence of dorsal column participation in the descending SCEPs was presented [9, 14, 18, 23]. However, it has been reported that the secondary descending peak is often difficult to obtain even with epidural stimulation [9] and others have shown significant differences in SCEPs depending upon the location of the cord stimulating electrodes [18]. Also the later peaks of the descending SCEP reported here demonstrated synaptic mediation and from their latency may be expected to mask the dorsal column peaks seen at lower cord levels. From all of these considerations we conclude that extraspinal stimulation preferentially activates axons within the lateral and ventral funiculi of the cord. The prior direct cord stimulation at a lower level completely occluding the extraspinally elicited descending response duplicates the findings of others in which both responses were elicited by epidural stimulation [9]. We therefore conclude that the initial positive-negative complex of the descending response reported here is analogous to the initial peaks obtained by others with epidural stimulation.

The identity of the tracts mediating the primary peak complex of the descending SCEP remains unknown. While there is general agreement with the findings that a proportion of the response is mediated via the dorsolateral segment of the cord and a greater proportion in the ventral half of the cord [9, 14, 18, 23] no single pathway has been shown to primarily mediate the response. The measurement of a maximal peak response deep within the thoracic cord only indicates that the activity is dependent upon the prior stimulation at higher levels. The response could be from collaterals or higher order neurons synaptically activated by the descending response. That is perhaps why the focus of many lesions at these maximum peak responses was without effect on the responses recorded caudal to the lesions. On the other hand, many other radio-frequency lesions made in this way did reduce, but never abolish the descending response. Of special significance is the fact that these lesions had a wide distribution over the lateral and ventral segments of the cord. These findings fostered the view that the descending response reflects activity in multiple pathways throughout the cord. Similar conclusions have been reached by others in experiments employing the epidural stimulating and recording techniques [23].

It is also noteworthy that the effect on the ascending C_2 response to sciatic nerve stimulation of the first dorsal column transection as illustrated in Fig. 1 was virtually to abolish the late polyphasic components without affecting the earlier larger positive-negative peak. These results are consistent with those of Satomi and Nishimoto [18] in which surgical destruction of the dorsal columns produced the same differential effect on early and late components of the ascending SCEP following sciatic nerve stimulation. The recording of a similar waveform configuration with extraspinal recording electrodes in the monkey indicates that the same information on the ascending SCEP to sciatic nerve stimulation may be obtained as with epidural recording electrodes. It remains an attractive hypothesis that both dorsal column and lateral column components can be differentiated in the ascending C_2 SCEP response to sciatic nerve stimulation.

While the monkey experiments were done to test the practicality of percutaneous electrode placement in a subhuman primate they also demonstrated the significance of level of recording on response configuration and perhaps the significance of species differences. The thoracic SCEP to sciatic nerve stimulation contained two clearly defined biphasic early peaks having conduction velocities of 79 and 65 m/s. Although it remains to be tested, it is tempting to speculate that the earliest peak represented activity in pathways which terminate as one ascends the cord and may explain why such a high conduction velocity was not seen in the cat with recordings at high cervical levels. The large late negativity of the L_2 descending response in the monkey has been seen in the cat with recording at L_4 and attributed to synaptic activation via the dorsal columns [24]. Since we have not seen this component in the cat following extraspinal stimulation and it was readily elicitable in the monkey it may be related to the fact that some descending tracts (e.g. corticospinal) terminate directly on motor neurons in greater proportion as one ascends the phylogenetic scale.

From these results we are encouraged that the extraspinal technique can provide information on the functional status of tracts within all three funiculi of the spinal cord. We are hopeful that application of this virtually noninvasive technique will prove safe and effective when applied to the clinical setting.

References

1. Allen A, Starr A, Nudleman K (1981) Assessment of sensory function in the operating room utilizing cerebral evoked potentials: A study of fifty-six surgically anesthetized patients. Clin Neurosurg 28:457–481
2. Bennett MH (1983) Effects of compression and ischemia on spinal cord evoked potentials. Experimental Neurol 80:508–519
3. Bennett MH (1984) SEP Monitoring during spinal cord operations. In: Homma S, Tamaki T (eds) Fundamentals and clinical applications of spinal cord monitoring. Saikon Publishing, Tokyo, pp 385–396
4. Brown RH, Nash CL (1979) Current status of spinal cord monitoring. Spine 4:466–470
5. Bunch WH, Scarf TB, Trimble J (1983) Spinal cord monitoring. J Bone Joint Surg 65A:707–710
6. Cracco RQ (1973) Spinal evoked responses: Peripheral nerve stimulation in man. Electroenceph Clin Neurophysiol 35:379–386
7. Engler GL, Spielholz NI, Bernhard NN, Danzinger F, Merkin H, Wolff T (1978) Somatosensory evoked potentials during Harrington instrumentation for scoliosis. J Bone Joint Surg 60A:528–532
8. Grundy BL (1982) Monitoring of sensory evoked potentials during neurosurgical operations: Methods and applications. Neurosurg 11:556–575
9. Harada Y, Takemitsu Y, Atsuta Y, Imai M (1984) Determination of the pathways of ascending and descending conductive spinal cord evoked potentials (SCEP). In: Homma S, Tamaki T (eds) Fundamentals and clinical applications of spinal cord monitoring. Saikon Publishing, Tokyo, pp 33–43
10. Homma S, Tamaki T (eds) (1984) Fundamentals and clinical applications of spinal cord monitoring. Saikon Publishing, Tokyo, p 396
11. Imai T (1976) Human electrospinogram evoked by direct stimulation of the spinal cord through epidural space. J Jpn Orthop Soc 50:1037–1056 (in Japanese)
12. Kurokawa T (1972) Spinal cord action potentials evoked by epidural stimulation of cord: A report of human and animal recordings. Jpn J Electroenceph Electromy 1:64–66 (in Japanese)
13. Levy WJ, York DH (1983) Evoked potentials from the motor tracts in humans. Neurosurg 12: 422–429
14. Maruyama Y, Shimoji K (1979) Origin of the human evoked electrospinogram. Adv Neurol Sci 23:397–408 (in Japanese)

15. Nordwall A, Axelgaard J, Harada Y, Valencia P, McNeal DR, Brown JC (1979) Spinal cord monitoring using evoked potentials recorded from feline vertebral bone. Spine 4:486–494
16. Raudzens PA (1982) Intraoperative monitoring of evoked potentials. Ann NY Acad Sci 388: 308–326
17. Saimada N, Matsuda H, Hirose T, Kondo M, Hashimoto T, Nakajima K, Abe J, Shimazu A (1984) Intraoperative evoked spinal cord potential for diagnosing the level of cervical myelopathy—Recording through the disc space and the yellow ligament. In: Homma S, Tamaki T (eds) Fundamentals and clinical applications of spinal cord monitoring. Saikon Publishing, Tokyo, pp 287–297
18. Satomi K, Nishimoto GI (1984) Effects of selective spinal cord transections on evoked spinal potentials in cats. In: Homma S, Tamaki T (eds) Fundamentals and clinical applications of spinal cord monitoring. Saikon Publishing, Tokyo, pp 87–98
19. Shimizu H, Shimoji K, Maruyama Y, Sato Y, Kuribayashi H (1979) Interaction between human evoked electrospinogram elicited by segmental and descending volleys. Experientia 35: 1199–1200
20. Shimoji K, Higashi H, Kano T (1971) Epidural recording of spinal electrogram in man. Electroenceph Clin Neurophysiol 30:236–239
21. Spielholz NI, Benjamin MV, Engler GL, Ransohoff J (1979) Somatosensory evoked potentials during compression and stabilization of the spine. Spine 4:500–505
22. Tamaki T, Tsuji H, Inoue S, Kobayashi H (1981) The prevention of iatrogenic spinal cord injury utilizing the evoked spinal cord potential. Int Orthop 4:313–317
23. Toyoda A, Kanda K (1984) Origins of spinal cord potentials evoked by stimulation of the cat spinal cord. In: Homma S, Tamaki T (eds) Fundamentals and clinical applications of spinal cord monitoring. Saikon Publishing, Tokyo, pp 99–111
24. Tsuyama N, Tsuzuki N, Kurokawa T, Imai T (1978) Clinical application of spinal cord evoked potential measurement. Int Orthop 2:39–46
25. Turnbull IM (1972) Blood supply of the spinal cord: Normal and pathological considerations. Clin Neurosurg 20:56–83
26. Worth RM, Markand ON, Derosa GP, Warren CH (1982) Intraoperative somatosensory evoked response monitoring during spinal cord surgery. In: Courjon J, Maugieure F, Revol M (eds) Clinical applications of evoked potentials in neurology. Raven Press, New York, pp 367–373

Scalp Distribution of Dermatomal Evoked Potentials

J. R. Toleikis[1], T. Sloan, S. Schrader, and A. Koht

Summary

Dermatomal somatosensory evoked potentials were recorded in 10 normal subjects, in order to optimize scalp recording sites for evaluation of spinal cord injuries. The results from 19 dermatomes studied showed a tendency for upper cervical responses to be maximal on the midline at CZ, lower cervical on the contralateral scalp and thoracic and lumbar responses once again on the midline.

Introduction

Somatosensory evoked potentials (SSEPs) have proven to be a valuable complement to the clinical examination in the evaluation of spinal cord injured patients [2, 6–9, 13]. Somatosensory evoked potentials provide an objective means for non-invasively evaluating inaccessible neural structures. However, conventional SSEP testing has several limitations. Somatosensory evoked potentials can suggest normality when definite abnormalities exist. For instance, SSEPs are useful for detecting lesions in peripheral nerves, the spinal cord and the brain but are oftentimes minimally affected by single spinal root lesions. On the other hand, SSEPs may be absent and suggest total loss of neural function when in fact some neural function may remain. Finally, certain areas of the trunk are not well suited for conventional SSEP testing because no easily accessible major nerves are present for direct stimulation (i.e. thoracic and sacral regions). To address these weaknesses, a diagnostic technique known as segmental or dermatomal somatosensory evoked potentials (DSSEPs) has been developed. Dermatomal somatosensory evoked potentials have demonstrated their usefulness in cases of spinal cord lesions such as tumors and spinal root entrapment problems as seen with herniated discs [1, 3–5, 10–12]. However, to date no one has reported on their use in evaluating spinal cord injured patients. Although well defined dermatomal responses are easily obtained from normal subjects, the size of these responses in spinal cord injured patients is often very small and they may go undetected if optimal recording placements are not utilized. This is particularly true in the case of sacral sparing. A means for optimizing data acquisition is therefore very important.

[1]Department of Anesthesia, Northwestern University Medical School, Chicago, IL, USA

Spinal Cord Monitoring
Ed. by J. Schramm and S. J. Jones

Methods

Optimizing the data acquisition technique consisted of finding the recording sites at which a maximum amplitude response occurred for each dermatomal level tested. Ten normal subjects ranging in age from 23 to 41 were used for this study. A four-channel Nicolet Pathfinder II and standard evoked potential recording techniques were used to acquire all data. Filter settings were set to pass frequencies between 5 and 250 Hz. Stimulation consisted of 300 microsecond squarewave pulses delivered at rates between 5.7 and 8.7 Hz. The current stimulus was applied at an intensity level which the subject described as feeling strong but not painful. Using the EEG international 10-20 system, active recording electrodes were carefully placed at C1′, C2′, C3′, C4′ and CZ′; 2 centimeters behind C1, C2, C3, C4 and CZ. A reference electrode was placed at FZ and a ground electrode was placed on the subject's shoulder. These active electrode positions were selected because they lie very close to the recording sites normally used to record upper and lower extremity SSEPs, are relatively easy to locate, and are situated such that the tongs used for traction in cervical spine injuries do not interfere with the easy placement of the recording electrodes. Five recording sites were delineated but because the Pathfinder II had only

Table 1. Latencies for the maximum amplitude responses at each dermatomal level have been analysed for a sample size of 20 (2 for each normal subject). Relative recording sites may differ between subjects or even in the same subject

	Latency (mean ± SD)	Latency δ L-R (mean ± SD)
C_4	21.26 ± 2.18	1.90 ± 1.12
C_5	24.70 ± 2.49	1.51 ± 1.53
C_6	27.96 ± 2.10	1.45 ± 1.63
C_7	29.17 ± 2.15	1.24 ± 1.50
C_8	29.04 ± 1.87	1.16 ± 1.06
T_1	25.82 ± 2.64	2.24 ± 1.63
T_4	26.57 ± 3.30	2.52 ± 2.76
T_6	29.23 ± 3.08	2.53 ± 2.25
T_8	29.79 ± 3.94	2.27 ± 1.85
T_{10}	30.77 ± 2.98	2.38 ± 2.32
L_1	30.84 ± 2.71	2.49 ± 1.92
L_2	34.55 ± 3.54	3.78 ± 2.64
L_3	38.87 ± 3.31	3.33 ± 1.66
L_4	44.14 ± 4.69	3.36 ± 1.93
L_5	50.62 ± 4.51	3.47 ± 1.37
S_1	49.58 ± 4.60	3.80 ± 2.56
S_2	35.52 ± 3.82	3.60 ± 1.67
S_3	35.28 ± 2.81	2.11 ± 1.50
S_4	33.51 ± 3.20	2.37 ± 1.77

four recording channels, only four recording sites could be used to acquire data at any given time. Therefore, C3′, C1′, CZ and C2′ were used for right sided stimulation and C4′, C2′, CZ′ and C2′ were used when the left side was stimulated. Care was taken in the placement of the stimulation electrodes so that each subject received stimulation at standard body surface dermatomal locations. Nineteen dermatomal levels from C_4 to S_4 were tested in each normal subject using this technique. An analysis time of 50 milliseconds was used for testing spinal levels from C_4 to T_{10} and 100 milliseconds was used for testing spinal levels below T10. The recorded data was analysed for each recording site in terms of latencies and amplitudes of the responses.

Results

For any given dermatome and any set of recording sites, the morphology of the responses was found to differ slightly. By using the peaks of the responses, this resulted in the measured latencies being slightly different at each recording site. Latencies of the maximum responses for all the dermatomes that were tested are shown in Table 1.

The amplitudes of the cervical dermatomal responses tended to be the largest whereas the thoracic and low sacral responses had the smallest amplitude. The

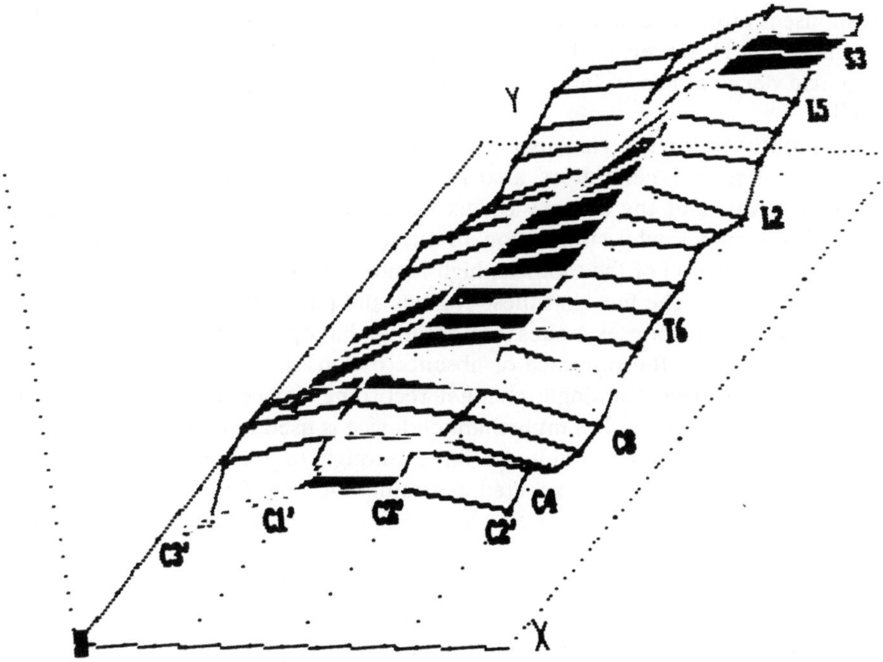

Fig. 1. Contour map illustrating the recording sites which provide the largest amplitude responses from stimulation of each dermatome. Data has been normalized for each dermatomal level so that comparisons of relative amplitudes of the responses from differing dermatomal levels cannot be made. Data is from right sided stimulation. An almost identical map was obtained for left sided stimulation

optimal recording electrode placement based on the maximal amplitude of the response for a given dermatome was not always consistent between subjects or even in the same subject. Because of changes in the morphologies of the measured waveforms at different recording sites, optimal recording sites for a given dermatomal level in the same individual might have differing peak latencies. However, the general tendency was for the high cervical (C_4 and C_5) dermatomal stimulation to produce a maximum response on the midline at CZ, whereas stimulation of the lower cervical dermatomes produced the maximal response on the contralateral side of the scalp near C1′ or C2′ rather than C3′ or C4′ (Fig. 1). Thoracic and lumbar stimulation tended to produce a maximum response along the midline of the scalp. Sacral dermatomal stimulation did not result in maximal values being along the midline as expected. Instead, maximal responses occurred off the midline on the ipsilateral side of the scalp near C1′ or C2′. These findings were generally true for both left and right sided stimulation.

Discussion

The findings regarding latencies and amplitudes of the dermatomal responses confirm those of other investigators [1, 3–5, 10–12]. Certainly factors of height, age and conduction velocity contribute to the variability in the latency values for specific dermatomes between individual subjects. In general, latency of a dermatomal response is directly related to the distance of the dermatome from the recording sites on the scalp. However, in the case of the C_6, C_7, and C_8 dermatomes versus the midthoracic dermatomes, relative conduction velocities apparently come into play as well.

Amplitudes for the various dermatomal responses are probably related to the number of neural generators associated with the various stimulation sites. The relative amount of the sensory cortex devoted to sensation in the hands and feet probably explains why stimulation of these areas produces the largest responses whereas stimulation of the thoracic, high lumbar and sacral regions produces relatively small responses. The fact that these regions and particularly the sacral area produce relatively small responses makes it imperative that optimal recording sites be defined to properly assess the presence or absence of sensory function in these areas. This is particularly true if a single channel recording device is being used to monitor dermatomal responses. If a multichannel device is used to monitor this activity, then the recording montage should be large enough to include sites where maximal activity may not always be expected. Certainly the orientation as well as the location of the neural generators dictates where the optimal locations for recording will be located.

To date, the dermatomal approach has been used to objectively evaluate over 300 spinal cord injured patients. A protocol for testing these patients has been established which includes testing by traditional SSEP techniques using major nerve stimulation. The use of the dermatomal technique has proven to be a valuable adjunct to these techniques. Many areas of the body cannot be tested using conventional SSEP testing techniques. These include the sacral area in particular. Dermatomal SSEPs have been used successfully to detect cases of sacral sparing and have provided a means for assessing patients on a single spinal level basis. It is believed that the results of this study will aid in improving the accuracy of the DSSEP technique.

References

1. Baust W, Ilsen HW, Jörg W, Wambach G (1972) A neurophysiological method for the localization of transverse lesions of the spinal cord (Abstract). Acta Neurochir 26:352–353
2. Dorfman LJ, Perkash I, Bosley TM, Cummins KL (1980) Use of cerebral evoked potentials to evaluate spinal somatosensory function in patients with traumatic and surgical myelopathies. J Neurosurg 52:654–660
3. Green J, Gildemeister R, Hazelwood C (1983) Dermatomally stimulated somatosensory cerebral evoked potentials in the clinical diagnosis of lumbar disc disease. Clin Electroencephalog 14:152–160
4. Jörg J (1977) Die elektrosensible Diagnostik in der Neurologie. Springer, Berlin
5. Jörg J, Düllberg W, Koeppen S (1982) Diagnostic value of segmental somatosensory evoked potentials in cases with chronic progressive para- or tetraspastic syndromes. In: Courjon J et al. (eds) Clinical applications of evoked potentials in neurology. Raven Press, New York, pp 347–358
6. Perot PL Jr (1972) The clinical use of somatosensory evoked potentials in spinal cord injury. Clin Neurosurg 20:367–381
7. Perot PL Jr (1976) Somatosensory evoked potentials in the evaluation of patients with spinal cord injury. In: Morley TP (ed) Current controversies in neurosurgery. Saunders, Philadelphia, pp 160–167
8. Perot PL Jr, Vera CL (1982) Scalp-recorded somatosensory evoked potentials to stimulation of nerves in the lower extremities and evaluation of patients with spinal cord trauma. Ann NY Acad Sci 388:359–368
9. Rowed DW, McLean JAG, Tator CH (1978) Somatosensory evoked potentials in acute spinal cord injury: Prognostic value. Surg Neurol 9:203–210
10. Scarff TB, Toleikis JR, Bunch WH, Parrish S (1979) Dermatomal somatosensory evoked potentials in children with myelomeningocele. Z Kinderchir 28:384–387
11. Scarff TB, Dallmann DE, Tokeikis JR, Brunch WH (1981) Dermatomal somatosensory evoked potentials in the diagnosis of lumbar root entrapment. Surgical Forum 32:489–491
12. Schramm J, Oettle GJ, Pichert T (1970) Clinical application of segmental somatosensory evoked potentials (SEP)-experience in patients with non-space occupying lesions. In: Barber C (ed) Evoked potentials. MTP-Press, Leicester, pp 455–465
13. Spielholz NI, Benjamin MV, Engler G, Ransohoff J (1979) Somatosensory evoked potentials and clinical outcome in spinal cord injury. In: Popp AJ (ed) Neural trauma. Raven Press, New York, pp 217–222

Technique of Evoked Potential Recordings from a Radiofrequency-Transmitted Spinal Cord Stimulating System

B. KEPPLINGER[1] and K. WALLNER

Summary

Radiofrequency (RF) transmitted spinal cord stimulation, as it has been used for clinical purposes (Long 1981; Meglio et al. 1981; Read et al. 1980) for more than a decade, offers the possibility of recording cortical (and peripheral) responses to spinally-delivered impulses, not just during the test stimulation period. Recordings can be obtained using conventional EMG averager devices, with the EMG machine triggering the RF stimulator unit or vice versa. For that purpose a home made electronic interface was developed. Registration of evoked responses using this method permits fast assessment of neurologic status. Comparison of components recorded with the methods described and by direct stimulation of the spinal cord during the test stimulation period showed no statistically significant changes using Student's *t*-test for paired data.

The techniques described allow for continuing recordings to be made for clinical and neurophysiological purposes in the great majority of patients who have an RF-transmitted stimulating system implanted.

Methods and Material

For recording purposes a DISA Neuromatic 1500 device was used, as well as a Neuromatic 2000. The averager unit was triggered externally from the Spinal Cord RF stimulating system (Medtronic Pisces-Selectr$_X$, Avery Percutaneous Electrical Nerve Stimulator). This was performed by bringing a second external RF receiver of the same type in close proximity to the antenna, which is located over the patient's subcutaneous implanted receiver. The electrical current, also produced in the external receiver, is transformed into a constant signal which triggers the averager unit. The transformer is equipped with a threshold value control, to adjust the system to various input levels, and a 5 volt TTL output as a safeguard for signals entering the DISA device (Fig. 1). The signal enters at the external trigger input accessory terminal. By this arrangement spinally-evoked responses can be recorded as with usually described methods. The number of averaged sweeps can be set from the EMG device while the stimulus frequency and duration, as well as the stimulus intensity is set from the RF transmitter. Recordings of RF-transmitted spinally-evoked responses also can be obtained when the EMG stimulator output current is transformed to an

[1]Neurologische Abteilung, NÖ LKH Mauer, A-3362 Mauer/Amstetten

Spinal Cord Monitoring
Ed. by J. Schramm and S. J. Jones

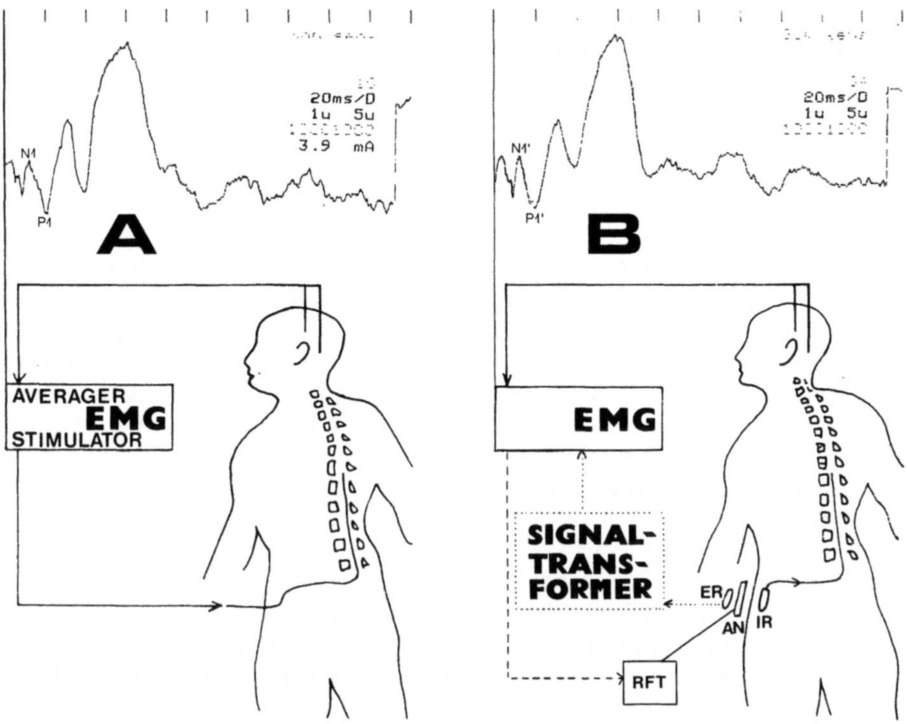

Fig. 1A, B. Tracings of scalp recordings and diagrams of spinal cord stimulation techniques. **A** with direct stimulation from the EMG stimulator, **B** with RF-transmitted stimulation. The different circuitry is indicated with dotted and broken lines. The scalp tracings resulted from stimulation at D10 spinal level. *RFT,* Radiofrequency transmitter; *AN,* antenna; *ER,* external receiver; *IR,* internal receiver

adequate RF signal (460 kHz for Medtronic-, 2.05 MHz for Avery receivers) or when the EMG stimulator output is used to trigger the usual RF transmitter (Fig. 1).

In twenty-two patients with implanted epidural leads, with electrode outlets at the C5, D3, D9, D10, D11 and D12 vertebral levels, recordings of scalp responses were obtained by delivering electrical current directly from the EMG stimulator unit to the epidural lead (Fig. 1A), as well as by the method described above after a receiver had been implanted (Fig. 1B). In either case recordings were obtained using a linked mastoid reference and stimuli were delivered at 1-s intervals, at an intensity which caused the patient to experience a non-painful sensation in one or both lower extremities. Care was taken to avoid motor rootlet stimulation.

Results

In fourteen patients without central nervous system lesions, in whom epidural leads had been implanted with the electrode outlet between the spinal levels of D9 and D12, the mean value of N1 latency in scalp recordings was 13.7 ms (SD 2.3 ms) and the mean value of P1 latency was 20.1 ms (SD 2.9 ms). In twenty-two patients tracings obtained by direct stimulation with the EMG device and with RF-mediated stimulation were compared. Even at a significance level of $P \leq 0.25$ no difference between compared values of N1 or P1 latency could be found using a paired *t*-test.

Discussion

The techniques described allow for repeated recordings of evoked responses to be obtained for clinical and neurophysiological purposes in the great majority of patients who have a remotely energized stimulating system implanted (Kepplinger 1983). In addition to the technique of spinal cord stimulation this would include stimulation of different brain regions, of the Gasserian Ganglion, the pelvic floor, etc. Our investigations concerning possible differencies in latency values of scalp responses evoked by direct stimulation and by the remote method demonstrated no statistically significant change. The disadvantage of these methods at the present time is that the exact current of the RF-delivered impulses cannot be measured directly, and therefore an exact comparison between stimulus intensities in consecutive trials is impossible. Secondly, patients with a totally implanted spinal cord stimulator (powered by an internal battery) cannot be evaluated by the techniques described. This could be implemented by a similar circuit from the EMG device to the programmer (coil), provided an adequate reed switch operation causes only a single output from the stimulator.

References

Kepplinger B (1983) Evoked potentials by means of temporary and permanent implanted spinal cord stimulating systems—preliminary report. Bull Soc Sci Med 1:59–64

Long MD (1981) Patient selection and the result of spinal cord stimulation for chronic pain. In: Hosobuchi Y, Corbin T (eds) Indications for spinal cord stimulation. Excerpta Medica, Amsterdam Geneva Princeton Tokyo, pp 1–12

Meglio M, Cioni B, Dal Lago A, De Santis M, Pola P, Serrichio M (1981) Pain control and improvement of peripheral blood flow following epidural spinal cord stimulation. J Neurosurg 54: 821–823

Read DJ, Matthews WB, Higson RH (1980) The effect of spinal cord stimulation on function in patients with multiple sclerosis. Brain 103:803–833

Pharmacological Influences

Anesthetic Technique for Spinal Cord Monitoring

W. T. Frazier[1], S. H. Odom, and B. D. Biggs

Summary

We have evaluated an anesthetic plan using a combination of nitrous oxide, Fentanyl and relaxants, with respect to its influence on per-operatively recorded somatosensory evoked potentials. Although this combination is suitable for cases involving SCM, an improved percentage of acceptable records might be obtained with some other agents and by improvement of electrophysiological techniques.

I. Introduction

This report describes a *retrospective analysis* of anesthetic technique and results obtained in *156* consecutive cases involving the use of Spinal Cord Monitoring (SCM) during surgery of the spine. SCM was accomplished via scalp-recorded somatosensory evoked potentials resultant from wrist or ankle stimulation. This series began in early 1979, extends through mid-1984 and includes surgery performed by several orthopedic (2) and neurological (4) surgeons. Ninety percent of the anesthetics were directed by one anesthesiologist. The utilization of this electrophysiological monitoring technique was requested by the surgeons on the basis of reports available in the literature at that time (Nash et al. 1977; Engler et al. 1978). The 156 cases for which scalp-recorded somatosensory evoked potentials (EPs) were monitored were not selected in any way and no patients were rejected for monitoring. Data such as drug doses, $PaCO_2$, etc., were collected from review of charts available (142 to 156, depending on particular items). Medical records missing from the series to date do not seem to represent any particular type of case or technique. The case mix for a majority sample of the cases (144 of 156) gives the following distribution: 47% Harrington rod; 25% anterior procedures (e.g., corpectomy and fusion); 6% anterior procedures with instrumentation (i.e., Dunn or Dwyer device); 11% cervical/high thoracic procedures and 11% posterior (low) decompression and/or fusion.

II. Choice of Anesthetic Technique

From the outset, it was determined that the principal anesthetic plan would be of the Nitrous Oxide-Analgesic-Relaxant (NAR) type. The intravenous analgesic was to be

[1]Department of Anesthesiology, School of Medicine, Emory University, Atlanta, GA 30322, USA

Spinal Cord Monitoring
Ed. by J. Schramm and S. J. Jones

given by continuous infusion (after a loading dose) in order to avoid abrupt changes in concentration (and possibly CNS effects) associated with bolus injections. No plans were made to rely heavily on Halogenated Volatile Anesthetics (HVAs), except possibly for the use of supplemental Halothane during intra-operative asthma. Some use of Ethrane was anticipated since it was relatively new and no negative reports were available. Forane became available during the series. In all cases, it was assumed that N_2O would be used as a basal anesthetic in concentration of 50–70%, depending on concurrent use of other anesthetics and the patient's PaO_2.

Choosing the NAR anesthetic plan for the majority of the cases was based on: (1) previous positive experience with such an approach on a wide variety of other types of cases; and (2) negative reports with the main available alternative (i.e., Halothane), (e.g., Engler et al. 1978).

Our experience had indicated that there were other potential benefits of the NAR approach: (a) adjustable CNS effects are relatively independent of effects on circulation, (b) intravenous analgesics can be administered independently of ventilatory requirements, (c) "baseline" analgesia obtained with i.v. analgesics can be maintained during "wake-up" (WU) tests and/or in the immediate post-operative period, (d) an i.v. analgesic combined with N_2O allows for the main kinetic requirement for changing sedation levels (i.e., at wake-up test or on emergence) to depend on N_2O (a very insoluble gas), thus resulting in transition from "asleep" to a level of "awakeness" permitting voluntary cooperation in 2–3 min, (e) cardiovascular control can be done relatively independently of sedation/analgesic agents with i.v. administration of such drugs as Inderal and Nitroglycerine.

Choice of the NAR technique was made knowing that it was not an ideal plan and would introduce some potential complexities such as: 1) the need to rely on other drugs for blood pressure control; 2) the relative unavailability of HVA for treatment of intra-operative bronchospasm; 3) the relative unavailability of increasing an HVA in compensation for decreased N_2O in cases requiring higher FIO_2.

Choice of an anesthetic plan for an open-ended series of such cases was with the knowledge that the ideal anesthetic is not likely to have yet been found and that the plan would evolve as experience was gained. It was also anticipated that, as permitted by circumstances, use of other agents (e.g., low dose Ethrane, Thiopental drip, etc.) would be tried.

Inherent in the plan to use moderate/intermediate dose Fentanyl was a commitment to relatively high dose N_2O (60–70%) to assure an overall adequate level of total anesthetic depth (i.e., to prevent awareness/recall), given that Fentanyl levels were to be maintained at a level infrequently requiring reversal (e.g., with Narcan) and/or ventilatory support at the end of the case (i.e., not a "high dose" Fentanyl technique). Attendant also to a N_2O/Fentanyl plan was the use of muscle relaxants. The most likely choice was Pavulon, which is known to have tachycardic effects especially in the presence of a rather "light" anesthetic. It was anticipated that some use of Inderal would be necessary to assist in prevention of tachycardia and hypertension. The level of relaxation was such that reversal was not absolutely necessary for a "wake-up" test. The choice of an i.v. analgesic was easy for us in that our experience indicated Fentanyl was superior to other choices (e.g., Morphine) due to: (a) shorter duration; (b) more cardiovascular stability; (c) less histamine release and; (d) less sedation for an apparent equal level of analgesia. On the basis of previous clinical experience, the loading dose and infusion rate were chosen to be 5 μg/kg and 2 μg/kg/h respectively.

The choices made for premedication were: Cimetidine, Benedryl, Glycopyrolate, Largon (related to Phenergan, later replaced with Haldol). Effects of these drugs in moderate doses via these routes (p.o. and IM) were not known at that time to have large effects on EPs.

Reports were already available in the literature regarding the effects of several anesthetics on EEG and evoked potentials (Abrahamian et al. 1963; Clark and Rosner 1973). A more extensive and recent review is now also available (Grundy 1983).

III. Neurophysiological Technique

After use of several types of equipment and techniques during the first year, *recording techniques* have been relatively conventional and include: Nicolet pre-amplifiers (gain of 10,000, high-pass of 0.5 Hz), ORTEC intermediate amplifiers (gain of 0.1–10; DC to 0.2, 0.5 or 3 kHz) and Nicolet MED-80 averager. Stimuli were presented in a pseudo-random fashion (inter-stimulus interval of 250–500 ms), a 2 kHz sample rate was used, and an epoch of 250 ms was analyzed with 200 stimuli in each run.

Stimulation techniques have been somewhat different from others reported in the literature in that an attempt has been made to increase the size of the EP by stimulating a larger number of fibers from each extremity. For hand stimulation both ulnar and median nerves are used, and for leg stimulation all major nerves are stimulated by a circumferential cathode just above the ankle, with a large anode on the dorsum of the foot. This stimulation technique necessitates large surface area electrodes and stimulus isolation units of high current capacity (80 ma, but low current density at skin). Such electrodes allow detection of sensory and motor thresholds without pain. The isolation units are of constant voltage type. Current (200 μs pulses) is monitored and set at approximately twice the motor threshold. Stimulation can be at 2 sites simultaneously (e.g., both ankles). Stimulation of another extremity (e.g., the hand) is used to control for drug level effects, etc.

IV. General Monitoring/Physiological Control

Conventional anesthetic monitoring has been used along with inhaled/exhaled gases (via mass-spectrometer), and frequent blood gases.

Blood pressure has been targeted at only moderate hypotension (e.g., 100/70 for a patient usually 120/80), with systolic pressure (arterial and cuff cross-checked) to be kept above 85 torr. Only moderate control of temperature has been accomplished, with almost all patients having an oesophageal temperature of 34.5°–36.5°C at the end of the operation.

V. Results

A. Anesthetic Management

A conventional induction technique has been used (Thiopental ~3 mg/kg), with a loading dose of Fentanyl (over the first 10–15 min) of 5 μg/kg. Blood pressure and heart rate control have been achieved with: (1) moderate intermittent vasodilator drip (e.g., nitroglycerine, up to approximately 100 μgm/min); and (2) i.v. Inderal (0.5–2 mg per case). Vasodilators were used in 65% of the cases and Inderal in 47%. The muscle relaxant most commonly used was Pavulon (75%), with some more recent use of Curare (15%, with Benadryl pretreatment for histamine block), or Atracurium (10%). The method of choice has been a loading dose followed by continuous infusion. As Atracurium has become available, it has become the best choice (as a drip) owing to its relative lack of cardiovascular effects and its short duration.

Awakening (at the time of a "wake-up" test or at the end of the case) has been in 2–3 min. Analgesic "reversal" with Narcan has been used in 10 instances (probably needed in only 3–4 of these). Analgesics have been needed in probably less than one half of the patients in their first hour of recovery room care (with intercostal blocks used at closing of thoracotomy cases). Ability to communicate immediately post-op has been very good (e.g., for neurological examination). Elevated $PaCO_2$ post-operatively has only been a problem in 5 patients with pre-existing pulmonary disease and has been relatively easy to manage with conventional techniques.

B. Neurological Events Detected

During the 156 cases, there were *12 significant neurological events:* (a) 9 "negative events" (EP change, or positive "wake-up" test, or post-operative neurological deficit), and (b) 3 "positive events". The 9 negative events were in the 149 "back" cases and the 3 positive events were in the 7 "neck" cases. Of the nine negative events (summarized in Table 1), there were two cases with temporary (2–4 week) weakness in one leg, six cases with intra-operative EP changes (2 "wake-up" tests done, one positive; no detectable post-operative neurological deficits), and one case with permanent one leg paralysis (plus Brown-Sequard picture, but with intact posterior columns). Of the 2 cases with temporary post-operative leg weakness: one had minimal-to-moderate EP changes (not clear initially), the deficit being first discovered on conventional "wake-up" test; in the other, the EP change was detected first and the deficit confirmed by "wake-up" test. The one case of permanent paralysis occurred during anterior corpectomy (pre-terminal metastasis at T-9 and T-10) from the left side and was accompanied by minimal but definite EP changes (mainly on the left) which partially resolved by the end of the case. Post-operative left leg paralysis, sensory (electrical) thresholds and post-operative EPs were consistent with a left hemi-cord lesion sparing the posterior column.

Of the 9 intra-operative negative events, all occurred at a time of known high risk. That is: 3 occurred at the time the spine was being "stretched"; 3 occurred when there was dissection very near a root/nerve; one occurred with necessary dissection near a critical vessel (T-9 on left); and 2 occurred when an instrument slipped accidentally into the spinal canal. In terms of *surgical consequences:* (a) 2 cases where Harrington instrumentation was planned were converted to fusion only (the 2 cases

Table 1. Negative intra-operative neurological events

Case	Date	Event detected first by			Wake-up test		Intra-op. EP recovery	Post-op. motor deficit	Post-op. sensory deficit	Change in surg. plan
		Decreased evoked potentials	Surgical event	Wake-up test	Done?	Motor deficit?				
1	12/81	A RL	C	√	√	↓ RL	±R	Mild (RL) temporary		√
2	1/82	√ LL	B		√	↓ LL	±L	Mild (LL) temporary		√
3	6/82	√	C				√			√
4	12/82	√ RL LL	B				±R>L	Complete (LL) permanent	↓ RL	√
5	1/83	√ LL	B				±L			
6	9/83	√ RL	B				±R			
7	5/84	√ RL	B				√			√
8	5/84	√ LL	B		√		√			
9	7/84	√ RL	C		√	↓ RL	±R			√

A, evolved later; B, correlated with surgical event; C, near time of H-rod "stretch"; ±, partial recovery; RL, right leg; LL, left leg with temporary paralysis and positive "wake-up" test); (b) in 2 cases with EP changes due to "stretch" (Harrington rod), distraction was reduced and EP changes reversed; (c) in one case of permanent one-leg paralysis, notification of EP changes resulted in a more conservative resection of the metastatic lesion; (d) in one case where the EP was lost on one side due to local compression (to achieve hemostasis), the time of compression was limited and the signal returned; (e) in 2 cases of local root/nerve trauma and in one case of accidental entry of an instrument into the canal, no changes in surgical plan were indicated (and no post-operative deficits occurred).

In the 7 "neck" cases, the 3 instances of "positive events" (i.e., EPs got better) were associated with release of bony/fibrous compression of the cervical cord.

C. Evoked Potential Grading

During the collection of the 156 cases of SCM, we have become increasingly impressed by: (1) the variability of success in obtaining EP signals (improving over the first 1–2 years, then levelling off); (2) the greater ease of obtaining adequate signals from the hand as compared with the foot; and (3) the greater difficulty in obtaining adequate leg signals in the presence of pre-existing neurological deficits. In an attempt to summarize our experience, it was decided to retrospectively grade the records on an A, B, C, D scale (converted to a numerical score as A+ = 10, A = 9, A− = 8, B+ = 7 ... C− = 2, D = 1, and recordings not attempted = 0). Although highly subjective and open to many sources of bias, such a grading scale has proven to be useful in tracing progress and highlighting areas which need attention and debate. The following is a *key to the grading scale adopted:*

A+ A A−	10 9 8	Signals available for essentially the entire case. Quality good enough to assure minimal ambiguity.
B+ B B−	7 6 5	Significant gaps in quantity or quality, but most of case covered with predominantly good signals.
C+ C C−	4 3 2	Significant gaps in quantity or quality; less than one-half of the case monitored and/or moderately poor quality signals.
D	1	Signals of no value (equipment? noise? physiology?)
---	0	No recordings attempted (e.g., leg signals only)

Graphs of typical results for the 32 cases monitored in 1982 are given in Fig. 1. EP grades for arm and leg signals are plotted in each case. Percentages of acceptability are calculated by computing the total of A grades and (A + B) grades and dividing by the total number of cases. A summary of the 1979–84 experience is given in Fig. 2 where it can be seen that: (1) the grades for arm/hand signals were generally better than for leg signals; and (2) the grades for leg/foot signals reached a maximum in

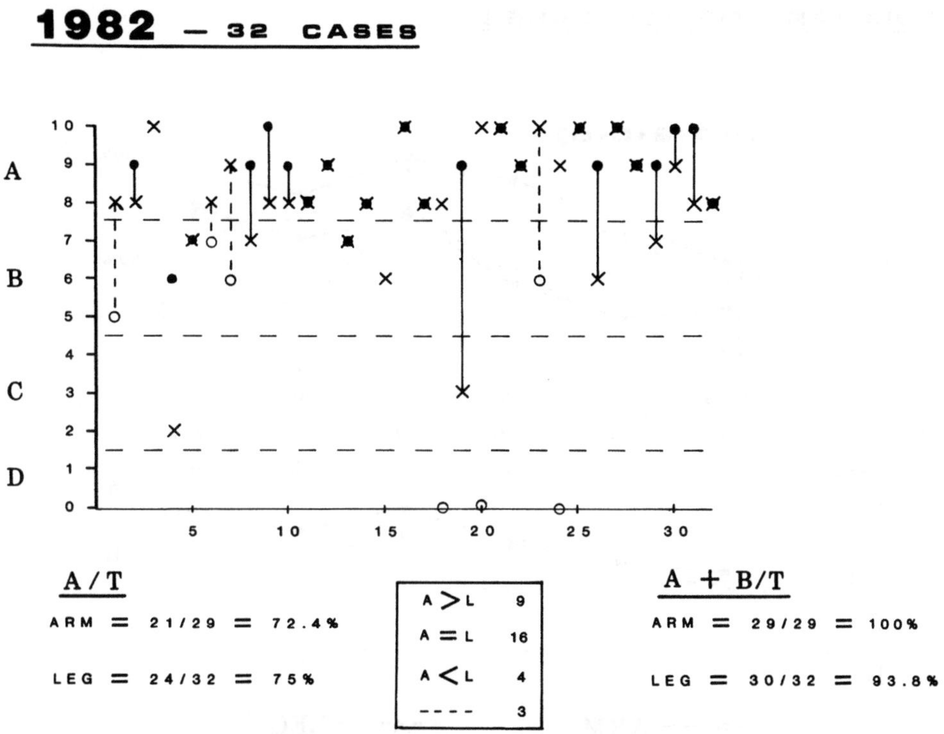

Fig. 1. The graph is a plot of grades assigned to Evoked Potentials (EPs) from the arm *(circles)* and from the leg *(crosses)*. The horizontal axis represents each patient in succession, and the vertical axis the grade of the signals for that patient (A+ = 10 .... D = 1; 0 = signal not attempted). Arm EPs are represented by filled circles when they are better than the leg signals for that patient, and by open circles when they are worse. Superimposed symbols indicate signals of equal quality. Symbols joined by a solid vertical line indicate that the arm EPs are better; when joined by a dashed vertical line, leg EPs were better. A/T is the sum of all A grades divided by the total number of cases (T), and A+B/T is the sum of all A and B grades divided by T. A > L etc. refers to the number of cases where arm EP grades were better than, equal to, or less than leg EP grades. — — — refers to cases where only leg signals were recorded (i.e., open circles at grade 0 level). See text for definitions of ABCD grades

1982 and then declined; this is probably explained by the fact that there was a change to using mainly unilateral stimulation (from simultaneous right and left leg stimulation) which yields a consistently smaller signal. In many instances, the bilateral leg signals were graded A or B, and the unilateral leg signals lower.

If one neglects the first learning year of the series (1979, 18 cases), the *overall results for leg/foot signals* are such that 69.5% of the signals are graded a B or better, with 77.5% B− or better, 15.9% are graded C and 6.5% D (i.e., 22.5% of signals were below acceptable level) (Fig. 3).

D. Use of Different Anesthetics

Of the 156 cases, 152 employed N_2O plus Fentanyl with or without other supplementary anaesthetic agents and 4 used mainly Ethrane (+N_2O). In the Fentanyl "drip"

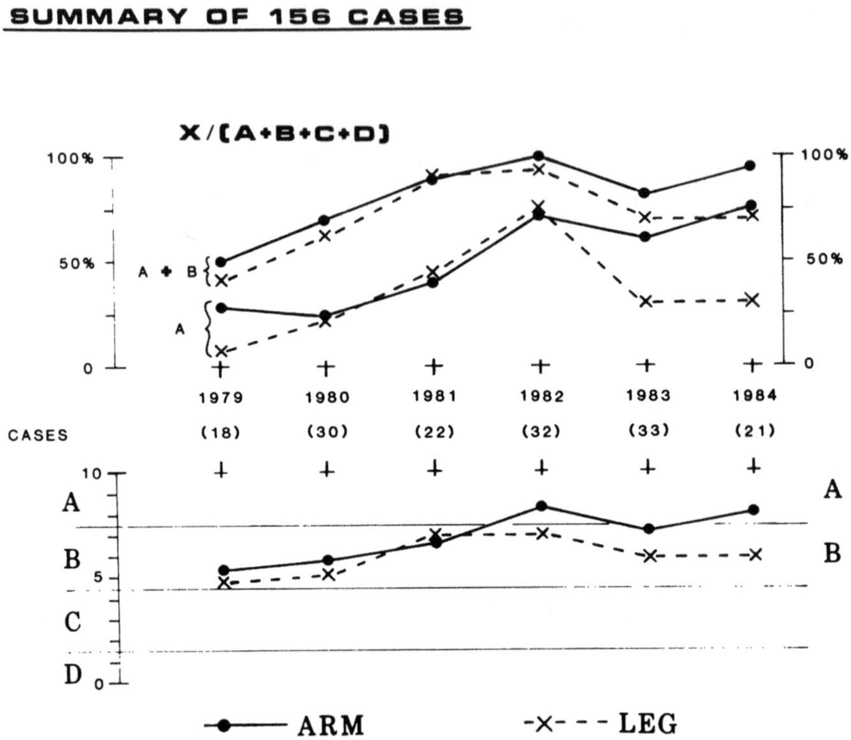

Fig. 2. *Upper half:* Percentage of acceptable Evoked Potential grades for the years 1979–1984. Solid lines connect arm (open circle) signal grades and dashed lines connect leg (cross) grades. Percentages are given for two levels of acceptability (A grades only, or A+B grades). *Lower half:* The average grades for arm EPs and for leg EPs. See text for definition of grades (A, B, etc. and number equivalent)

cases, *125 employed Fentanyl plus N_2O alone,* and *27 had other drugs added:* (1) *2 cases* had significant low dose *Halothane;* its effect on EPs was not clear in one case (poor technique) and in the other it caused a moderate amplitude decrease; (2) *10 cases* had significant low dose *Ethrane;* in only one case were conditions such that the effects were unambiguous; ¼% Ethrane caused a moderate and easily reversible decrease (~30%) in amplitude; (3) *4 cases* had significant *Forane;* in only one case were conditions such that interpretation was unambiguous; in that case, prolonged addition of ⅛–¼% Forane had minimal to no effect; (4) *11 cases* had the addition of *Thiopental* (continuous infusion; 1–2 mg/kg/h), with a coincident reduction in N_2O from 70% to 50–60%; signals were no worse and possibly better than with Fentanyl plus 70% N_2O.

It is important to note that: (a) the cases employing *mainly Ethrane* (N_2O) were performed in the first year with signals of marginal quality; thus by virtue of number and quality no firm conclusion can be drawn, but the results probably indicate that Ethrane (>0.5%) plus 60–70% N_2O is not a good plan; (b) other uses of Halothane, Ethrane and Forane were all "low dose" (<½%) and administered *in addition* to the regular (Fentanyl plus N_2O) plan; for a fairer evaluation, a coincident decrease in

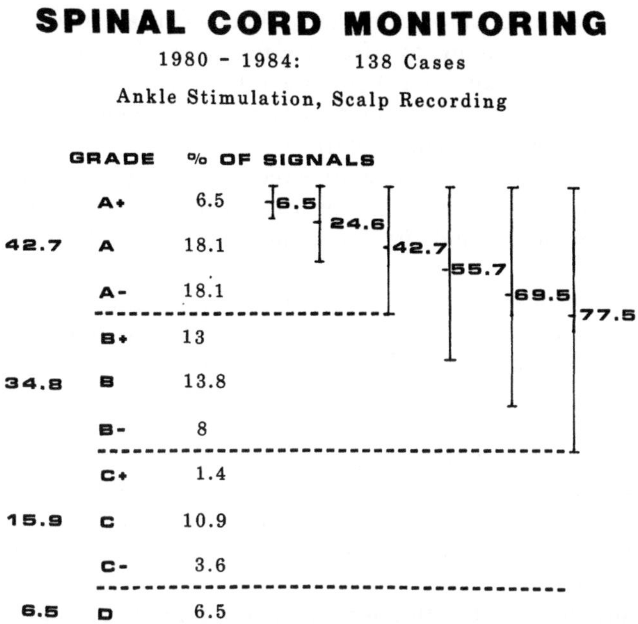

Fig. 3. A summary of grading results (leg signals) for the years 1980–84, omitting first "learning" year (1979). Percentages are given for each grade (A+, A, etc.), for combined A, B, etc. grades, and for successive cumulative grade groups (e.g., 77.5% of all A+ through B− grades, our current level of "acceptability")

one of the other agents (e.g., N_2O) would be necessary; and (c) when using Thiopental "drip", a reduction in the percentage of N_2O was made.

VI. Discussion

A. Anesthetic Technique

From a purely anesthetic management point of view, we were generally pleased. Use of multiple i.v. "drips" (e.g., Fentanyl, Atracurium, with or without nitroglycerine) was compensated for by not having to use predominantly one agent for multiple (and sometimes conflicting) purposes. With evolution of better "hardware" (i.e., small, cheap, "smart" i.v. pumps), the logistics of multiple continuous i.v. infusions will become much easier.

The Fentanyl drip rate target was always at least 2 μg/kg/h (on the basis of previous experience with frequent i.v. bolus technique) and was increased to as high as 5 μg/kg/h to compensate for analgesic tolerance (e.g., where there was a history of chronic analgesic use) or in response to intra-operative signs of need (e.g., increased BP accompanied by pupillary size consistent with underdose). All but 4 cases were below 4 μg/kg/min. In only 1 of these was it clear that the need was overestimated. In the 6 cases with the highest post-operative $PaCO_2$ level ($>\sim 51$ torr), there was clear evidence of predisposing pulmonary problems in 5, and 1 was due to an overestimated dose in compensation for chronic drug use. Although the anesthetists were given limited freedom to titrate Fentanyl to effect, the few cases with doses as low as 1–1.5 $\mu g/kg/min$ were most likely due to: (1) dose recording errors (i.e., estimate of amount gone from i.v. bag); and (2) deliberate cuts in dose to compensate for addition of other drugs (e.g., Ethrane or Thiopental).

There was only one instance of possible intra-operative recall. When the teenage patient was interviewed with the mother, all reports of events remembered were consistent with events in OR before induction and after awakening (e.g., faces, clanging of instruments, etc.).

The use of vasodilators and low dose Inderal was only moderate in incidence and amount. Conditions for the "wake-up" test and post-operative neurological examination and communication were excellent (i.e., fast recovery, no vigorous limb movements, etc.).

In the last 15 cases approximately, the usual premedication with Haldol (p.o.) and Benadryl (p.o.) was changed to i.v. Fentanyl (to effect) in the holding area. This change assures that no premedicant is being absorbed during the operation (and might thus be responsible for slow "drift"), keeps the number of drugs lower (and thus the interpretation of drug effects clearer) and allows an estimate of that patient's dose response before induction.

The availability of Atracurium as a substitute for Pavulon has made control of muscle relaxation much better and has reduced need for vasodilators.

The most difficult drug dose decisions were associated with adjusting the Fentanyl dose for patients with pre- or intra-operative evidence of tolerance to analgesics (e.g., due to use of unknown amounts of chronic pain medications).

B. Stimulating/Recording Technique

It is possible that better results could be obtained by (1) better pre-amplifiers (e.g., higher common-mode rejection ratio), (2) raising the low frequency filter (from 0.5 Hz to some higher level), (3) another type of ankle stimulation, (4) different location of recording electrodes, (5) faster signal acquisition by using a shorter epoch (down to 150 ms), and (6) use of more modern signal processing techniques (e.g., "smart" digital filters).

C. Grading of EP Records

Certainly an attempt should be made to refine and make the grading as objective as possible. For example, individual leg and dual leg signals should be graded separately and the percentage time of the operation covered by adequate monitoring should be quantified. In terms of correlations, better data is needed on pre-operative neurological deficits.

D. Influence of Anesthetics

Our experience indicates that, in this population of patients, the N_2O-Fentanyl-Relaxant anesthetic plan does allow moderate success in spinal cord monitoring. Since adopting this particular approach, several reports have appeared reporting a preference for such a plan (e.g., Worth et al. 1982; Pathak et al. 1983).

Regarding the use of Halogenated Volatile Anesthetics (HVA), our data are insufficient to say anything firm. It seems that most investigators have already eliminated Halothane from their choice list as a primary agent, but it is likely that low dose Halothane supplement (e.g., ¼%) may be used when indicated (e.g., for intra-operative asthma) with some acceptable reduction in signal amplitude. Similar low dose supplements of Ethrane or Forane may also be possible with similar moderately reduced signals (for Ethrane) and perhaps minimal reduction with Forane. One report has appeared (McPherson 1983) pointing out the virtues of Ethrane supplement as well as the depressant effects of N_2O (an agent probably universally used in most studies to date).

As a modification of a "pure" Fentanyl/N_2O technique, our preliminary experience with 11 cases would indicate that addition of a Thiopental continuous infusion (with reduction of N_2O concentration) seems promising. For example, there is a reduced need for high concentration of N_2O, and there is theoretically better protection of compromised neural tissue. Our experience would indicate that there are no clear disadvantages to this plan from a purely anesthetic point of view (except with poor cardiac status) and experience with this approach in our other neurosurgical patients suggests that it is potentially very useful.

E. Possible Reasons for Some Inadequate Leg EPs

Our results indicate that there may be a significant proportion of patients in whom good scalp-recorded single leg evoked potentials cannot be achieved. At our present level of acceptance (grades A + B), this proportion is probably near 20–25%. If we consistently used simultaneous bilateral stimulation, this would undoubtedly be lower. It is our impression that the proportion would also be lower if one chose patients without significant pre-operative neurological deficits and eliminated the kneeling position (i.e., possible nerve compression).

Reports in the literature vary quite widely in their success rate for SCM under anesthesia. Raudzens (1982) reported inadequate signals in 6 of 31 patients (19.4%). Allen et al. (1980) reported "satisfactory" records in 77% of mixed patients (visual, auditory and somatosensory EPs), with the qualification that leg EPs were the most difficult to define (and presumably had lower success rate). However, two other studies (Engler et al. 1978; Worth et al. 1982) report success in 54 of 55 and 12 of 12 patients respectively. The latter report selected the 12 out of a total of 21 cases, but only on the basis of completeness of "stored data", not on the basis of results. Although it is likely that acceptability criteria vary with different institutions, it remains to be seen why there is such a wide variation in reported results.

Although our results are comparable with those of Raudzens (1982) and Allen et al. (1980), particularly if both legs are stimulated, there may be other factors related to neurophysiological technique that allow some clinicians to achieve a higher success rate (e.g., location and level of leg/ankle stimulation). It is relatively clear that pre-operative neurological deficits do have a marked effect on results. Although our audit is not yet finished in this regard, our patient population may be biased in that direction due to a surgical referral pattern emphasizing patients with deficits and/or patients who have had multiple procedures elsewhere prior to referral.

Other factors which could be called on to explain poorer results for leg signals than arm signals are: (1) temporal dispersion of the peripheral nerve/posterior column afferent volley; (2) the leg region of sensory cortex is further away from the scalp recording electrodes (medial vs. lateral cortical surface); (3) for the leg signals, other non-stimulated cortex is interposed between the leg area and the recording electrodes (generating background EEG "noise"); (4) the amount of cortex responding to ankle stimulation may be less than that responding to hand stimulation; and (5) the orientation of the neural current generators for the leg area may be less advantageous (i.e., more iso-potential at the scalp surface).

It is also possible that stimulus conditions used (trials repeated every 3–6 min in many instances) may cause some sort of habituation at the cord or higher level.

VII. Conclusion

An anesthetic plan of N_2O/continuous Fentanyl infusion/relaxant is suitable for cases involving SCM. In our opinion, it is not yet clear if it is the best available technique. Questions remain in our minds about the utility of agents such as Forane (as a low dose supplement). Substitution of low dose continuously infused Thiopental (with a decrease in N_2O) may give results as good as with N_2O/Fentanyl alone and may offer some advantages.

Neurophysiological monitoring *not* dependent upon posterior column function may be needed in some types of cases. One of our cases had permanent one leg paralysis with evidence of sparing of posterior columns. Since EP changes during the operation were not dramatic, there is the distinct possibility that anterior cord damage could occur and be missed by the usual kind of SCM. Results consistent with this possibility have been reported (Halliday and Wakefield 1963; Szilagyi et al. 1978; Carlson 1984).

We feel that if our acceptability percentage (~70%) cannot be improved upon by better application of available neurophysiological and anesthetic techniques, then better SCM (in our hands) may await newer anesthetic drugs or the use of other neurophysiological techniques (i.e., spinal cord stimulation and/or recording, e.g. Shimoji et al. 1978).

References

Abrahamian HA, Allison T, William RG, Rosner BS (1963) Effects of thiopental on human cerebral evoked responses. Anesthesiology 24:650–657

Allen A, Starr A, Nudleman K (1981) Assessment of sensory function in the operating room utilizing cerebral evoked potentials: A study of fifty-six surgically anesthetized patients. Clin Neurosurg 28:457–481

Carlson CA (1984) Somatosensory evoked potential monitoring in aortic coarction surgery: Failure to predict adverse neurologic outcome. Tenth Annual Gulf/Atlantic Resident's Conference, Atlanta, Georgia 1984

Clark LC, Rosner BS (1973) Neurophysiologic effects of general anesthetics: I. The electroencephalogram and sensory evoked responses in man. Anesthesiology 38:564–582

Engler GL, Spielholz NI, Bernhard WN, Danziger F, Merkin H, Wolff T (1978) Somatosensory evoked potentials during Harrington instrumentation for scoliosis. J Bone Joint Surg 60A: 528–532

Grundy BL (1983) Intraoperative monitoring of sensory evoked potentials. Anesthesiology 58: 72–87
Nash CL, Lorig RA, Schatzinger LA, Brown RH (1977) Spinal cord monitoring during operative treatment of the spine. Clin Orthop 126: 100–105
Pathak KS, Brown RH, Nash CL, Cascorbi HF (1983) Continuous opioid infusion for scoliosis fusion surgery. Anesth Analg 62: 841–845
Raudzens PA (1982) Intraoperative monitoring of evoked potentials. Ann NY Acad Sci 388: 308–326
Shimoji K, Shimizu H, Maruyama Y (1978) Origin of somatosensory evoked responses recorded from the cervical skin surface. J Neurosurg 48: 980–984
Szilagyi DE, Hageman JH, Smith RF, Elliott JP (1978) Spinal cord damage in surgery of the abdominal aorta. Surgery 83: 38–56
Worth RM, Markand ON, DeRosa GP, Warren CH (1982) Intraoperative somatosensory evoked response monitoring during spinal cord surgery. Adv Neurol 32: 367–373

Effects of Fentanyl and Enflurane on Cortical and Subcortical SEP During General Anesthesia in Man

F. THURNER[1], J. SCHRAMM, and J. ROMSTÖCK

Summary

The effects of fentanyl and enflurane on the amplitudes and latencies of cortical and subcortical somatosensory evoked potentials were tested in two groups of 10 patients. These anesthetics were given together with flunitrazepam and etomidate for induction and nitrous oxide for maintenance. Preoperative values of amplitude and latency were compared to intraoperative values obtained with three different doses of fentanyl and enflurane. Enflurane induces dose-dependent latency delay and amplitude attenuation in both subcortical and cortical somatosensory evoked potentials. Fentanyl-benzoediazepine anesthesia causes no apparent alteration of subcortical responses, whereas cortical potentials are attenuated in amplitude, but little affected in latency.

Introduction

Intraoperative recordings of somatosensory evoked potentials (SEP) are currently being tested as an instrument for monitoring neurophysiological functions. They are used during operations of spinal tumors, for Harrington instrumentation of scoliosis (Bunch et al. 1983; La Mont et al. 1983) as well as an indirect measure of cerebral blood flow (Okada et al. 1983) during carotid endarteriectomy and artificial hypotension during aneurysm clipping (McPherson et al. 1983; Grundy 1982). Several publications about the effect of different anesthetic agents on SEP already exist (Drummond et al. 1984; Pathak et al. 1984; Peterson et al. 1984; Schröder et al. 1984). Less precise data have been accumulated about the effects of anesthetic regimes compromising several drugs (Frazier et al. 1985). We wanted to investigate (1) how the combination of anesthetics commonly used by us would change the SEP, (2) which of 2 anesthetic regimes would give us the more reliable potentials and (3) whether the so-called subcortical potential would be affected by these anesthetic regimes in a different way than the cortical SEP.

For monitoring somatosensory function during surgery of the spinal cord and brain stem a different mode of action on different CNS-structures would have practical implications.

Material and Methods

20 patients, 29 to 51 years old, undergoing lumbar disc operations gave informed consent to have somatosensory evoked potentials tested intraoperatively. Preopera-

[1]Institute of Anesthesiology, University of Erlangen, Krankenhausstr. 12, D-8520 Erlangen

Spinal Cord Monitoring
Ed. by J. Schramm and S. J. Jones

Table 1. Drugs administered at various stages of procedures

Premedication	*Drugs investigated*	
0.015 mg/kg Flunitrazepam	Fentanyl	
0.5 mg Atropine	Step 1	1.8 μg/kg
Induction (Step 0)	Step 2	3.6 μg/kg
0.2 mg/kg Etomidate	Step 3	7.2 μg/kg
0.01 mg/kg Flunitrazepam	Enflurane	
0.1 mg/kg Pancuronium	Step 1	0.5%
66% N_2O	Step 2	1.0%
	Step 3	1.5%

tive recordings were taken the day before surgery. Median nerve evoked potentials with neck and scalp recordings were used. The evening before the operation they received 2 mg flunitrazepam orally. Before being brought to the theatre a premedication consisting of 0.5 mg Atropine and 0.015 mg/kg bodyweight flunitrazepam were administered intramuscularly. Anesthesia was induced by an intravenous dose of 0.01 mg/kg flunitrazepam, 0.2 mg/kg etomidate, 0.1 mg/kg pancuronium and 66% nitrous oxide (Table 1). Etomidate is a very short acting substance ($t\frac{1}{2} = 2.6$ min). Nitrous oxide reaches an equilibrium within 5 min. Flunitrazepam ($t\frac{1}{2} = 2.86$ h) and pancuronium are long-acting substances (Forth et al. 1977; Dudziak 1980). In this way we obtained a steady state of anesthesia lasting for the duration of all SEP recordings, since the action of etomidate will have disappeared, whereas the action of the other agents will have been approximately constant for this time.

20 min after induction of anesthesia the first SEP was recorded. As this was done at a base level of anesthesia and none of the more powerful anesthetics was yet added, we called it step 0.

Then either enflurane or fentanyl were administered in three different doses, increasing stepwise. Enflurane was given in doses of 0.5, 1.0, and 1.5 volume-% through a calibrated vaporizer (Dräger) in a half-open system. Ten minutes after a new dose was started the next SEP recording was performed. This finding was the same, both for cortical and subcortical potential recordings and in the fentanyl and enflurane-groups. Regardless of the type of premedication and anesthetic agent, the potentials seemed to be slightly faster and possibly better synchronized than in the awake patient. Therefore, to evaluate the influence of increasing doses of fentanyl and enflurane properly, the data obtained with increasing doses were compared to those obtained at step 0.

Fentanyl

After the administration of fentanyl there was no increase in the latency of cortical potentials affecting the early components (N20 and P25). Irrespective of the fentanyl concentration, the latency variations were within limits of measuring precision (Fig. 1). For components N35 and P45 there was a latency increase which was significant at the $P < 0.05$ level (Table 2). There was also a reduction of amplitude of

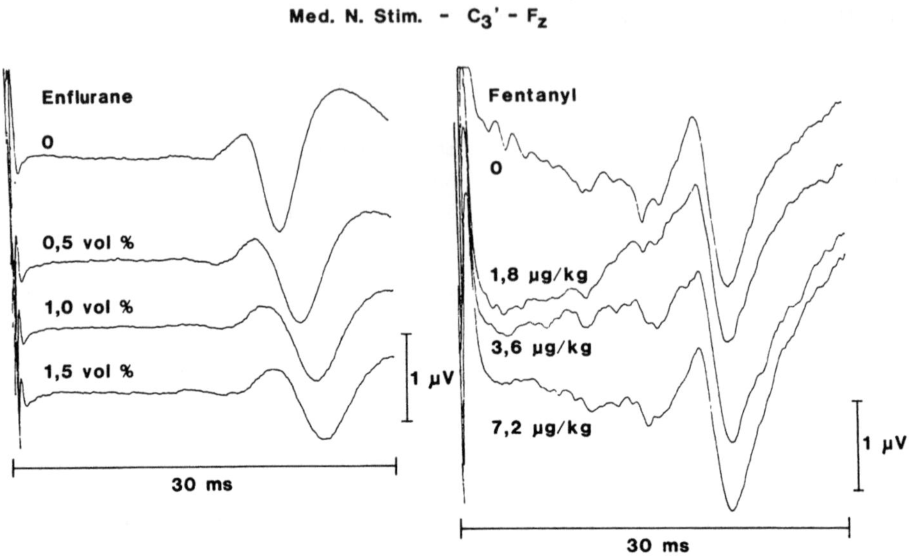

Fig. 1. Scalp SEP after median nerve stimulation with different concentrations of fentanyl and enflurane. 300 runs per average, 30–3000 Hz bandpass

Table 2. Fentanyl effects on cortical SEP

SEP	Fentanyl				
	N20	P25	N35	P45	N55
Preop.					
Lat	19.3	23.1	34.0	44.0	60.5 ms
Amp		1.69	2.16	2.22	3.62 µV
Step 3					
Lat	18.7*	23.1	36.8*	52.8*	64.7 ms
Amp		1.56	2.93*	2.48	1.30 µV

* $P<0.05$

Table 3. Fentanyl effects on subcortical SEP. *CCT,* central conduction time (N11–N20)

SEP latency	Fentanyl					
	N9	N11	N13	P15	N20	CCT
Preop.	10.14	12.38	14.11	16.14	19.84	7.46
Step 0	9.80	11.96	13.70	16.06	19.72	7.76
Step 1	9.90	12.01	13.80	16.19	19.71	7.70
Step 2	9.95	12.07	13.77	16.24	19.66	7.59
Step 3	9.97	11.96	13.73	16.24	19.76	7.80

No significant change

N55, whereas the amplitude of N35 was enhanced ($P<0.05$). This, however, might already be seen at step 0, where fentanyl was not yet applied, whereas in both stages flunitrazepam was given together with nitrous oxide. Therefore this effect was most likely due to the application of flunitrazepam. With regard to subcortical potentials, there was no significant change to be found after the application of fentanyl (Table 3). There were also no effects which might have been due to the application of flunitrazepam as described above.

Enflurane

During enflurane application there was a continuous increase in latency related to the enflurane concentration (Tables 4, 5). With regard to subcortical peaks (N9 and N13) the delay was significant ($P<0.01$) but not yet very large. N11 remained constant. For the P15-peak the latency increase was more marked and also statistically significant ($P<0.01$). The latencies of all cortical SEP components were increased by enflurane. Compared to step 0, P25 was already distinctly depressed in step 1 (Fig. 1). The N35 amplitude, which was initially enhanced by flunitrazepam,

Table 4. Enflurane effects on subcortical SEP. *CCT,* central conduction time (N11–N20)

SEP latency	Enflurane					
	N9	N11	N13	P15	N20	CCT
Preop.	9.32	11.79	13.62**	15.55**	19.30**	7.51*
Step 0	9.12	11.54	13.18	15.25	18.95	7.41
Step 1	9.22**	11.57	13.33**	15.70**	19.36**	7.79**
Step 2	9.23**	11.63	13.43**	16.21**	20.15**	8.52**
Step 3	9.31**	11.64	13.62**	16.45**	20.92**	9.28**

* $P<0.05$; ** $P<0.01$

Table 5. Enflurane effects on cortical SEP

SEP	Enflurane				
	N20	P25	N35	P45	N55
Preop.					
Lat	19.22**	22.78**	30.10	40.5**	55.2
Amp		2.28*	3.02	3.73	3.51*
Step 0					
Lat	18.42	21.44	29.00	42.8	54.9
Amp		3.89	4.41	3.32	1.97
Step 3					
Lat	20.48**	25.76**	34.90*	55.8**	77.8**
Amp		2.74*	2.84	2.14	0.88**

* $P<0.05$; ** $P<0.01$

was reduced to its former amplitude again and occurred 10 ms later (Table 5). The next positivity was found at 55.8 ms and the next negativity at 77.8 ms. The amplitude of both peaks was very low, the latter measuring only 0.88 μV on average in step 3. No later components were found in the somatosensory evoked response with an enflurane concentration as low as 0.5 volume-%. On the other hand it must be stressed that SEPs were never completely depressed (i.e. with no identifiable peaks) even if 1.5 volume-% of enflurane were administered.

Central Conduction Time

The latency difference between reproducible components of subcortical potentials recorded over the cervical spine and cortical components recorded over the contralateral cortex has been used to estimate the time needed for conduction of the afferent volley from the cervical spinal cord to cortical or subcortical structures (Desmedt and Brunko 1980). The central conduction time used in our study was determined by using peak N11 from subcortical potentials and peak N20 from the cortical recordings. N11 was chosen because it remained fairly constant in both types of anesthesia, whereas N20 was changed very markedly. At the same time both peaks have the advantage of being exactly measurable and well reproducible. The difference between the two latencies was termed "central conduction time". These values were submitted to linear regression analysis. An increase of latency of 1.27 ms/1 volume-% of enflurane was found in this way. The correlation coefficient for enflurane values was 0.98. The same regression analysis applied to fentanyl values showed a very small increase in latency (0.058 ms) and the correlation coefficient r^2 was only 0.038 (Fig. 2).

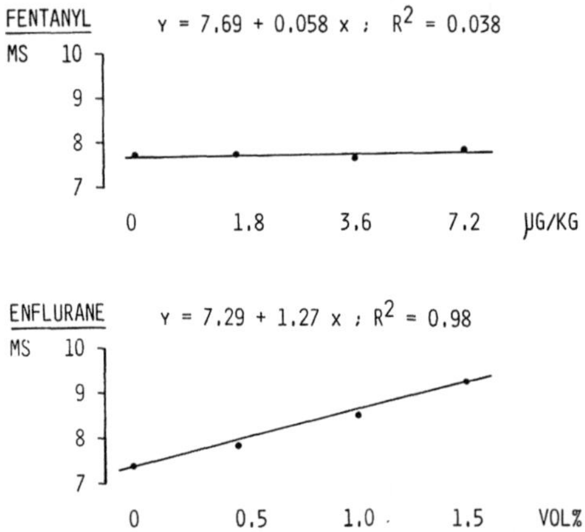

Fig. 2. Linear regression analysis of central conduction time (N11–N20) alteration with different concentrations of fentanyl and enflurane

Discussion

The pharmacological action of volatile anesthetics is unexplained in detail. In spite of their quite different chemical configuration, it is assumed that they act on synapses, because in lower concentrations they are antagonized by naloxone (Freye 1982). The increase in central conduction time allows some speculation about the pharmacological action of anesthetics. Fentanyl does not seem to affect spinal and peripheral neuronal structures from which the somatosensory responses are generated. According to autoradiographical studies (Pert et al. 1973; Kuhar et al. 1973; Simon et al. 1973), opiate receptors are located in the substantia gelatinosa, the spinothalamic tract, the formatio reticularis, the periaqueductal grey matter, the hypothalamus and the limbic system. Somatosensory evoked potentials, however, are most likely to be propagated and generated through the dorsal column and medial lemniscal system. Despite the large number of studies which have been devoted to the subcortical SEP, no definite conclusions are yet possible concerning the precise location of the generators of different cortical and subcortical SEP components. In this paper the generator discussion will not be taken up, but generally it seems acceptable to assume that N20 and all later components of the median nerve evoked potential are generated beyond the thalamus or in cortical structures.

Freye's hypothesis (1982), that volatile anesthetics mainly act on synapses, seems to be confirmed by our findings. Peripheral conduction velocity represented by latency values at Erb's point and the N11 peak latency remain virtually unchanged, whereas central conduction time is strongly influenced by enflurane.

The differences in latencies and amplitudes between recordings performed in the awake stage and in the stage of basic anesthesia (step 0) need to be discussed. In the awake patient we stimulated at submaximal motor response intensity levels. Following administration of anaesthetics and muscle relaxants, the motor response is suppressed and the degree of stimulation cannot be precisely assessed. Therefore, to be sure and to compensate for possible malalignment of stimulus electrodes and other negative influences which cannot be detected due to the influence of relaxants, we always used a stimulus intensity of 20 mA intraoperatively.

This means that stimulus intensities were different pre- and intraoperatively, and the latency decrease and increase in amplitudes seem to reflect the fact that supramaximal stimulation yields better responses due to recruitment of additional fibers and completeness of depolarization. This difference, however, could not be avoided, unless by abandoning the practice of applying the same stimulus intensity to all patients in all stages of anesthesia. To exclude this influence the effects of fentanyl and enflurane anesthesia were judged in comparison to the values obtained at stage 0, corresponding to basic anesthesia.

The changes affecting subcortical potentials are not easily understood. The latency changes between the preoperative and intraoperative recording at stage 0 may also be explained by the increased stimulus strength with larger numbers of fibers being depolarized in a more synchronized fashion. It remains unclear, however, why N11 should be affected less than N9, although occurring later. It may be that this potential reflects activity in another parallel pathway system conducting at a different velocity, rather than reflecting subsequent responses of a single volley travelling centrally.

The enhancement of a cortical component—namely N35—appearing at step 0 was still apparent when fentanyl was added in steps 1, 2 and 3 in the fentanyl-group, and was reversed when enflurane was added in the enflurane-group. At first glance it is surprising that in light anesthesia one is recording an increased amplitude of cortical SEP components, as cortical responses depend not only on neuronal conduction velocity but also on vigilance and attentiveness. The amplitude increase in N35 was most probably due to flunitrazepam, because a similar pattern of change was described for diazepam by Prevec (1980) after its administration in awake persons (enhancement of N35, delay of P45 and attenuation of N55). Both substances are benzodiazepines and similar in pharmacological and clinical properties. So the effect on evoked responses may justifiably be assumed to be similar.

The primary response N20–P25 is little affected by most kinds of anesthesia. On the other hand it is very sensitive to reduction of cerebral blood flow. This makes it a suitable instrument for monitoring cerebral blood flow. The N20–P25 component has been used by Grundy (1982, 1983) in monitoring during aneurysm surgery and by Russ and Krummholz (1984) during carotid endarteriectomy. In enflurane anesthesia, however, the latency of N20–P25 is increased approximately 3 ms by anesthesia alone, with the amplitude also affected. SEPs might therefore be affected not only by spinal or brain stem pathology disturbing the neuronal pathways themselves but also by factors influencing the cerebral blood flow.

The following conclusions may be drawn:

Enflurane induces dose-dependent latency delay and amplitude attenuation, in both subcortical and cortical somatosensory evoked potentials.

Fentanyl-benzodiazepine anesthesia causes no apparent alteration of subcortical responses. Cortical potentials are attenuated in amplitude but little affected in latency.

References

Bunch WH, Scarff TB, Trible J (1983) Spinal cord monitoring. J Bone Joint Surg 65A: 707–710

Desmedt JE, Brunko E (1980) Functional organisation of far field and cortical components of somatosensory evoked potentials in normal adults. Progr Clin Neurophysiol 7: 27–50

Drummond JC, Todd MM, U HS (1984) The effect of high dose sodium thiopental on brain stem auditory and median nerve somatosensory evoked responses. Anesthesiology 61/3A: 345

Dudziak R (1980) Lehrbuch der Anaesthesiologie. Schattauer, Stuttgart New York

Forth W, Henschler D, Rummel W (1977) Pharmakologie und Toxikologie. Bibliographisches Institut, Mannheim

Frazier WT, Odom SH, Biggs BD (1985) Anesthetic technique for spinal cord monitoring. In: Schramm J, Jones SJ (eds) Spinal cord monitoring. Springer, Berlin Heidelberg New York, pp 69–81

Freye E (1982) Opiatrezeptoren im Gehirn. peri-med, Erlangen

Grundy BL (1982) Monitoring of sensory evoked potentials during neurosurgical operations. Methods and applications. Neurosurgery 11: 556–575

Grundy BL (1983) Intraoperative monitoring of sensory evoked potentials. Anesthesiology 58: 72–87

Kuhar MJ, Pert CB, Snyder SH (1973) Regional distribution of opiate receptor binding in monkey and human brain. Nature 245: 447–450

La Mont RL, Wasson SL, Green MA (1983) Spinal cord monitoring during spinal surgery using somatosensory spinal evoked potentials. J Pediatr Orthop 3/1: 31–36

Okada Y, Shima T, Yamamoto M, Uozumi T (1983) Regional cerebral blood flow, sensory evoked potentials, and intracranial pressure in dogs with MCA occlusion by embolization or trapping. J Neurosurgery 58:500–507

Pathak KS, Brown RH, Cascorbi HF, Nash CL (1984) Effects of fentanyl and morphine on intraoperative somatosensory cortical-evoked potentials. Anesth Analg 63:833–837

Pathak KS, Amadio BS, Shaffer JW, Scoles PV (1984) Effect of halothane, nitrous oxide anesthesia on spinal versus cortical evoked potentials during spinal surgery. Anesthesiology 61/3A:346

Pert CB, Snyder SH (1973) Opiate receptors: demonstration in nervous tissue. Science 179:1011–1014

Pert CB, Paternak SH, Snyder SH (1976) Opiate receptors: autoradiographic localisation in rat brain. Proc Natl Acad Sci 73:3729–3733

Peterson DO, Drummond JC, Todd MM (1984) Effects of halothane and isoflurane on somatosensory evoked potentials in man. Anesthesiology 61/3A:344

McPherson RW, Niedermeyer EF, Otenasek RJ, Hanley DF (1983) Correlation of transient neurological deficit and somatosensory evoked potentials after intracranial aneurysm surgery. J Neurosurgery 59:146–149

Prevec T (1980) Effect of valium on the somatosensory evoked potentials. Prog Clin Neurophysiol 7:311–318

Russ W, Krummholz W (1984) Monitoring in der Carotischirurgie mit somatosensorisch evozierten Potentialen. Der Anesthesist 33/9:475

Schröder FW, Hyman AA, Powel RN, Boettner RB (1984) Spinal evoked potentials: response alterations to anesthesia, hypotension and hypothermia. Electroencephalogr Clin Neurophysiol 57:4P

Simon EJ, Hiller JM, Edelmann I (1973) Stereospecific binding of the potent narcotic analgesic etorphine to rat brain homogenate. Proc Natl Acad Sci 70:1947–1949

Symon L, Wang AD, Costa e Silva IE, Gentili F (1984) Perioperative use of somatosensory evoked responses in aneurysm surgery. J Neurosurg 60:269–275

Pentobarbital: The Influence on Somatosensory Conduction in the Rat

D. Claus[1], W. Weitbrecht, and B. Neundörfer

Summary

Somatosensory nerve conduction was measured in 13 male Wistar rats repeatedly while the animals were awaking from a Barbiturate narcosis. The peripheral, spinal and lemniscal sensory volley conduction was examined simultaneously at 20-min intervals. The Barbiturate narcosis did not influence the volley conduction in the peripheral nerve, but the spinal and intralemniscal sensory volley conduction was affected. The Barbiturate caused a conduction delay which was more pronounced in the central sections of the sensory pathway.

Introduction

By stimulating a hind paw, evoked potentials can be recorded in the rat over spinal roots, spinal medulla, brain stem and convexity of the skull (Fig. 1).

The L5-responses reflect activity in the cauda equina (Riffel and Stöhr 1983). The L1-potential is generated in the lumbosacral spinal medulla. As the comparison with

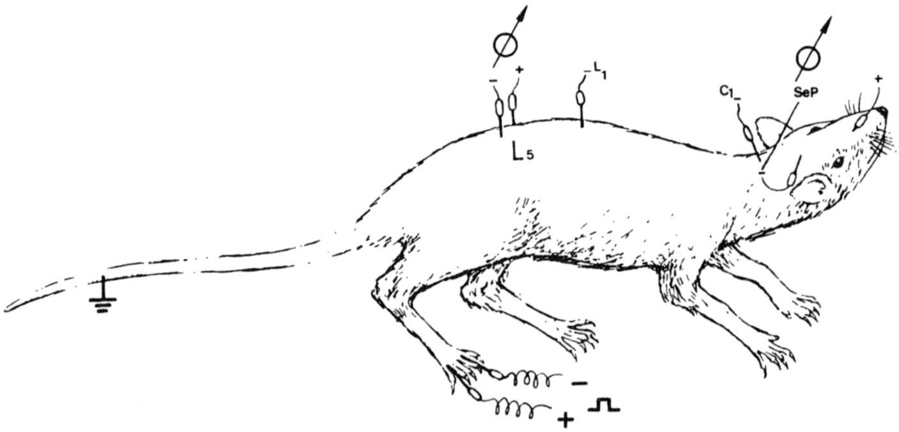

Fig. 1. The needle electrode positioning in the rat. Stimulation at the right hind paw. Recording over L5, L1, C1 and convexity of the skull

[1]Neurologische Klinik, Universität Erlangen, D-8520 Erlangen

Spinal Cord Monitoring
Ed. by J. Schramm and S. J. Jones

fore-paw evoked potentials has proven (Claus and Neundörfer 1983), the C1-response reflects conduction of the somatosensory volley in the brain stem. The first C1-component (CI) can be ascribed to the cervical dorsal tracts; CII to postsynaptic structures of the Nucleus gracilis and medial lemniscus; CIII probably to thalamic synapses and/or thalamocortical tracts.

The first cortical SEP component, a positivity P1 (Budnick et al. 1981; Shaw and Cant 1981), is followed by N1 and P2. With regard to these potentials, it is of interest to study the influence of Pentobarbital on peripheral and central volley conduction following a hind-paw stimulus in the rat.

Material and Methods

13 male Wistar rats, average weight 373 g, received 40 mg/kg Pentobarbital-Sodium intraperitoneally. Until a deep state of narcosis had been achieved (no pain reaction after a needle prick on the tail, no corneal reflex elicitable) the narcotic was reinjected in small doses as necessary. The examination started when a constant level of narcosis had been reached.

Responses were recorded following stimulation of the right hind paw with a 0.1 ms rectangular current (causing a slight paw twitch) at 20-min intervals. The signal was picked up by needles inserted over the spinal process L5 on both sides, over L1 on the right side paravertebrally and finally between the occipital squama and the first cervical vertebra paravertebrally on the right. The SEP-electrode was inserted subcutaneously over the bregma in the direction of the left posterior eye-corner. The reference-electrode for the pseudo-unipolar derivations L1, C1, was inserted subcutaneously over the tip of the nose. The sweep time amounted to 20 ms for the derivations over the spinal medulla and brain stem and 50 ms for the cortical SEP. For the L5- and L1-derivations 128 sweeps were averaged; for the cervical C1- and SEP-examinations 512 or 1024 sweeps were necessary. During the last examination, after 60 min, the animals were already showing signs of spontaneous activity.

Results and Discussion

The results are shown in tabular form (Table 1). The L5- and L1-latencies remained constant during recovery from anaesthesia. Consequently there was no evidence that peripheral and radicular nerve conduction was influenced by Barbiturate narcosis. The latencies of the three components of the cervical potential were minimally shortened. This difference was significant between initial and final recordings (Wilcoxon $P<0.05$). The difference increased from CI to CIII, i.e. the shortening of latency was more obvious for the postsynaptic potentials CII and CIII. Even more obviously, the latencies of the SEP-components P1, N1, P2 decreased significantly (Wilcoxon $P<0.01$). This shortening of latency was more pronounced for the P2 component.

Shaw and Cant (1981) were able to demonstrate an influence of Barbiturate on the early components of the cortical SEP in the rat after fore-paw stimulation. They discovered a significant shortening of the SEP-components III, P1, N1 and P2 be-

Table 1. Average values of radicular (L5), spinal (L1), brainstem (CI, II, III) and cortical potential (SEP) latencies, plus amplitude of P2 measured from N1. $N = 13$, $\bar{x}$ in ms

Min	L_5	L_1	C_I	C_{II}	C_{III}	SEP			Ampl. N_1/P_2 (μV)
						P_1	N_1	P_2	
0									
$\bar{x}$	3.05	3.98	5.02	6.86	8.12	10.34	11.41	14.48	11.54
SD	0.203	0.188	0.336	0.226	0.372	0.984	0.897	1.345	4.03
20									
$\bar{x}$	3.04	3.96	4.98	6.81	8.05	10.30	11.32	14.29	11.69
SD	0.198	0.202	0.346	0.266	0.463	1.098	1.039	1.539	4.01
40									
$\bar{x}$	3.04	3.98	4.92	6.76	7.99	10.11	11.21	14.06	11.46
SD	0.180	0.213	0.332	0.325	0.466	1.049	1.051	1.481	4.70
60									
$\bar{x}$	3.03	3.95	4.92	6.73	7.92	9.88	10.95	13.82	11.50
SD	0.184	0.222	0.332	0.287	0.460	1.021	1.077	1.411	4.96
P			<0.05	<0.01	<0.02	<0.01	<0.01	<0.01	

tween a deep narcosis and the waking phase. Therefore, it was suggested that sensory volley conduction in the medial lemniscus was delayed by barbiturate influence. The same result was obtained by Allison and Hume (1981) in the cat.

The conduction delay might be explained by a GABAergic inhibitory Barbiturate effect, predominantly on thalamic synapses. No Barbiturate influence was found for the early presynaptic spinal potentials.

The effect of Barbiturate on brain stem potentials has not previously been investigated. Allison and Hume (1981) reported that the more caudally the potentials were generated, the later they were depressed in deep narcosis. Shorvon and Reynolds (1982) found a reversible peripheral sensory conduction delay in a few patients under Barbiturate therapy. This cannot be explained by an inhibitory effect on synaptic transmission, but may be due to altered ionic permeability of the nerve membranes (Deupress 1980). Deupress (1980) described an inhibitory effect of Phenobarbital on spinal volley transmission.

With the aid of spinal, cervical and cortical recordings following hind-paw stimulation, conduction of the peripheral and spinal sensory volley can be examined in the rat. The peripheral and radicular conduction velocity (to L5 and L1) is not measurably altered by Barbiturate. The more centrally the potentials are generated, the more their latencies are influenced. The latency of the first cervical component CI decreased minimally from 5.0 to 4.9 ms during the waking phase (Fig. 2). This shortening was significant ($P<0.05$) and indicates a Barbiturate effect on conduction of the spinal volley.

The CII-latency, reflecting activity of postsynaptic structures of the Nucleus gracilis and medial lemniscus, decreased clearly and more significantly ($P<0.01$).

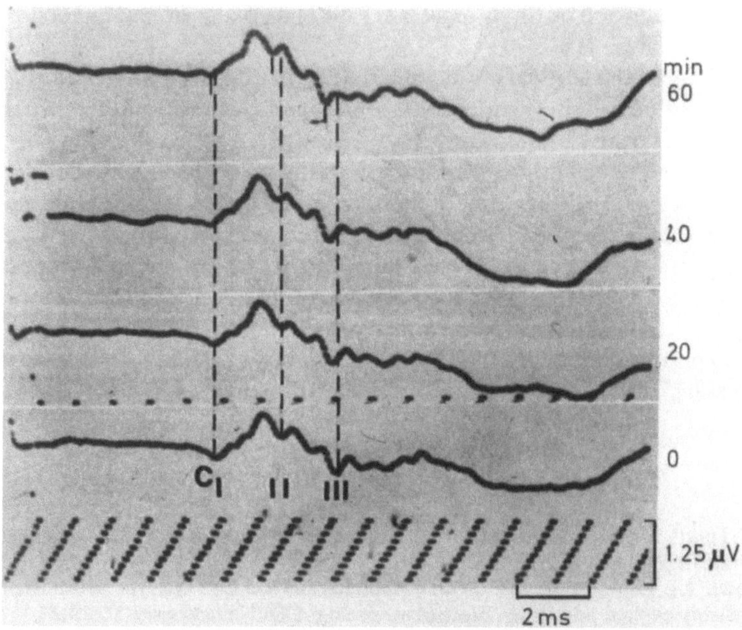

Fig. 2. Cervical recordings obtained 20, 40 and 60 min after constant narcosis. The latencies decreased minimally

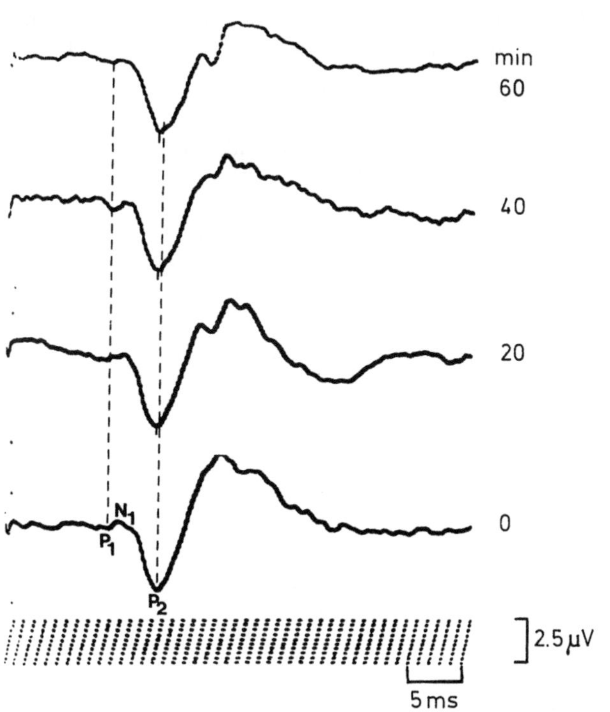

Fig. 3. SEP-examinations which show the shortening of latencies during the waking phase

Therefore, the results described by Shaw and Cant (1981) were confirmed by recording of brain stem potentials.

The clear shortening of the cortical potential latencies P1, N1 and P2 (Fig. 3) confirms the inhibitory influence of Pentobarbital on synaptic transmission through somatosensory structures of the brain stem.

While we did not observe an influence of Pentobarbital on sensory conduction of the peripheral nerve, a significant slowing of the intraspinal volley was measurable. This should be taken into account in animal experiments.

This delay is not provable because of the short intraspinal conduction distance after fore-paw stimulus. Therefore, Shaw and Cant (1981) found no intraspinal delay.

The obvious Barbiturate effect observed on central sensory pathway conduction confirms earlier results.

References

Allison T, Hume AL (1981) A comparative analysis of short-latency somatosensory evoked potentials in man, cat and rat. Exp Neurol 72:592–611

Budnick B, McKeown KL, Wiederholt WC (1981) Hypothermia-induced changes in rat short latency somatosensory evoked potentials. Electroencephalogr Clin Neurophysiol 51:19–31

Claus D, Neundörfer B (1983) Zur Untersuchung der peripheren und zentralen Nervenleitung bei der Ratte. Z EEG-EMG 14:160–163

Deupress JD (1980) Mode of action of anticonvulsant drugs. In: Tyrer JH (ed) The treatment of epilepsy. MTP Press, Lancaster, pp 1–28

Riffel B, Stöhr M (1983) Spinale SEP in der Diagnostik und Lokalisation von Rückenmarkserkrankungen. In: Struppler A (Hrsg) Elektrophysiologische Diagnostik in der Neurologie. Thieme, Stuttgart New York, S 166–167

Shaw NA, Cant BR (1981) The effect of pentobarbital on central somatosensory conduction time in the rat. Electroencephalogr Clin Neurophysiol 51:674–677

Shorvon SD, Reynolds EH (1982) Anticonvulsant peripheral neuropathy. J Neurol Neurosurg Psychiatr 45:620–626

Pharmacogenic Changes of Cortical Responses to Spinal Cord Stimulation

B. Kepplinger[1], H. Schmid, O. Schröttner, and R. Leuteritz

Summary

Scalp recordings of responses to spinally-delivered electrical stimuli were performed before and after epidural bupivacaine injection in nine patients. A marked decrease of evoked potential amplitude values to 30% of their initial level was observed within 35 min after epidural blockade. A wide reaching diffusion with involvement of spinal pathways is demonstrated by means of the described neurophysiological method, which may also be valuable in other investigations looking for a pharmacogenic change of spinal cord function.

Introduction

In 1875 Corning proposed that the spinal cord would absorb drugs injected close to it and that the spinal cord could be medicated in this way. This idea was rejected for a long time, and epidural anaesthesia was considered to be merely due to multiple paravertebral blockades.

Studies, as carried out by Wilkinson and Lund (1970), on cerebrospinal fluid (CSF) after epidural local anesthetic instillation showed, however, that there is a certain amount of perfusion through the dura. Beside others, Cohen (1968) and Bromage (1978) demonstrated in experiments on animals the regional distribution of 14C-labeled lidocaine in the various neighboring structures, e.g. external roots, dorsal root ganglia, dura, intradural roots, pia and spinal cord. Varying concentrations were even demonstrated between different structures of the spinal cord. Concentrations as measured in the cord were considered by Wall (1967) to be sufficient to cause blockade of some long pathways and modulation of inhibitory systems in the dorsal horn. A neurophysiological approach to investigation of the topic outlined above is illustrated by the following study.

Method and Material

The effect of epidural blockade with bupivacaine was demonstrated in nine patients, with respect to the influence on scalp potentials evoked by means of a spinal cord stimulating system (Kepplinger 1983; Kepplinger and Wallner 1985). These tests

[1]Neurologische Abteilung, NÖ LKH Mauer, A-3362 Mauer/Amstetten

Spinal Cord Monitoring
Ed. by J. Schramm and S. J. Jones

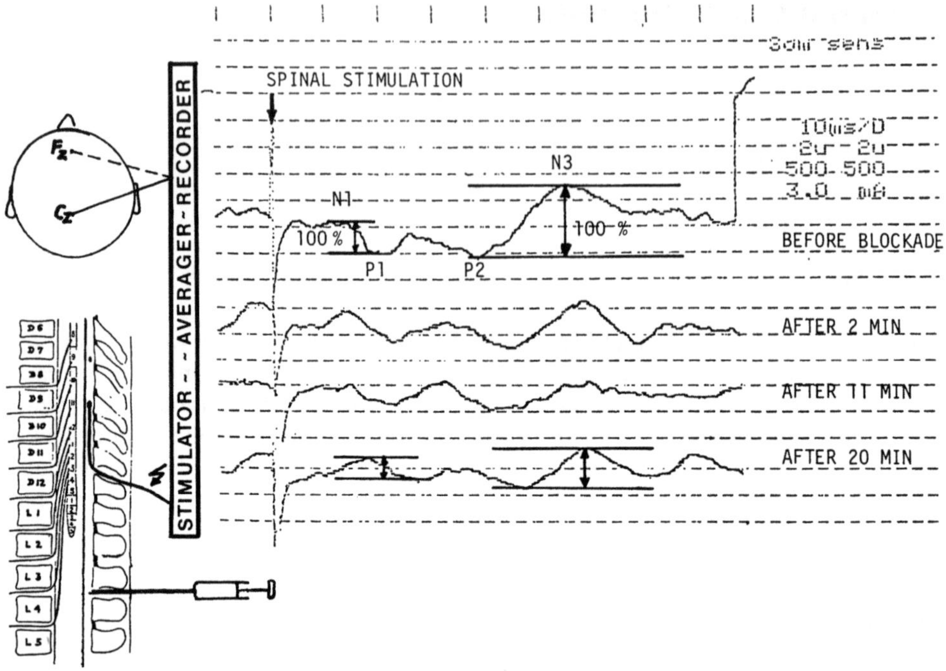

Fig. 1. Sites of injection, stimulation and scalp recordings. Spinally-evoked scalp responses before and after epidural blockade, showing N1–P1 and P2–N3 measured amplitudes for statistical evaluation

were carried out either immediately after introduction of a percutaneous epidural lead (for suture pain or original pain alleviation) or in patients with already implanted radiofrequency receivers, when the patients reported persisting pain. The stimulating electrodes were placed epidurally with the active tip at or above D9 and the lead outlet between the D9 and D12 vertebral levels. Scalp recordings were performed in referential fashion from the vertex (Cz) with the reference electrode at Fz. An epidural instillation of 5 ml 0.5% bupivacaine (Carbostesin) was performed at the L2/3, L3/4 or L4/5 lumbar spine levels, after having obtained initial recordings. For clinical and technical reasons the recordings were performed at varying intervals up to a maximum of 128 min after bupivacaine instillation.

From the recordings received we restricted this work only to measurement of the amplitude values N1–P1 and P2–N3 (as shown in Fig. 1). In order to be able to compare the interindividual amplitude values, with respect to their modifications due to bupivacaine instillation, we converted the absolute values into relative percentage values, setting the initial N1–P1 and P2–N3 amplitudes at 100 percent. For statistical evaluation (Wilcoxon matched pairs signed rank-test) the measurements up to 35 min were collected at 5-min intervals; measurements from 35 min to 75 min were obtained at intervals of 10 min. Two of the patients were excluded from the statistics because they did not achieve satisfactory pain alleviation. In addition to this their curves (Figs. 2, 3; dotted lines) were atypical. The rest of the patients reported pain alleviation, but did not show clinical signs of intrathecal local anesthesic application.

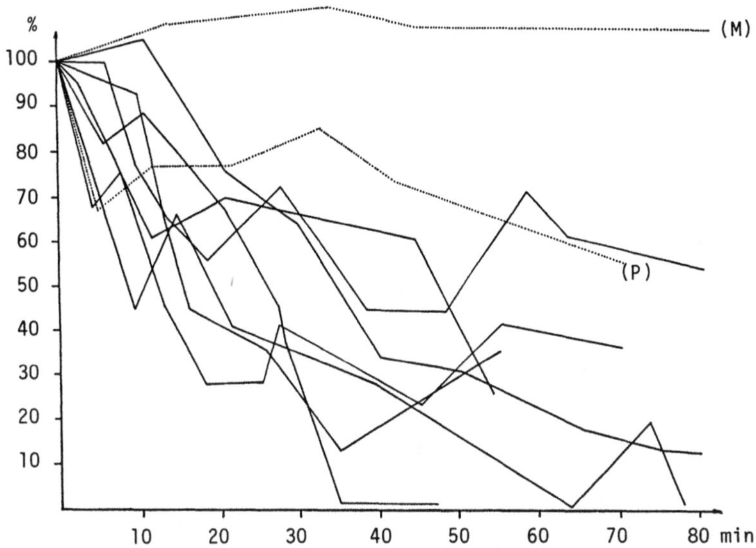

Fig. 2. Percentile N1–P1 amplitude values in the real time course up to 80 min following instillation of bupivacaine. Values of dotted lines *(M, P)* were excluded from the statistics

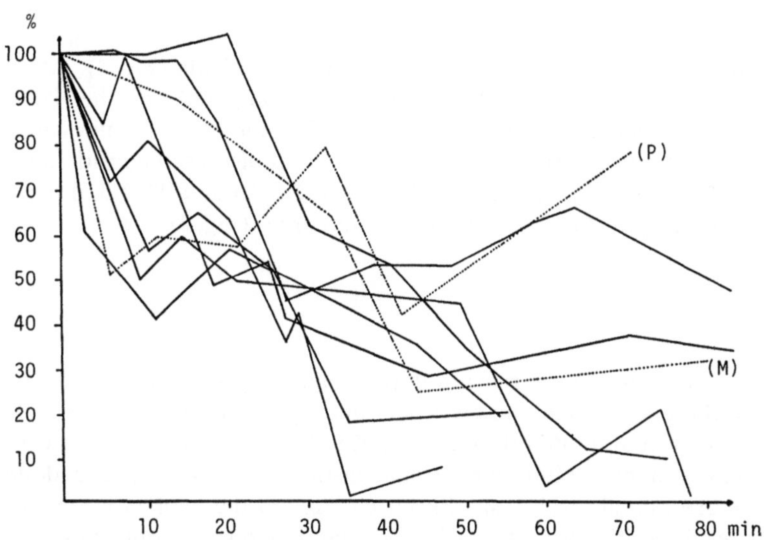

Fig. 3. Percentile P2–N3 amplitude values in the real time course up to 80 min following instillation of bupivacaine. Values of dotted lines *(M, P)* were excluded from the statistics

Results

In the 35 min after epidural instillation of bupivacaine the N1–P1 (Fig. 2) and P2–N3 (Fig. 3) amplitude values showed a progressive decrease. This drop in amplitude was statistically significant at the five percent level ($P \leq 0.05$) within the first 3 min,

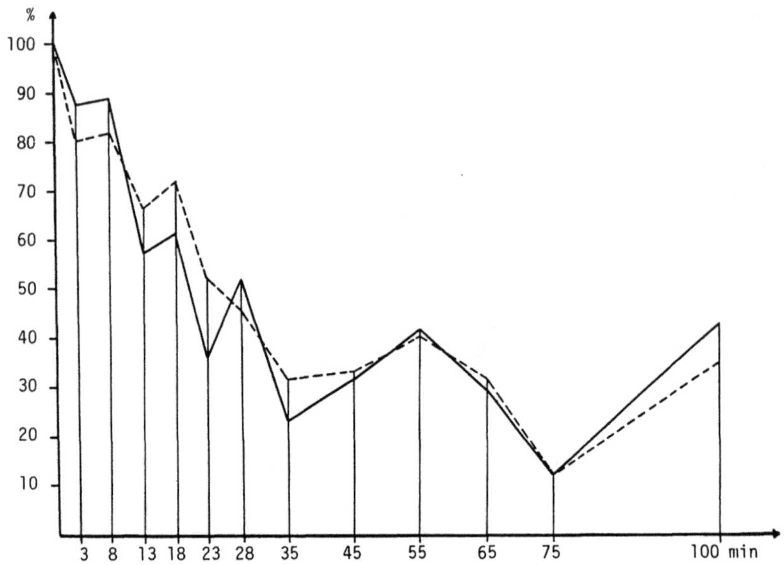

Fig. 4. Arithmetic means of percentile N1–P1 *(solid line)* and P2–N3 *(broken line)* amplitude values in the blocked time course up to 100 min after instillation of bupivacaine

between 8 and 13 min, between 18 and 23 min and from 28 to 35 min (Fig. 4). After 35 min the mean amplitude had dropped to a level of 30% compared with the 100% level of the initial amplitudes. The dispersion of N1–P1 and P2–N3 values was smallest in the period of about 30 min after instillation and increased considerably later on (Figs. 2, 3). The mean percentage amplitudes of N1–P1 and P2–N3 in the 5 and 10 min registration blocks showed fairly congruent curves (Fig. 4).

Discussion

Epidural spinal electrical stimulation reaches in particular the long pathways of the dorsal columns and therefore this method was denominated "Dorsal Column Stimulation" (DCS) by Shealy et al. (1967). Using this form of stimulation for the recording of evoked scalp responses after bupivacaine instillation (which is also done epidurally but at more caudal levels), significant changes in the evoked potentials were found. This result verifies Wall's (1967) aforementioned suggestion, in that it demonstrated at least a partial blockade of long ascending pathways of the dorsal columns. Apart from that, the result demonstrates neurophysiologically the far-reaching distribution and diffusion of a relatively small quantity of a local anesthesic agent instilled in the epidural space. In the investigated patients the distance from the site of instillation to the site of stimulation was from two to seven vertebral levels (four on average). This variation in distance might partially explain the wide range of interindividual potential (N1–P1, P2–N3) amplitude curves. The phenomenon described above could also be observed after instillation of lidocaine (Kepplinger et al. 1983) and procaine (Kepplinger, unpublished material). These results were not considered in the present study, because of the known differing clinical effect of the local anesthesics mentioned, but especially because of their differing biochemical properties with respect to penetration and capacity of binding to the structures of the spinal cord (Cohen 1968; Bromage 1978). In our opinion, however, this method would offer an approach to the investigation of the differing modes of action of anesthetic agents on the spinal cord. By this method also the effect of epidural morphine or systemic (enteral/parenteral) morphine, diezepam, baclofen, etc. application could be investigated.

Corning's idea of direct medication of the spinal cord (1885)—rather utopian at that time—has in our day become a reality, with the clinical use of epidural catheters with depot and pump systems in the fight against intractible pain conditions. A neurophysiological approach to further investigation of the mechanism of action of spinal medication is provided by this method.

References

Bromage PR (1978) Epidural analgesia. Mechanism of action. Saunders, Philadelphia Eastburne Toronto, Chap 4, pp 119–159

Cohen EN (1968) Distribution of local anesthetic agents in the neuraxis of the dog. Anesthesiology 29:1002

Corning JL (1885) Spinal anaesthesia and local medication of the cord. NY Med J 42:483

Kepplinger B (1983) Evoked potentials by means of temporary and permanent implanted spinal cord stimulating systems—preliminary report. Bull Soc Sci Med Grande Duche Luxemb 120: 59–64

Kepplinger B, Wallner K (1985) Technique of evoked potential recordings from a radiofrequency transmitted spinal cord stimulating system. In: Schramm J, Jones SJ (eds) Spinal cord monitoring. Springer, Berlin Heidelberg New York Tokyo, pp 64–66

Kepplinger B, Schmid H, Wallner K (1983) Samples of cortical responses in patients with a permanent cord stimulating system. Electroencephalogr Clin Neurophysiol 54:99

Shealy CN, Mortimer JT, Reswick JB (1967) Electrical inhibition of pain by stimulation of the dorsal columns. Anesth Analg 46:489–491

Wall PD (1967) The laminar organization of the dorsal horn and effects of descending impulses. J Physiol 188:402

Wilkinson GR, Lund PC (1970) Bupivacaine in plasma and cerebrospinal fluid following peridural administration. Anesthesiology 33:482–486

considered in the present study, because of the known differing clinical effect of the [illegible] investigation of the differing modes of action of anesthetic agents on the spinal cord [illegible] the effect of [illegible]

[illegible]

[illegible]

[illegible]

Traumatic Cord Lesions

Acute and Chronic Pathomorphological Reactions to Cord Injury

H. SCHNEIDER[1]

Summary

The complex pathomorphological aspects of spinal cord injury suggest the contribution of different and overlapping factors. The different aspects of the vulnerability of the spinal cord are summarized. It seems to be justified to differentiate between arterial, capillary and venous lesions. The traumatic myelopathy is characterized by the microcirculatory failure at the injured segment and increased permeability and impaired drainage of the postcapillary and venous vessels within and beyond the impact site.

The primary traumatic lesions consist of direct damage to all structures of the injured level of the spinal cord. The autodestructive and progressive involvement of the microvasculature in the central gray matter is characteristic of the early posttraumatic lesions, leading to central hemorrhagic necrosis and focal ischemia. The edema formation and the following edematous necrosis of the central white matter (predominantly in the ventral parts of the dorsal columns) indicates impaired venous drainage as the decisive pathological factor in the delayed posttraumatic phase.

The acute and chronic morphological sequelae of spinal cord injury as the cause of conduction failure for somatosensory evoked potentials involve at least two general problems of spinal cord pathology: first the pathological reactions and the lesion patterns of the white matter, second the vascular factors contributing to the traumatic malacia of the cord.

The essential anatomical picture of spinal cord injury is that of contusion, laceration and compression, either alone or in combination. The morphological changes in the cord after blunt injury depend on the type, extent and dynamic of trauma. The cord suffers first from the direct injury to axons, myelin sheaths, neurons and blood vessels. These primary lesions are followed by a chain of events leading to focal ischemia, breakdown of the microcirculation, edema and, finally, hemorrhagic necrosis of the cord. The reparative phase varies considerably. The posttraumatic myelopathies like arachnoiditis or syringomyelia become manifest months and years after the incident.

1 The Vulnerability of the Spinal Cord

For the better understanding of the traumatic myelopathy some remarks on the vulnerability of the spinal cord may be made in advance, since some rather stereo-

[1]Zentralinstitut für Seelische Gesundheit, I 5, D-6800 Mannheim 1

Spinal Cord Monitoring
Ed. by J. Schramm and S. J. Jones

typic concepts about watersheds and susceptibility to lack of oxygen are incompatible with newer morphological and pathophysiological findings in the spinal cord.

The vulnerability of the spinal cord was first discussed by Adamkiewicz (1881 and 1882), an anatomist in Krakow, considering a vascular factor in the pathogenesis of the luetic myelopathy. His post-mortem examinations of the human spinal cord vascularity by injection of red and blue dyes into the extramedullary arteries and veins led him to the concept of vascular territories in the spinal cord, separated from each other, and to the hypothesis that segments with minor vascular supply (e.g. the middle part of the thoracic cord) may exhibit a special susceptibility under conditions of reduced blood supply.

In an coincident study Kadyi (1889), also a Polish anatomist from Lemberg, denied categorically the existence of such vascular territories. Later authors tended to affirm the interpretation of Adamkiewicz. Suh and Alexander (1939) introduced the term "watershed" into the discussion and supposed that the upper thoracic segments are poorly vascularized and therefore at risk. Herren and Alexander (1939) stated: "The thoracic region has the poorest blood supply; viewed from this standpoint it appears to be the most vulnerable". Bolton (1939) postulated another watershed in the spinal cross section between the anterior and posterior horns.

This concept of vulnerability has been refuted by several authors (Gillilan 1958; Corbin 1961; Turnbull 1971). Corbin, reviewing the localization of spinal cord infarcts, found that the cervical and lumbar enlargements, which receive the main arterial tributaries, are affected by ischemic lesions more often than the so-called "unprovided" thoracic segments. This was confirmed by Jellinger (1966) and Schneider (1980a, b). Gillilan (1958) stated, from the anatomist's standpoint, that "the blood supply to the thoracic cord is entirely adequate for the volume of gray matter present, and it is relatively as good as for any other cord segment".

This anatomical statement arguing for the equal and adequate blood supply for every spinal segment has recently been confirmed in functional studies. Hayashi et al. (1983a) have demonstrated that local blood flow and local tissue oxygen metabolism are maintained at constant values within the various spinal cord levels. This means that the vascular supply and blood flow corresponds closely to the needs of the tissue. The same authors (Hayashi et al. 1983b) proved that vascular density, metabolic activity and local blood flow are highest in those regions of the cross section previously mentioned as poorly supplied. The vulnerability of laminae VII and VIII seems to correspond to the high metabolic activity and therefore high oxygen consumption.

In many reports on vascular spinal pathology the role of the arterial blood supply has largely been overestimated. The term ischemia is widely used meaning, on the one hand, pure arterial occlusion and, on the other, circulatory arrest due to microcirculatory failure after impact injury or cord compression. In most cases with a syndrome of the anterior spinal artery, the occlusion of the spinal artery cannot be proven (Schneider 1980b). Many studies have shown that the experimental occlusion of radicular arteries or even the anterior spinal artery does not regularly provoke ischemic lesions in the cord (Killen and Forster 1966; Tveten and Loeken 1975; Fried and Aparicio 1973; Shimomura et al. 1968; Hukuda and Wilson 1972). Meanwhile spinal arteries, even the artery of Adamkiewicz (a. radicularis magna), can be ligated in the treatment of arteriovenous malformations without obligatory neurological deficits (Decker et al. 1975).

2 The Vascular Factor in Myelopathy

If we look at the three sections of the intramedullary spinal cord vessels (arteries, capillaries and veins), it is to assume that the different organization of each vessel type (wall, branching, reactivity) would imply a different pathology with characteristic reactions of the injured tissue. The question arises whether it is possible to characterize spinal cord lesions corresponding to the affected vessel type. In fact, some spinal lesions allow clear attribution to the basic vascular process, i.e. arterial occlusion (pure ischemia), microcirculatory failure or impairment of the venous drainage.

2.1 Ischemic Myelopathy

The ischemic myelopathy is seen after cardiac arrest, after arterial occlusion due to arteriosclerosis, clamping or dissecting aneurysm of the aorta or following surgery with resection of segmental arteries.

Experimental spinal cord ischemia represents a classical neuropathological model leading to bilateral gray matter necroses in different animal species, similar to findings in human neuropathology (for literature see Schneider 1980a).

The ischemic period necessary to produce necroses ranges from 15 min in cats and rabbits through 18–21 min in rats to 40 min in dogs.

Ischemic lesions may be discrete, with single neuronal necroses in the intermediate gray (Rexed's laminae VI and VII, Fig. 1a, b). In the progressive vascular myelopathy of the elderly, rarefaction of neurons and neuropil occurs in the central gray matter. Severe ischemic lesions are characterized by bilateral central necroses extending to the laminae VIII and IX and into the posterior horns (Figs. 1c, 2). Only marginal structures of the gray matter are likely to be preserved.

Occlusions of single spinal arteries are not so rare, but spinal cord infarcts are not obligatory when *extra*medullary arteries are obstructed (Zuber et al. 1970; Schneider 1980b). On the other hand, the occlusion of *intra*medullary arteries, which are true endarteries, regularly provokes ischemic infarcts in the cord. Embolic occlusion occurs more often than arteriosclerosis of spinal cord arteries. When multiple small emboli occlude several sulcal and central arteries the resulting infarct may correspond to the territory of the anterior spinal artery including the anterior horns and the ventral and ventrolateral white matter (Fig. 1e). The obstruction of a single pial artery causes a wedge-shaped necrosis of the peripheral white matter (Fig. 1h).

2.2 Microcirculatory Failure

The bilateral hemorrhagic necrosis of the central gray matter represents the typical result of breakdown of the microcirculation.

There are two prototypes of hemorrhagic gray matter necrosis: trauma and embolic (or vasotoxic) damage to the microvasculature. Figure 3 demonstrates schematically 3 stages of the change after blunt spinal cord injury, beginning with centrally located multifocal hemorrhages (Fig. 3a), progressing to subtotal hemorrhagic infarction of the gray matter (Fig. 3b) and to necrosis of the transverse section (Fig. 3c). The other type of microcirculatory disturbance induced by multiple embolization into the intramedullary endarteries or by vasotoxic agents (e.g. the

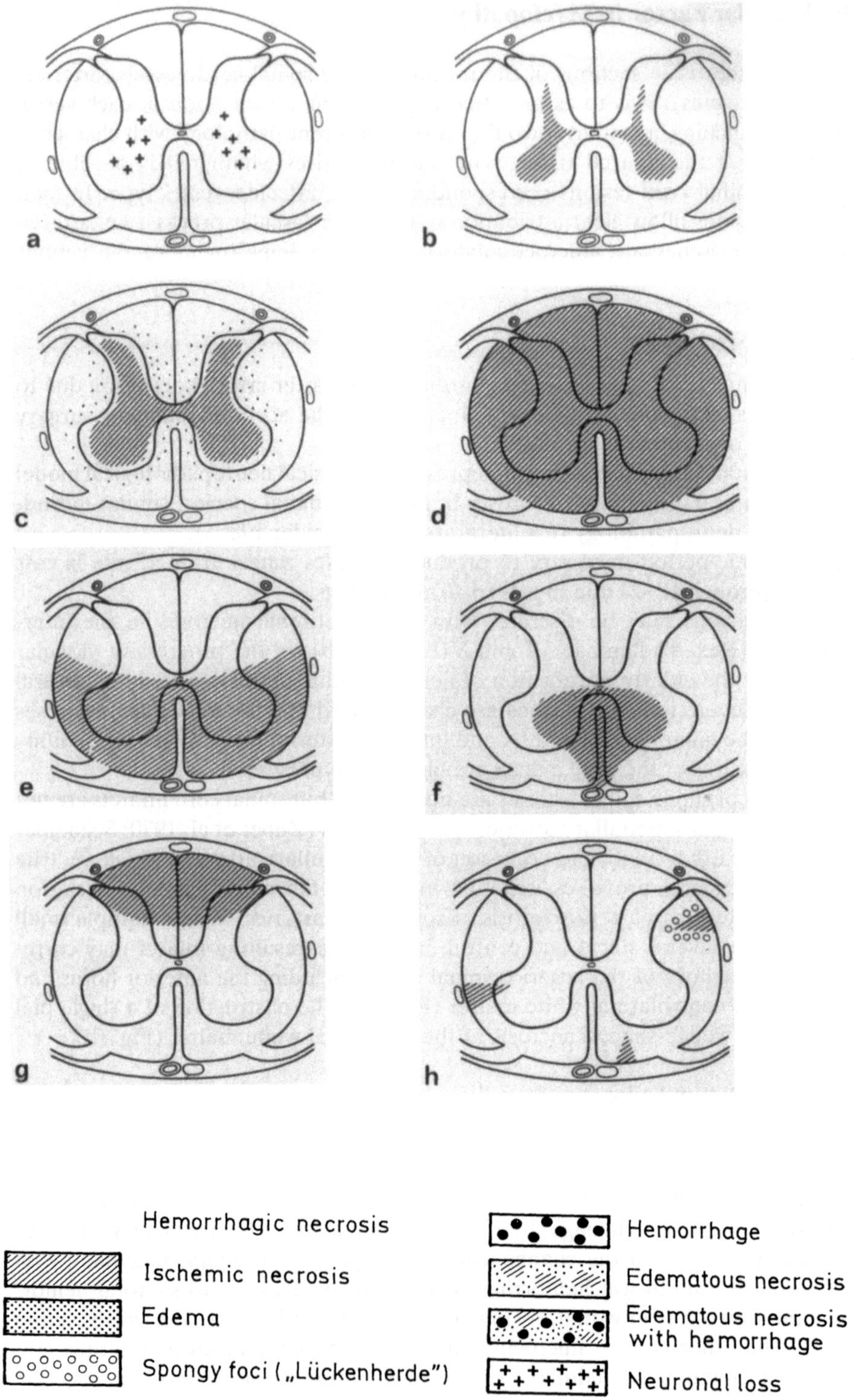
a
b
c
d
e
f
g
h
Hemorrhagic necrosis
Ischemic necrosis
Edema
Spongy foci („Lückenherde")
Hemorrhage
Edematous necrosis
Edematous necrosis with hemorrhage
Neuronal loss

formerly used contrast medium Na-acetrizoate) is represented in Fig. 4. There is a multifocal process with confluent hemorrhages and ischemic necroses in the gray matter, spreading centrifugally into the white matter.

The primary traumatic damage to the microvasculature, mainly capillaries and venules, the accumulation of toxic metabolites or the direct action of vasotoxic substances seem to introduce an autodestructive process in the highly capillarized gray matter. The temporary cessation of perfusion followed by recirculation into the defective vascular bed may reinforce this process. The focal and, later on, confluent necroses spread centrifugally in the adjacent gray and white matter. The peripheral white matter remains intact, but total necrosis of the transverse section may occur.

2.3 Impairment of the Venous Drainage

The pathogenetic principle of impaired venous drainage of the spinal cord appears to be less well defined and perhaps controversial. The venous pathology, however, plays an decisive role in the pathogenesis of white matter lesions (Jellinger 1966; Gillilan 1970; Schneider 1980a, b).

There are only four reports on the morphology of intra- and extramedullary veins (Gillilan 1970; Clemens 1961). Gillilan noted that the venous capillaries in the spinal gray matter can be traced directly from the capillary network into the postcapillary venules. The capillaries sometimes open abruptly into the venules of comparatively greater diameter at the border of the gray matter. This transition characterizes the beginning of the collecting and capacitance veins with low resistance. The larger venules and veins are formed by two venules coming together in a very narrow angle. At this point some authors supposed the existence of valves preventing retrograde filling of the capillary bed. The veins then course radially through the white matter to the pial surface, or can be followed ventrally as central and sulcal veins.

Although the capillary plexus within the spinal cord represents an anastomotic system, intrinsic venous anastomoses occur only exceptionally. The important venous anastomoses are extramedullary.

The drainage via the central and sulcal veins and the anterior medullary vein is less important than the centrifugal system of radially passing veins (vasocorona), emptying in the pial veins and in the dorsal and dorsolateral anastomotic chains. The extramedullary venous network and radicular branches converge to a coarse plexus surrounding the spinal roots and communicate with the internal (epidural) venous plexus. From the epidural plexus there are segmental communications with the external (paravertebral) plexus and from there with the venae cavae and the azygos and hemiazygos veins. The medullary and radicular veins are also interconnected through the vertebral vein system with the intracranial veins and with the veins of the thoracic, abdominal and pelvic cavities. Since the volume of this venous pool ("venous lake"—Batson 1957) is about 150–200 ml—twenty times that of the arterial sector—the circulation is rather slow. Clemens (1961) described the presence of valves in the radicular veins at their junction with the vertebral plexus, which probably limit the backflow of venous blood into the spinal cord.

Fig. 1a–h. Ischemic lesions of the spinal cord (cardiac arrest, anoxia, arterial occlusion). **a–c** Graded gray matter lesions ranging from single neuronal necroses to complete necrosis; **d** complete ischemic myelomalacia; **e–h** types of incomplete myelomalacia after occlusion of the anterior spinal artery (**e**), central artery (**f**), dorsolateral arteries (**g**) or pial arteries (**h**)

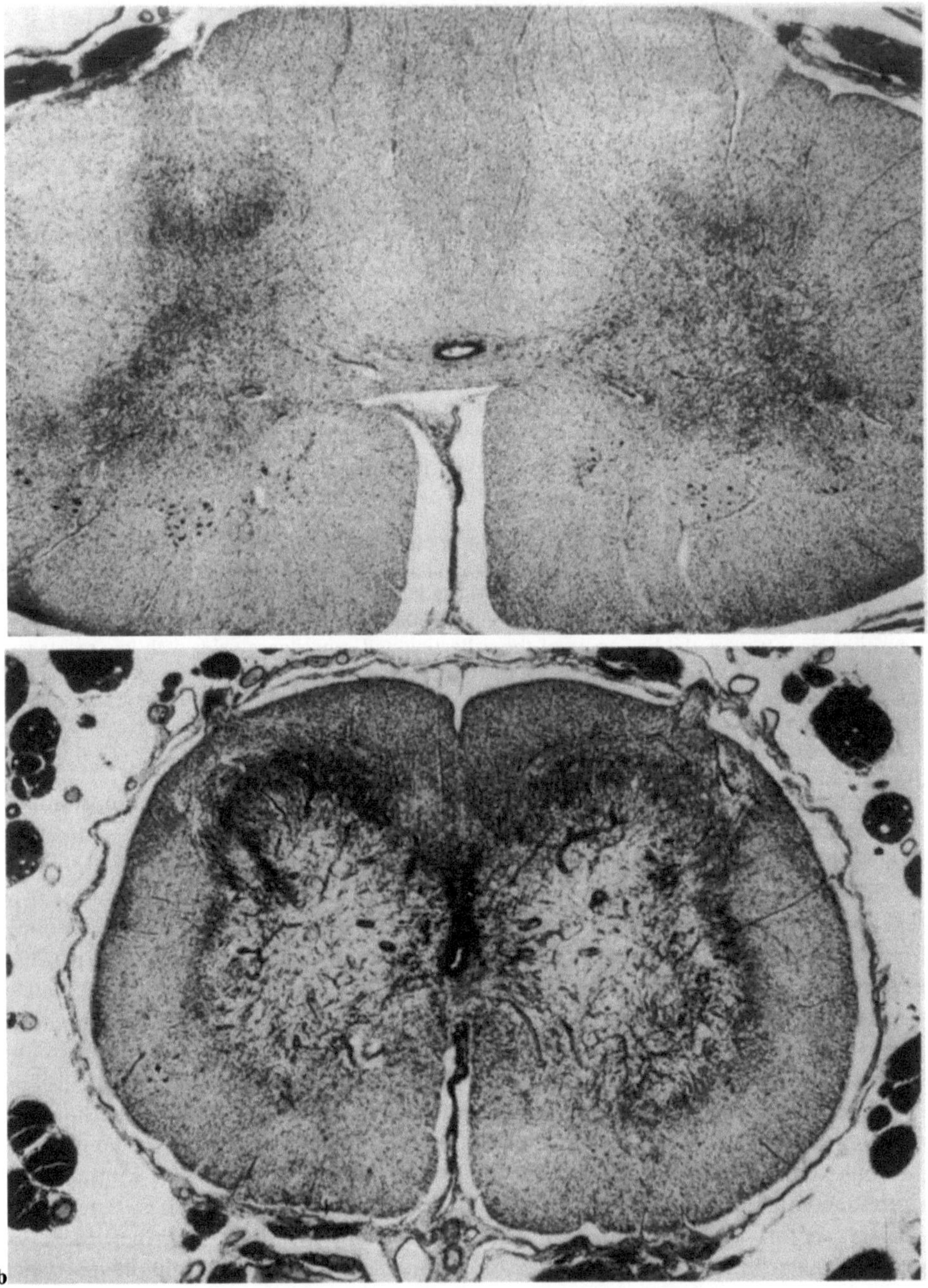

a

b

Fig. 2a, b. Ischemic lesions of the spinal gray matter after temporary cardiac arrest of a newborn infant; **a** cervical and **b** lumbo-sacral cord (Nissl, ×25). Compare with Fig. 1c. Microcirculatory failure

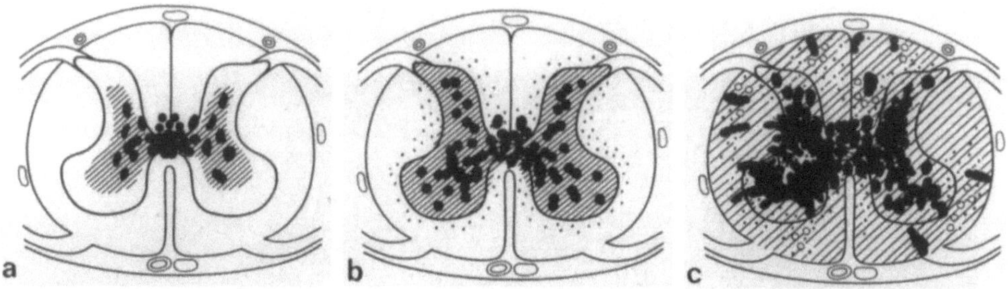

Fig. 3a–c. Stages of the progressive destruction of the terminal vascular bed after blunt spinal cord injury (see also Fig. 9). The centrifugal expansion of the central hemorrhagic necrosis involves primarily the gray matter, later the whole transverse section

One must expect that there will be a wide spectrum of lesions considering the great anatomical and functional differences of the described venous structures. Are there any common pathomorphological patterns which are characteristic of impairment of the venous drainage?

There are indeed some criteria indicating impairment or obstruction of the venous drainage (Fig. 5):

1) Sparing of the gray matter primarily.
2) Increased vascular permeability and edema formation occurring mainly in the white matter,
a) the tendency to "edematous necroses" in the deep white structures, often found as pencil-like necroses or cavitations in the ventral part of the posterior horns ("venous swamp"; Fig. 5a, b),
b) the development of perivenous hemorrhages and hemorrhagic necroses—either in the peripheral white matter (vasocorona, Fig. 5c) or throughout the white matter (Figs. 6 and 7),
c) the combination of central necrosis and peripheral hemorrhages (Fig. 6), and finally edematous malacia of the transverse section (Fig. 7).

Involvement of the venous compartment of the spinal cord has been observed in several experimental conditions. After *impact injury* to the spinal cord, sludging and stasis of the extramedullary veins can be seen followed by a prolonged traumatic dilatation (Ducker 1976). Intramedullary hemorrhages by rupture of vessels also occur first in the venules (Dohrmann et al. 1971; Sasaki et al. 1978). The injection of vasotoxic agents (e.g. the formerly used contrast medium Na-acetrizoate) induces in the vessels of the medullary surface a transient circulatory arrest, followed by prolonged venous dilatation, congestion and extravasation of indicator substances (Margolis et al. 1959). In experimental compression of the spinal cord by a balloon catheter, Griffiths (1975) described the small veins in the intermediate gray matter as the first target of increased vascular permeability. From here the leakage of Evans' blue albumin spread into the dorsolateral white matter. It should be remembered that the posttraumatic hyperemia in the white matter also promotes the increasing venous permeability.

The crucial role of impaired venous drainage has been demonstrated in the model of spinal cord compression. After implantation of an epidural neoplasm (Ushio et al.

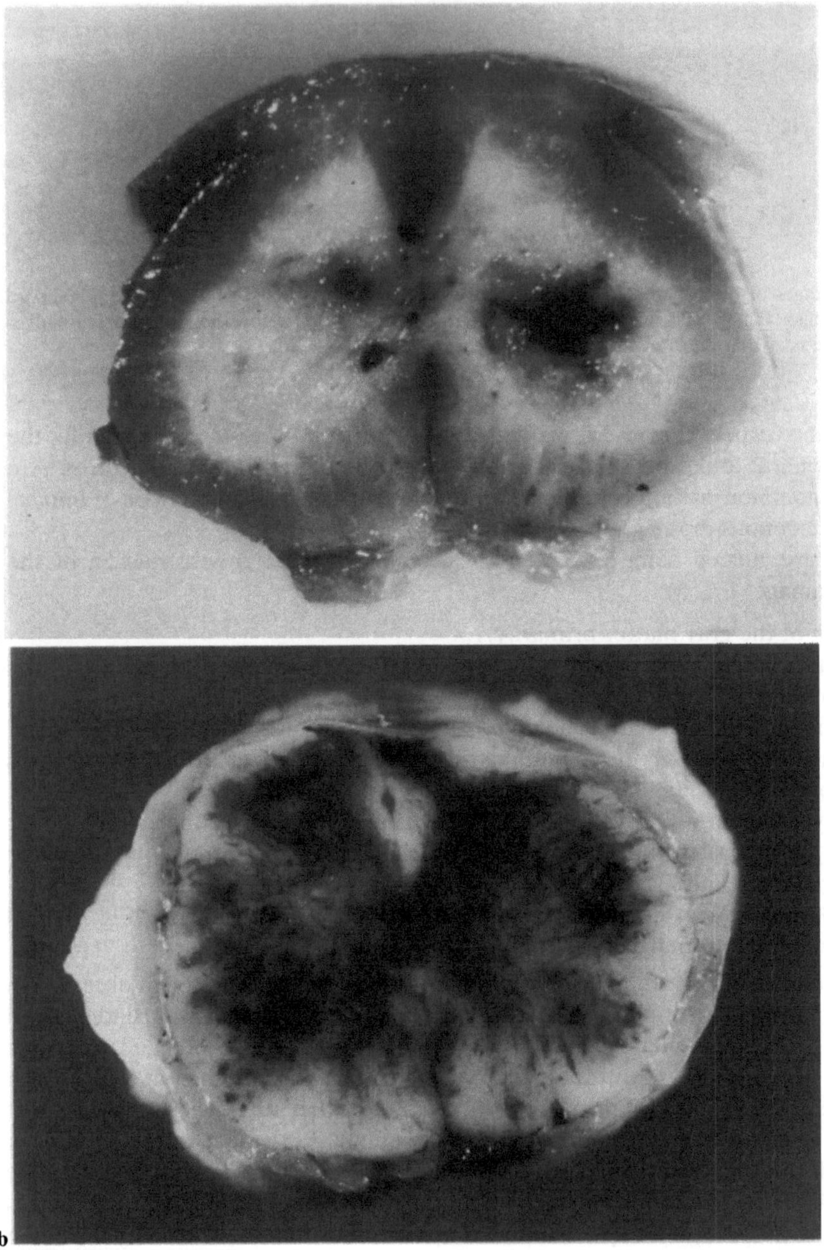

a

b

Fig. 4. a, b. Centrifugal hemorrhagic necrosis induced by the vasotoxic action of Na-acetrizoate (formerly used as angiographic agent). The early stage is characterized by scattered and confluent hemorrhagic necroses within the gray matter. Lumbar spinal cord of the dog after aortography

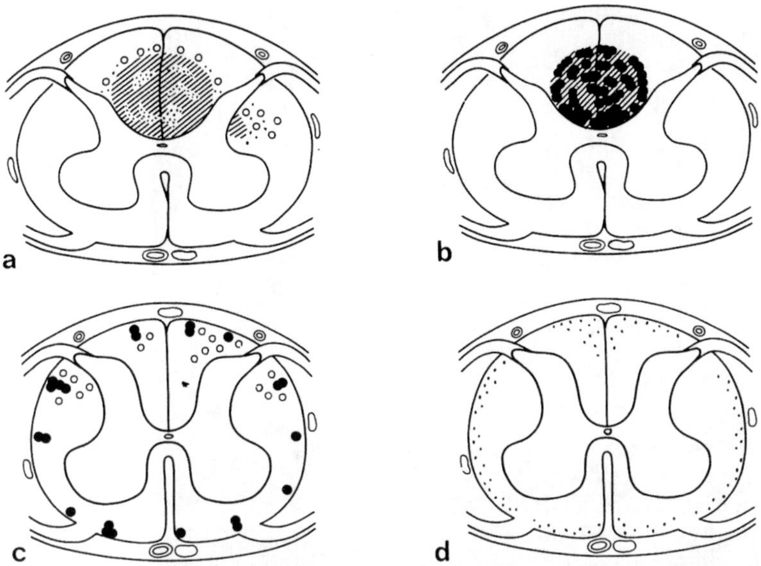

Fig. 5a–d. Spinal lesions of venous type involving mainly the white matter. **a** Edematous necroses in the dorsal and lateral columns; **b** hemorrhagic edematous necrosis; **c** peripheral hemorrhages and focal edema ("Lückenherde") due to the obstruction of pial veins (see also Fig. 6); **d** marginal demyelination during chronic meningitic reactions

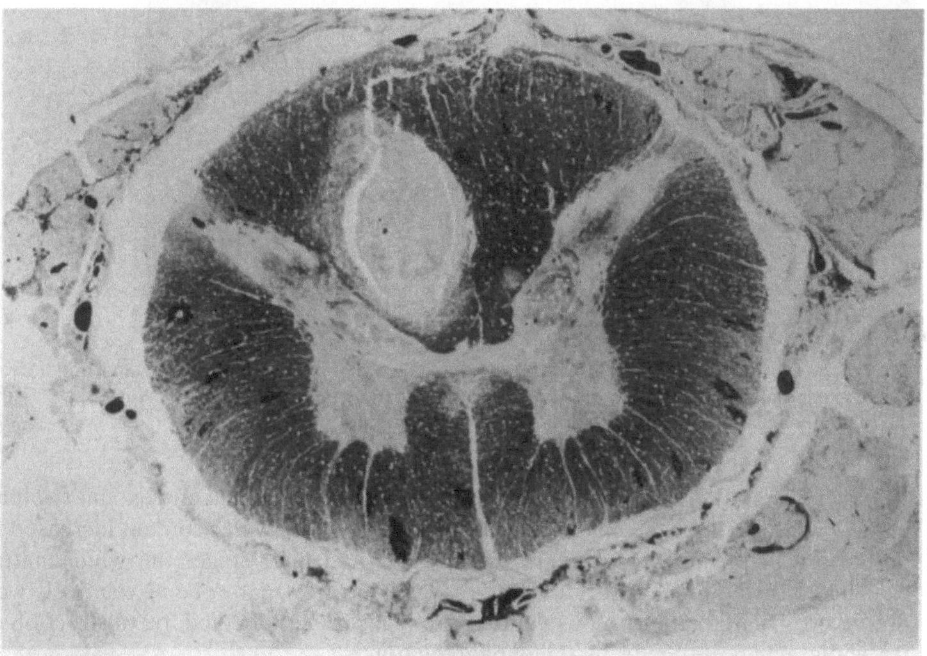

Fig. 6. Multiple hemorrhages in the vasocorona and edematous necrosis in the posterior columns in a case with purulent meningitis. Upper thoracic cord (Heidenhain-Woelcke, × 30)

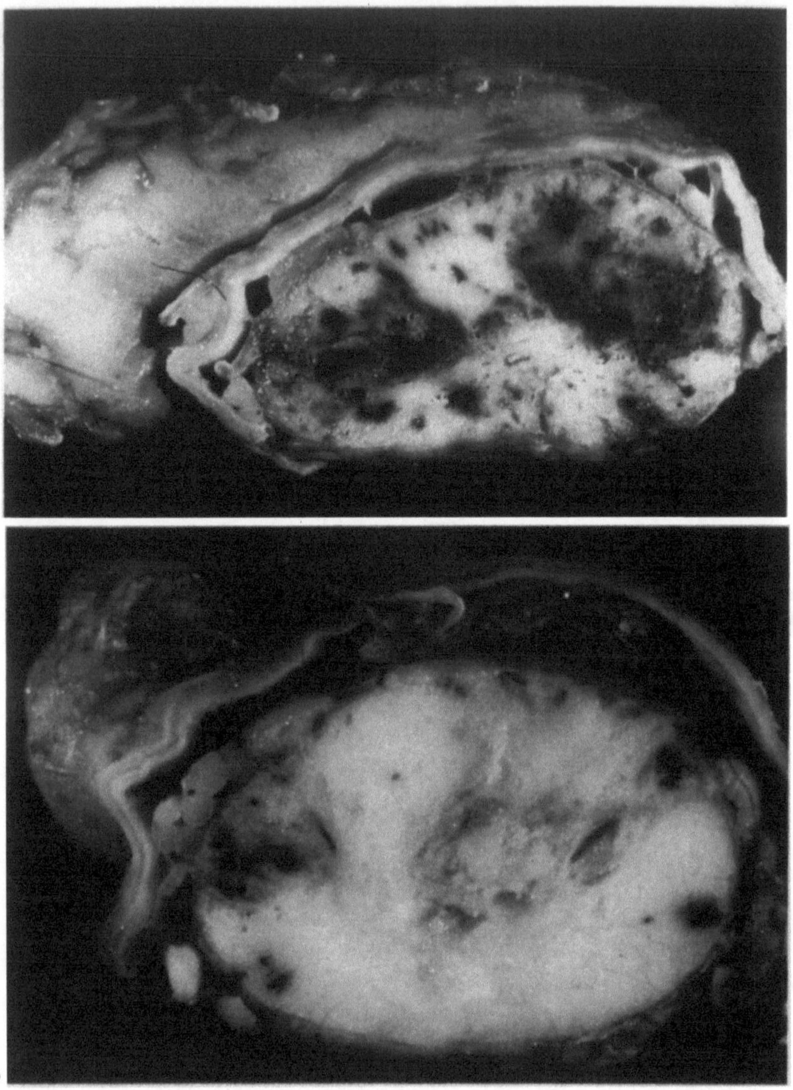

a

b

Fig. 7. a, b. Two different levels (T9 and L1) of a spinal cord of a patient showing edema and hemorrhagic myelomalacia due to compression by a carcinomatous metastasis in the epidural space. Note the predominant involvement of the peripheral white matter

1977a; Ikeda et al. 1980) the first histological signs were edema and axonal swelling in the white matter of the compressed cord segment. The water content increased in proportion to the degree of neurological deficit. In later stages the white matter changes consisted of necrosis and hemorrhages, the gray matter being relatively well preserved. Dilated and tortuous veins at the pial surface reflected the obstruction of the epidural venous plexus by the expanding tumor.

In experiments with antero-posterior compression of the cord it has been shown by angiography (Ramsey and Doppmann 1973; Doppmann 1975; Doppmann and

Girton 1976) that the anterior spinal artery remains intact, whereas the sulcal arteries may exhibit stretching, kinking or obstruction. But only the filling of the dorsal spinal vein is representative for the intramedullary hemodynamics. Paraplegia develops simultaneously with the appearance of filling defects of the dorsolateral extramedullary veins.

Consequently, it seems to be necessary to stress the relevance of the venous sector, whose reactivity and drainage capacity is still poorly understood. The pathological changes of blood volume in brain and spinal cord concern essentially the venous pool. When the postcapillary and collecting venules are damaged the sequelae may closely correspond to the microcirculatory failure described above. The consequence of the rising venous congestion for the antecedent microvasculature seems to be evident, but it can be difficult to differentiate the pathogenetic factors in the vicious circle. Intramedullary edema leading to increased intraspinal hydrostatic pressure impairs capillary and venous perfusion equally.

Besides the effects of vertebral mechanisms, the medullary veins can be affected directly in meningitic processes. In purulent meningitis the accompanying phlebitis reaches from the subarachnoid space radially into the peripheral white matter leading either to hemorrhages or to foci of spongy demyelination ("Lückenherde", Fig. 6). The venous factor in the chronic obliterating arachnoiditis is still unclear, but the marginal demyelination and the development of central cavitating necroses ("syringomyelia") can also be interpreted—at least partly—as the result of the severely impaired venous drainage.

3 The Morphology of the Spinal Cord After Impact Injury

3.1 The Acute Traumatic Changes

The basic pattern of early traumatic change is the *central hemorrhagic necrosis* involving the central, intermediate and ventral gray matter, to a lesser degree also the posterior horns. The hemorrhagic type of central necrosis corresponds closely to that occurring after rapid compression of the cord by a balloon catheter (Griffiths 1975). One of the first abnormalities seen immediately after injury is a perivenous diapedesis on the cord surface (Ducker 1976). These changes are already present in the gray matter 90 s after impact injury, indicating that the capillary and venous hemorrhages are partly caused by rupture of the vessel wall (Figs. 8, 9; Sasaki et al. 1978). This was shown in our experiments with rats using the weight dropping technique (80 g.cm) on the lumbosacral cord, and has been described in other species (Ducker et al. 1971; Wagner et al. 1971; Dohrmann et al. 1971; Fairholm and Turnbull 1971). Some of the venular defects are closed by an accumulation of thrombocytes, whereas the densely packed erythrocytes outside indicate the primary expulsion through the bursted vessel wall (Fig. 8).

Other venules show distension and marked regressive changes of the endothelial cells, and partial loss of the interendothelial tight junctions (Sasaki et al. 1978; Dohrmann et al. 1971).

Other primary lesions are single neuronal necroses and shearing damage to axons and myelin sheaths. In the neuropil there are swollen glial processes and myelin sheaths with enlarged hydropic periaxonal spaces and intramyelinic vacuoles.

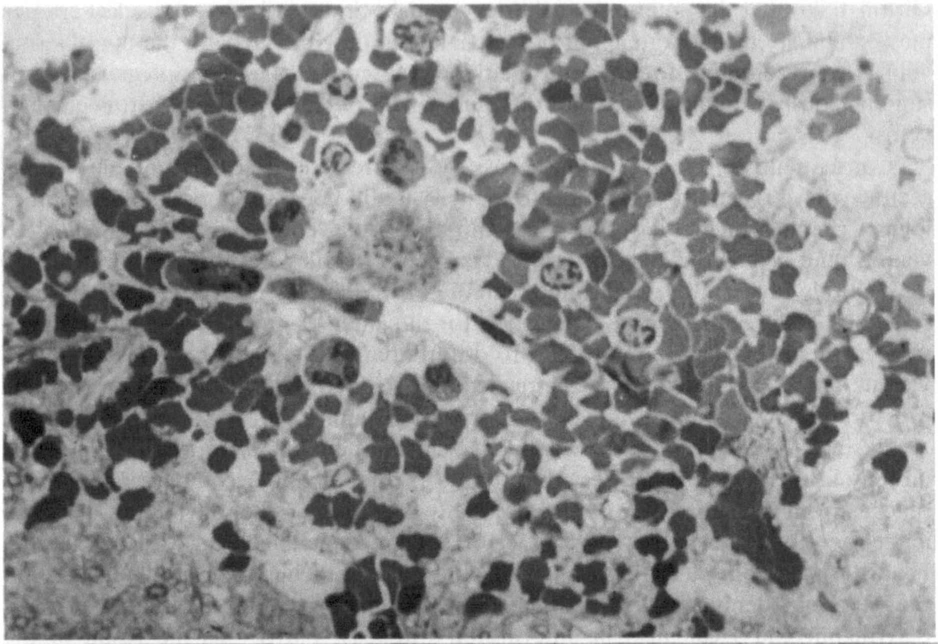

a

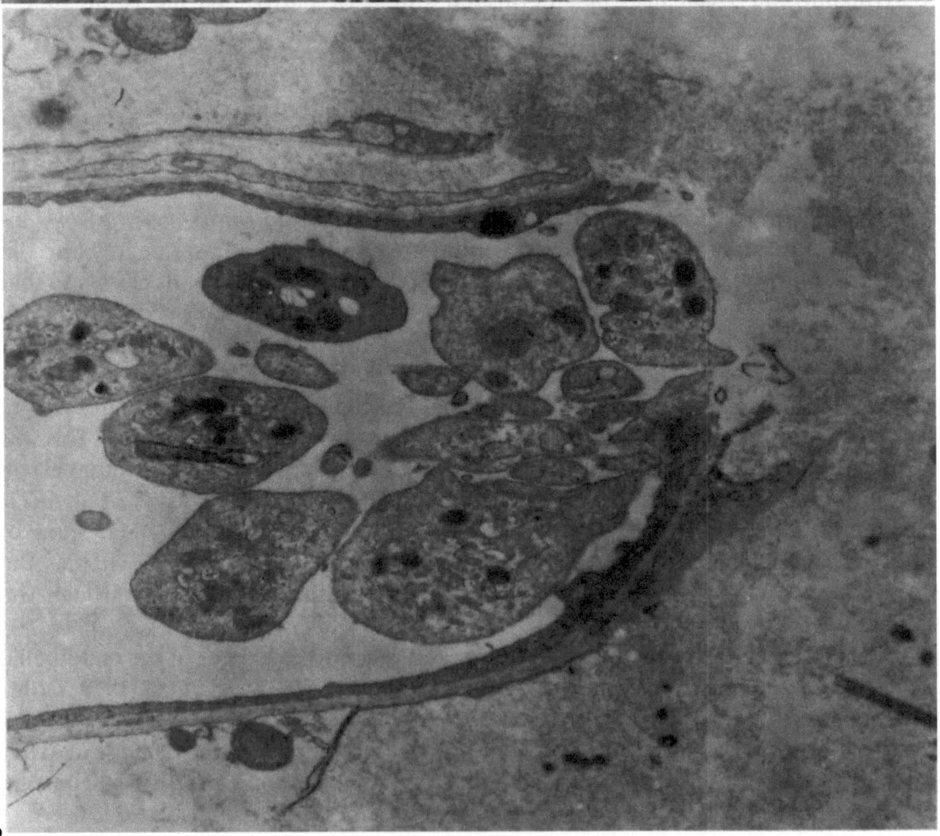

b

3.2 Secondary Posttraumatic Changes

The microangiographic studies in rabbits and dogs of Fairholm and Turnbull (1971) showed the progressive nonfilling of small vessels in the central gray matter and the ventral and posterior horns 2–4 h after impact injury using 200–400 g.cm. The area of nonfilling was proportional to the injury force and the time elapsed following trauma, and showed extravasations of the barium perfusate and necrosis of all elements including capillaries.

Figure 9 demonstrates the rapid progression of the traumatic necrosis in the rat (80 g.cm weight dropping technique) in transverse and longitudinal direction. There are large extravasations of Evans blue injected 5 min before sacrifice by perfusion fixation. Immediately after trauma (90 s, Fig. 9a) there are multifocal extravasations within and outside the traumatic zone. Five minutes after trauma the increased vascular permeability becomes visible in the enlarged peritraumatic zone, involving also the central and peripheral white matter. Sixty minutes after trauma leakage of the tracer is found mainly in the periphery, whereas the necrotic and edematous centre remains unstained due to local cessation of perfusion.

The following hours are characterized by enlargement and confluence of the central necrosis, shifting into the ventral and dorsal horns and into the adjacent white matter. The velocity and degree of this process are by and large proportional to the severity of trauma. The temporal sequence appears rather uniform. Within hours the transverse lesion may be complete, leading to necrosis of the cross section. The peripheral white matter, however, may be spared. Experimental data have shown that the traumatic destruction of the microvessels in the gray matter is followed by leakage of blood cells and plasma, space-occupying edema, impaired and finally halted circulation. The leakage is presumably due to direct tissue injury and secondary release of endogenous biogenic amines which may contribute to continuing permeability and progressive microcirculatory failure. The traumatic edema shows ultrastructurally both vasogenic and cytotoxic criteria (Goodman et al. 1976). Increased vascular permeability is first seen in the intermediate gray matter. Presumably the progressive edema leads to a further impairment of the venous outflow. Since the venous drainage is from the inside of the cord to the periphery, the outflow insufficiency endangers primarily the central white matter of the posterior columns (Jellinger 1976).

3.3 Late and Defective Stages

The *reparative phase* is characterized by resorption and liquefaction of the necrotic masses with the tendency of cyst formation above and below the injured segment. Structures of the gray matter with incomplete necrosis develop intensive astrocytic reaction and vascular proliferation. The terminal vascular bed of the injured gray matter is replaced by coarse, tortuous and thick-walled vessels.

Fig. 8a, b. Ruptured venules in the central gray matter in the rat lumbar cord, 90 s after blunt trauma (80 g.cm). **a** Semithin section, × 440). **b** Leaking venule with exudation of a flocculent (proteinaceous) material in the enlarged extracellular space. The vascular defect is occluded by adherent thrombocytes (ultrathin section, × 16,000)

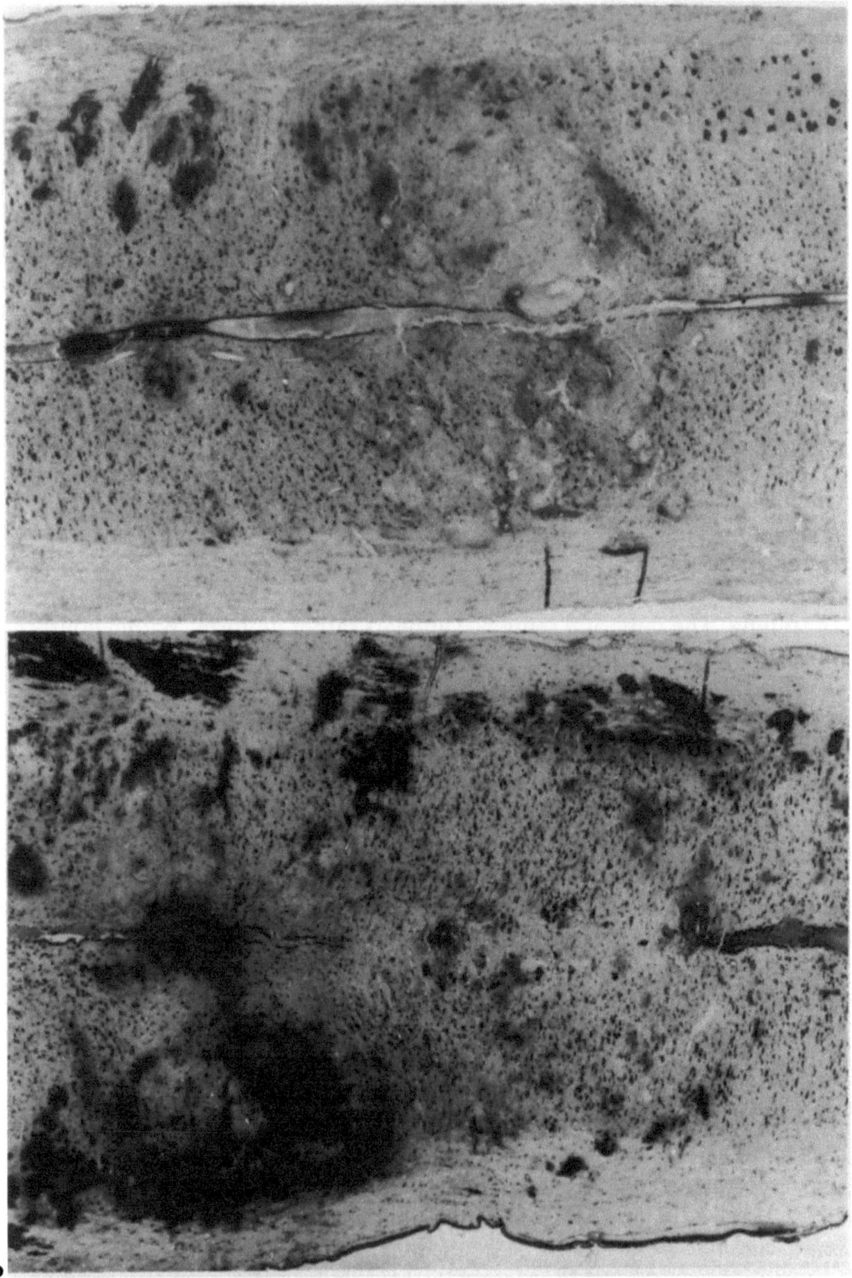

a

b

Fig. 9a–c. Rat lumbar cord 90 s (**a**), 5 min (**b**) and 60 min (**c**) after blunt trauma (80 g.cm). The dark spots represent extravasations of Evans blue injected 5 min before death. The time sequence demonstrates the growing central necrosis (Fig. 3) surrounded by the zone with increased vascular permeability and shifting edema

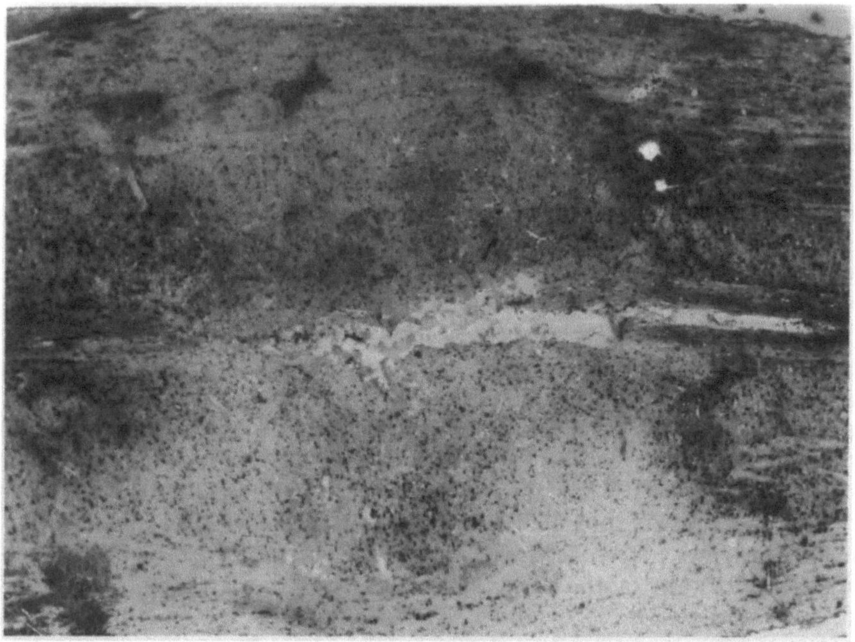

c

The organization and resorption of the necrotic structures, however, is delayed due to the disappearance of the neuroglia and destruction of the mesenchyma. Larger necroses cannot be replaced by glial or mesenchymal scars. The leptomeninges become thickened and fibrotic from residual hemorrhages, sometimes with obliteration of the subarachnoid space. The cicatrication of the mesenchymal structures and the transformation of the remaining medullary tissue may induce different types of posttraumatic myelopathies, for example posttraumatic arachnopathy, cystic degeneration of the cord ("traumatic syringomyelia") and cervical spondylotic myelopathy.

3.4 The Traumatic Myelin Changes in the Spinal Cord

In animals with transient traumatic paraparesis the primary central gray lesion and the white matter edema are less severe. Permanent traumatic paraplegia is the result of the progressive damage to the white matter. The question arises whether therapeutic factors can stop the autodestructive microcirculatory failure and the spreading edema.

Since morphological changes require time to become visible, it is difficult to prove myelin damage of primary traumatic origin. Presumably there are acute shearing forces damaging myelin sheaths and axons of the spinal white matter at the level of contusion, comparable to the diffuse white matter shearing injury of the brain (Adams 1982). Up until now it is not clear, however, whether there is a functional block or a definite morphological lesion when paralysis and loss of evoked potential transmission are present immediately after acute spinal cord trauma. In any case, a vascular factor such as ischemia can be excluded as the cause of the acute functional loss, since the perfusion rate diminishes much later. Kobrine (1980) has demonstrated that long tract neural conduction in the spinal cord seems to be quite resistant to the effects of ischemia alone, disappearing only after 8–18 min of absolute ischemia. Later on, 1–4 h after cord injury, the microvascular reactions and the spreading edema overlap with the hypothetical primary changes, leading to necrosis of the central white matter.

Griffiths and McCulloch (1983) described (in the cat, 90 min after impact injury) the early edema in the base of the dorsal columns, which was replaced by a cystic area 7 days later. The cats showed varying degrees of paraparesis which usually improved over the first week. The early myelin changes were enlarged periaxonal spaces and intramyelinic vacuoles. The necrotizing process, however, with destruction of the neuroglial and mesenchymal elements, allowed no further examination. In the transitional zone of damage, between center and cord periphery, the changes were more discrete, consisting of paranodal changes and retraction of lateral loops of the sheath as signs of partial demyelination. Some axons showed swelling and accumulation of organelles. Griffiths and McCulloch suggested that the earliest paranodal changes may reflect the direct mechanical injury. The partial demyelination increases in severity in the intermediate dorsal column 24 h after trauma. After 3 days many demyelinated fibres are present, intermingled with normal and degenerating axons. The majority of naked axons, however, appear morphologically normal and undergo remyelination 10–14 days after trauma.

Data regarding the posttraumatic changes of the blood flow in the spinal white matter are contradictory. Some authors noted a marked decrease, others an increase of white matter perfusion (Griffiths 1975; Dohrmann et al. 1971; Kobrine et al. 1975). Sandler and Tator (1976a, c) stated that there was first a decrease of the white matter blood flow and after 24 h an hyperemic perfusion. Local hyperemia may be followed by dilatation, prolonged circulation time and increased permeability of capillaries and veins. Studies with Evans blue-albumin showed that leakage of the marker occurs centrally, spreading peripherally and also in a longitudinal direction. In a first step the protein rich fluid passes through the defective vessel wall in the traumatized regions. When the perfusion ceases within the central necrosis increasing permeability develops in the surrounding white matter and in the gray matter rostral and caudal to the level of injury (Nemecek et al. 1977; Sasaki et al. 1978). The spreading vasogenic edema is followed by marginal hemorrhages from leaking veins (Fig. 9) and transforms the ventral part of the dorsal columns into a more or less circumscribed cylindrical necrosis, which may be plurisegmental. Jellinger (1966) first insisted on the intimate connexion between impaired venous drainage and the pathogenesis of the pencil-like (cavitating) necroses. This type of edematous necrosis originates usually from myelomalacic processes or hemorrhages, and is generally situated in the ventral part of the dorsal columns. The cylindrical necrosis may have an hemorrhagic component when the adjacent veins are damaged.

The liquefaction and resorption of the necrotic masses are delayed and incomplete, since the glial and mesenchymal elements in the vicinity undergo regressive changes and do not participate in the degradation of the necrotic white matter. Vascular proliferation is generally missed. The cavitating plurisegmental necroses are not specific for spinal cord trauma, but they are also observed in spinal infarct hemorrhages or in combination with extramedullary space occupying processes (Fig. 7) or meningitis (Fig. 6).

References

Adamkiewicz A (1881) Die Blutgefäße des menschlichen Rückenmarkes. I. Die Gefäße der Rückenmarkssubstanz. S-B Akad Wiss Wien, Math-Nat Kl Abt 3, 84:469–502

Adamkiewicz A (1882) Die Blutgefäße des menschlichen Rückenmarkes. II. Die Gefäße der Rückenmarksoberfläche. S-B Akad Wiss Wien, Math-Nat Kl Abt 3, 85:101–130

Adams H (1982) Diffuse axonal injury due to nonmissile head injury in humans. Ann Neurol 12: 557–562

Batson OV (1957) The vertebral vein system. Am J Roentgenol 78:195–212

Bolton B (1932) The blood supply of the human spinal cord. J Neurol Psychiatry 2:137–148

Bresnahan J, King JS, Martin GF, Yashon D (1976) A neuroanatomical analysis of spinal cord injury in the rhesus monkey. J Neurol Sci 28:521–542

Clemens HJ (1961) Die Venensysteme der menschlichen Wirbelsäule. De Gruyter, Berlin

Corbin JL (1961) Anatomie et pathologie artérielle de la moëlle. Masson, Paris

Decker RE, Stein HL, Epstein JA (1975) Complete embolization of artery of Adamkiewicz to obliterate an intramedullary arteriovenous aneurysm. J Neurosurg 43:486–489

Di Chiro G, Wener L (1973) Angiography of the spinal cord. A review of contemporary techniques and applications. J Neurosurg 39:1–29

Dohrmann GJ, Wagner FC Jr, Bucy PC (1971) The microvasculature in transitory traumatic paraplegia. An electron microscopic study in the monkey. J Neurosurg 35:263–271

Doppmann JL (1975) The mechanism of ischemia in anteroposterior compression of the spinal cord. Invest Radiol 10:543–551

Doppmann JL, Girton M (1976) Angiographic study of the effect of laminectomy in the presence of acute anterior epidural masses. J Neurosurg 45:195–202

Ducker T (1976) Experimental injury of the spinal cord. In: Handbook of clinical neurology, vol 25 (Injuries of the spine and spinal cord). North-Holland, Amsterdam Oxford New York, pp 9–26

Ducker T, Kindt G, Kempe G (1971) Pathological findings in acute experimental spinal cord trauma. J Neurosurg 25:700–707

Fairholm DJ, Turnbull IM (1971) Microangiographic study of experimental spinal cord injuries. J Neurosurg 35:277–286

Fried LC, Aparicio O (1973) Experimental ischemia of the spinal cord. Histologic studies after anterior spinal artery occlusion. Neurology (Minneap) 23:289–293

Gillilan LA (1958) The arterial blood supply of the human spinal cord. J Comp Neurol 110:75–103

Gillilan LA (1970) Veins of the spinal cord. Neurology (Minneap) 20:860–868

Goodman JH, Bingham WG Jr, Hunt WE (1976) Ultrastructural blood-brain barrier alterations and edema formation in acute spinal cord trauma. J Neurosurg 44:418–423

Griffiths IR (1975) Vasogenic edema following acute and chronic spinal cord compression in the dog. J Neurosurg 42:155–165

Griffiths IR, McCulloch MC (1983) Nerve fibres in spinal cord impact injuries. Part 1. Changes in the myelin sheath during the initial 5 weeks. J Neurol Sci 58:335–349

Griffiths IR, Miller R (1974) Vascular permeability to protein and vasogenic edema in experimental concussive injuries to the canine spinal cord. J Neurol Sci 22:291–304

Hayashi N, Green BA, Gonzalez-Carvajal M, Mora J, Veraar P (1983a) Local blood flow, oxygen tension, and oxygen consumption in the rat spinal cord. Part 1: Oxygen metabolism and neuronal function. J Neurosurg 58:516–525

Hayashi N, Green BA, Gonzalez-Carvajal M, Mora J, Veraar P (1983b) Local blood flow, oxygen tension, and oxygen consumption in the rat spinal cord. Part 2: Relation to segmental level. J Neurosurg 58:526–530

Herren RY, Alexander L (1939) Sulcal and intrinsic blood vessels of human spinal cord. Arch Neurol Psychiat 41:678–691

Hukuda S, Wilson CB (1972) Experimental cervical myelopathy: Effects of compression and ischemia on the canine cervical cord. J Neurosurg 37:631–652

Ikeda H, Ushio Y, Hayakawa T, Mogami H (1980) Edema and circulatory disturbance in the spinal cord compressed by epidural neoplasms in rabbits. J Neurosurg 52:203–209

Jellinger K (1966) Zur Orthologie und Pathologie der Rückenmarksdurchblutung. Springer, Wien New York

Jellinger K (1967) Spinal cord arteriosclerosis and progressive vascular myelopathy. J Neurol Neurosurg Psychiatry 30:195–206
Jellinger K (1976) Neuropathology of spinal cord injuries. Hand Clin Neurology 25:43–121
Kadyi H (1889) Über die Blutgefäße des menschlichen Rückenmarkes. Gubrynowicz und Schmidt, Lemberg
Killen DA, Forster JH (1966) Spinal cord injury as a complication of contrast angiography. Surgery 59:969–981
Kobrine AI, Doyle TF, Martins AN (1975) Local spinal cord blood flow in experimental traumatic myelopathy. J Neurosurg 42:144–149
Kobrine AI (1980) Relative vulnerability of the brain and spinal cord to ischemia. J Neurol Sci 45:65–72
Margolis G, Griffin AT, Kenan PD, Tindall GT, Riggins R, Fort L (1959) Contrast medium injury to the spinal cord. The role of altered circulatory dynamics. J Neurosurg 16:390–406
Nemecek S, Petr R, Suba P, Roszival V, Melka O (1977) Longitudinal extension of oedema in experimental spinal cord injury. Acta Neurochir 37:7–16
Ramsey R, Doppmann JL (1973) The effects of epidural masses on spinal cord blood flow. Radiology 107:99–103
Sandler AN, Tator CH (1976a) Review of the effect of spinal cord trauma on the vessels and blood flow in the spinal cord. J Neurosurg 45:638–646
Sandler AN, Tator CH (1976c) Effect of acute spinal cord compression injury on regional spinal cord blood flow in primates. J Neurosurg 45:660–676
Sasaki S, Schneider H, Renz S (1978) Microcirculatory disturbances during the early phase following experimental spinal cord trauma in the rat. Adv Neurol 20:423–431
Schneider H (1980a) Kreislaufstörungen und Gefäßprozesse des Rückenmarks. In: Doerr W, Seifert S, Uehlinger E (Hrsg) Spezielle pathologische Anatomie, Vol 13/I: Pathologie des Nervensystems I. Springer, Berlin Heidelberg New York
Schneider H (1980b) Das sogenannte Verschlußsyndrom der Arteria spinalis anterior. Verh Dtsch Ges Neurol 1:667–670
Schneider H, Dralle J (1973) Ultrastructural changes in the rat spinal cord after temporary occlusion of the thoracic aorta. Acta Neuropathol (Berl) 26:301–315
Schneider H, Renz S, Stoltenburg G, Sasaki S (1974) Microcirculatory disturbances in the canine spinal cord produced by contrast media in aortography. In: Cervós-Navarro J (ed) Pathology of cerebral microcirculation. De Gruyter, Berlin, pp 256–266
Shimomura Y, Hukuda S, Mizuno S (1968) Experimental study of ischemic damage to the cervical spinal cord. J Neurosurg 28:565–581
Suh TH, Alexander L (1939) Vascular system of the human spinal cord. Arch Neurol Psychiatry (Chic) 41/4:659–677
Turnbull IM (1971) Microvasculature of the human spinal cord. J Neurosurg 35:141–147
Turnbull IM (1972) Blood supply of the spinal cord. In: Vinken PJ, Bruyn GW (eds) Handbook of clinical neurology, vol 12. Elsevier, Amsterdam Oxford New York, pp 478–491
Tveten L (1976e) Spinal cord vascularity. V. The venous drainage of the spinal cord in the rat. Acta Radiol (Diagn) (Stockh) 17:653–662
Tveten L, Loeken AC (1975) Spinal cord vascularity. A histopathological and angiographic study of the effects of thoracic-lumbar aortic mobilization in the rat. Neuropathol Appl Neurobiol 1:379–395
Ushio Y, Posner R, Shapiro WR (1977a) Experimental spinal cord compression by epidural neoplasms. Neurology (Chic) 27:422–429
Wagner FC, Dohrmann G, Bucy P (1971) Histopathology of transitory traumatic paraplegia in the monkey. J Neurosurg 35:272–276
Zuber WF, Gaspar HR, Rothschild PD (1970) The anterior spinal artery syndrome—a complication of abdominal aortic surgery: Report of five cases and review of the literature. Ann Surg 172:909–915

Evoked Potentials in Experimental Spinal Cord Injury

T. B. Ducker[1]

Summary

Somatosensory evoked potentials, a reflection of spinal cord conduction, are altered after spinal cord injury in this experimental model. In those animals which recover their motor function, sensory conduction through the cord returns rapidly, often before motor function, and is sustained. In those animals who are rendered paralyzed for at least one week, the initial tendency for a return of evoked potentials in the majority of animals is not sustained after the first few days or one week. The sequential changes in the evoked potentials in this latter group provides further supportive evidence that the ongoing pathophysiological process in certain animals is progressive in nature, so that the initial reparative processes fail along with axonal conduction.

Introduction

In spinal cord injury in the monkey using the contusion model, somatosensory evoked potentials were recorded in an effort to understand better the pathophysiological events as they relate to electrical conduction, blood flow, clinical function, and histopathology. The results of these studies have been published (Ducker et al. 1971, 1978a, b, c). This report includes not only the animals in that publication, but additional animals utilized to further expand the sequential changes observed in cord injured animals where the force of impact was just sufficient to cause either transient paraparesis or transient paraplegia for at least one week. As indicated in our earlier publication, as many as one half of the animals that were paralyzed for one week had the capability of conducting an evoked potential across the lesion site at 2–3 h post injury (Ducker et al. 1978b). Unfortunately that conduction capability disappeared with time. This study amplifies the previous publication, with additional animals, and reproduces the tracings of those animals which best illustrate the changes that occur in evolving pathophysiology after cord injury.

Methods

Rhesus monkeys were used in this experiment and were denied food for 24 h prior to the study. For the initial surgery and experiment, the animals were sedated with sodium thiamytal (15–20 mg/kg) and with phencyclidine hydrochloride (1 mg/kg). Arterial and venous lines were put up to monitor the blood pressure and central

[1]Neurosurgery, University of Maryland, 100 Cathedral Street, Annapolis, MD 21401, USA

Spinal Cord Monitoring
Ed. by J. Schramm and S. J. Jones

venous pressure, and for sampling of blood gases. Other monitoring included, in the majority of animals, determinations of spinal blood flow in spinal cord tissue cases as described in our earlier publication. Once the animal was stabilized and placed on a special animal operating table, the back was shaved and a dorsal lumbar junction two space laminectomy was carried out down to the dura mater. Before injury, immediately after trauma, at 1 h, 2 h, 3 h, 1 day and 1 week, somatosensory evoked potential recordings were made in the animals. The trauma consisted of 500–800 gcm energy injuries. In the 500 gcm injury, 50 g was dropped through a tube of 10 cm. In 800 gcm the weight was dropped from 16 cm. The animals were treated with methylprednisolone, 2 mg/kg at the time of injury and 1 mg/kg every four hours after that for eight doses, as this was a part of the protocol from the previous studies. At 36 h after injury, a special suspension of methylprednisolone was given as a single deep intramuscular injection to provide for a tapering glucocorticoid medication dosage. The injury inflicted rendered well over half of the animals completely paraplegic for one week. The total that remained paralyzed was 15. The number of animals that showed significant recovery was 13. More complete lesions were noted with the higher energy injury. In the majority of the closures, the spinous processes above and below the laminectomy site had immobilizing heavy suture material inserted to keep the animal from flexing and bending the injured cord. One week after the trauma, a second operative procedure was carried out to measure the evoked potentials, various gases, flow, etc. The animals were also sacrificed at that time for pathological study.

Somatosensory evoked potentials were recorded according to the technique previously described (Perot 1973). A Grass Stimulator, Model S44 isolation unit, provided an electrical stimulus to the sural and median nerves. The recordings represented in this study were from the sural nerves. The recording electrodes were placed on the scalp, midway between the inion and nasion, at the midline for the hindlimb potentials. These potentials were transmitted to a four channel EEG preamplifying unit and averaged with the Fabritek, Model FT1074, summating responses to 128 repetitions of the stimulus. The wave forms were displayed on the oscilloscope, recorded on a Polaroid camera and then copied onto illustration paper. The animals were sedated for all the recordings, not only at the time of the surgery, but also one day after injury in order to have a cooperative animal. These recordings were later copied and transposed for grouping and further analysis.

Results

While in the overall experimental design, several grades of neurologic recovery were recorded. This data is available in our previous publications (Ducker et al. 1978a, b). Reviewed and added to at this time are only those animals who remained paraplegic for one week, or who were at least grade 3 or grade 4 one week after injury.

Grade 0 signifies those animals which were completely paraplegic. Grade 3 animals were able to stand and move about their cage but did have some obvious difficulty in their lower extremities. Grade 4 animals had no obvious neurologic deficit in their lower extremities. As recorded in our earlier data, the animals that were completely paraplegic had a progressive drop off in spinal cord blood flow along with diminished tissue oxygen. In animals that recovered from the cord trauma, after a very transient initial diminution of spinal cord blood flow, there was a hyperemic response which was often maintained throughout the week after injury (Ducker et al. 1978a). In the first two hours after injury, the change in spinal cord blood flow was a better predictor of how the animal would eventually do. At one day, the spinal cord evoked potential was a fairly accurate predictor of the neurologic status one week after injury.

In the animals that recovered, to walk about their cage, within one to two hours, more than 90% recovered and maintained a healthy appearing evoked potential. While the amplitude was usually less, the initial N1 and P1 were clearly recognized. These evoked potentials in the animals that recovered appeared before the animal had good motor function. Thus, as represented in Fig. 1, the early return and sustaining of a clearly recognizable evoked potential after experimental cord injury is a good predictor that the animal will have significant motor function.

In the paraplegic animals an entirely different pattern evolved. As seen in Fig. 1, the initial recording after the injury was lost in more than 90% of the animals. Where perceivable evoked potentials were present, in less than 10% of the animals, they were quite abnormal. Within an hour, nearly 20% of animals had a return of their evoked potential. By two hours, this figure rose to 60%, only for it to dwindle to just below 50% within three hours. The seven animals who had clearly recognizable evoked potentials three hours after injury are all plotted together in Fig. 2. Looking at this composite overlay, the pretrauma evoked potential is fairly consistently reproduced between animals. Immediately post trauma, there is no recognizable evoked potential. By one hour, there seem to be some potentials in a few animals, but certainly no primary P1 or N1 wave. By two to three hours many of these animals had an evoked potential, which was diminished in amplitude and on occasion lacked

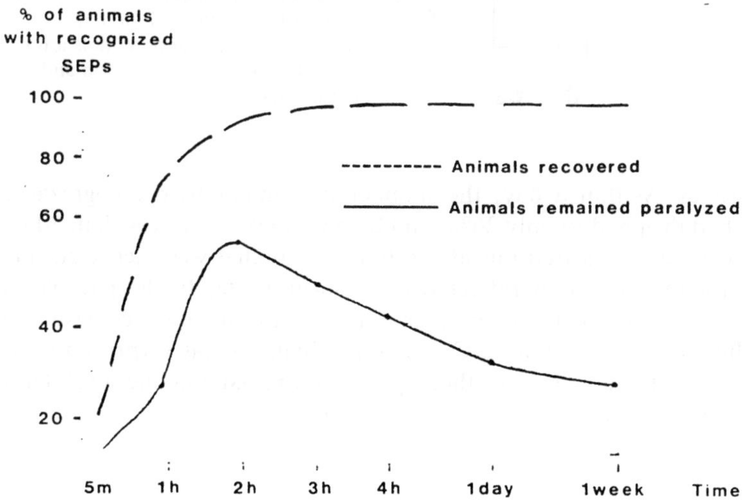

Fig. 1. Graph showing proportion of animals with recognizable somatosensory evoked potentials at various times after trauma. A total of 13 animals showed significant recovery while 15 remained paralysed

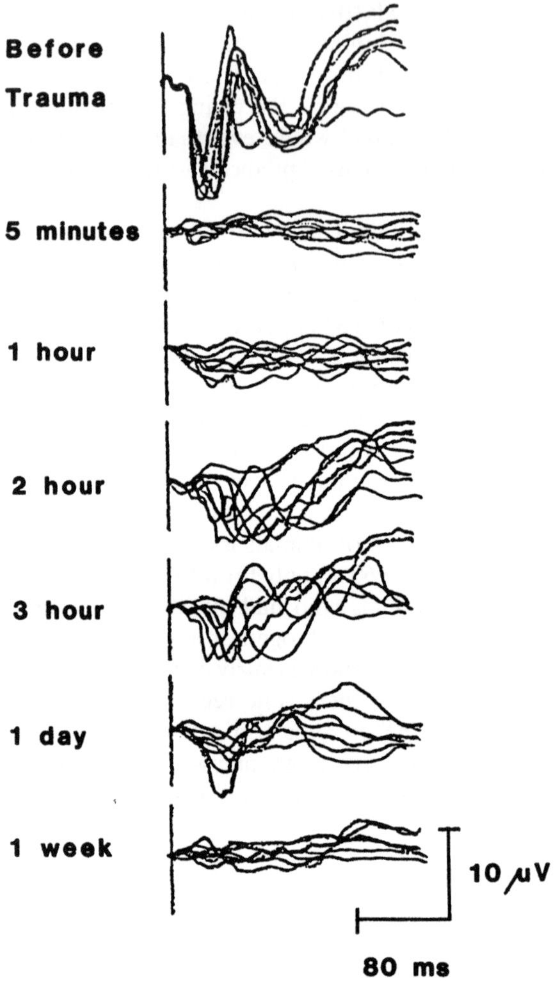

Fig. 2. Superimposed somatosensory evoked potentials recorded at various times after trauma in 7 animals, showing recovery of potentials after 3 h but eventual deterioration and paraplegia

the initial early waves. Within a day, the number of animals with recognizable evoked potentials had dropped to only 26%, and by one week it was less than 20%.

Histopathological studies carried out at the time the animal was sacrificed one week after injury for the most part reflected the clinical grade. In the paraplegic animals, the proportion of viable tissue in the major area of trauma was consistently less than 5% of the cross-sectional area. On the other hand, it was surprising to us that an animal with as little as 8–10% of their spinal cord remaining one week later could ambulate and had fairly normal looking evoked potentials.

Discussion

The prognosis of a spinal cord injured animal soon after injury can be predicted by the evoked potentials. In animals wherein there will be significant recovery, the evoked potential tends to return early, rapidly develop its healthy looking pattern with respectable amplitude, and sustain itself in the initial days after the trauma. On the other hand, an animal with a threshold injury that remains paralyzed in one week may have sequential changes within the evoked potentials themselves, but they are rarely sustained beyond the first day or two. Coupled with spinal cord blood flow determinations, the two factors in the laboratory have a very high predictive capability for the prognosis of the animal.

The sequential changes observed in the somatosensory evoked potential in animals that remain paralyzed for at least one week is one of many pieces of evidence suggesting that the pathology after cord injury evolves. The initial concussive effect, which is associated with a transient and overall ischemia (Dohrmann et al. 1973; Kobrine et al. 1979; Fairholm and Turnbull 1971), obviously adversely effects spinal cord conduction. This is true even in animals with complete recovery. In the majority of animals which remain paralyzed, there clearly appears to be an attempt for conduction to return, only for the spinal cord conduction to fail within the first few days after injury. In previous publications, we have suggested that this may be linked to the vascular support of the spinal cord (Ducker et al. 1971, 1978b). There are now further investigational findings supporting this conclusion (Dohrmann et al. 1973; Fairholm and Turnbull 1971; Sandler and Tator 1976). Maintaining a viable spinal cord by vascular manipulation has thus far failed in all controlled experimental studies. Therefore, whether the vascular observations are epiphenomena or represent true cause and effect is not fully determined (Kobrine et al. 1979).

The nature of the evoked potential waveform abnormality seen after injury is primarily a loss of amplitude of the response. A true shift in latency is difficult, if not impossible, to prove. It appears that various positive and negative components call on different neuronal pools. While the damage may be mostly in posterior columns which will effect the early components, it appears that the anterior columns contribute to the later component. The actual evoked potential may, in its final shape, be more reflective of which axonal pathways are called upon to conduct the response which we see (Perot and Vera 1982; Rowed et al. 1978; Young 1982). If the entire cord is damaged so that there is no motor function, the sensory response will clearly be absent in the majority of animals within the first few days, and definitely by one week.

Clinical material gathered from many centers has confirmed the first report from Rowed et al. (1978) that the early return of an evoked potential is a favorable prognostic sign. Evoked potentials carried out weeks or months after spinal cord injury are not nearly as clinically valuable (Perot and Vera 1982; Young 1982). What the patient states is often more important. If anything, the evoked potential may give some false hope in the subacute phases after injury (Young 1982).

References

Dohrmann GH, Wick KM, Bucy PC (1973) Spinal cord blood flow patterns in experimental traumatic paraplegia. J Neurosurg 38:52–58

Ducker TB, Kindt GW, Kempe LG (1971) Pathological findings in acute spinal cord trauma. J Neurosurg 35:700–708

Ducker TB, Salcman M, Perot PL Jr, Ballantine D (1978a) Experimental spinal cord trauma, I: Correlation of blood flow, tissue oxygen and neurologic status in the dog. Surg Neurol 10: 60–63

Ducker TB, Salcman M, Lucas JT, Garrison WB, Perot PL Jr (1978b) Experimental spinal cord trauma, II: Blood flow, tissue oxygen, evoked potentials in both paretic and plegic monkeys. Surg Neurol 10:64–70

Ducker TB, Salcman M, Daniell H (1978c) Experimental spinal cord trauma, III: Therapeutic effect of immobilization and pharmacologic agents. Surg Neurol 10:73–78

Fairholm DJ, Turnbull IM (1971) Microangiographic study of experimental spinal cord injuries. J Neurosurg 35:277–286

Kobrine AI, Evans DE, Rizzoli HV (1979) The effects of ischemia on long-tract neural conduction in the spinal cord. J Neurosurg 50:639–644

Perot PL Jr (1973) The clinical use of somatosensory evoked potentials in spinal cord injury. Clin Neurosurg 28:367–381

Perot PL Jr, Vera CL (1982) Scalp-recorded somatosensory evoked potentials to stimulation of nerves in the lower extremities and evaluation of patients with spinal cord trauma. Ann NY Acad Sci 388:359–368

Rowed DW, McLean JAG, Tator CH (1978) Somatosensory evoked potentials in acute spinal cord injury: prognostic value. Surg Neurol 9:203–210

Sandler AN, Tator CH (1976) Effect of acute spinal cord compression injury on regional spinal cord blood flow in primates. J Neurosurg 45:660–676

Young W (1982) Correlation of somatosensory evoked potentials and neurological findings in spinal cord injury. In: Tator CH (ed) Early management of acute spinal cord injury. Raven Press, New York, pp 153–165

Somatosensory Evoked Potentials (SEPs) in Spinal Cord Injury

W. YOUNG[1]

Cortical somatosensory evoked potentials (SEPs) were described in 1947 by Dawson. Recorded from scalp and activated by peripheral nerve stimulation, SEPs are noninvasive, robust, and reliable measures of somatosensory function in patients (Allison 1962, 1982; Allison and Hume 1981; Arezzo et al. 1979, 1980, 1981, 1982; Desmedt and Cheron 1982; Larson et al. 1966; Nakanishi et al. 1978; Perot 1973; Perot and Vera 1982). They have been extensively used in spinal cord injury (SCI) (Croft et al. 1972; Cusick et al. 1979; D'Angelo et al. 1973 Dimitrijevic et al. 1978; Dorfman et al. 1978, 1980; Dorfman 1977; Donaghy and Numoto 1969; Ducker et al. 1978; Eisen and Nudleman 1979; Engler et al. 1978; Ertekin 1978; Jones 1977, 1982; Jones and Small 1978; Katz et al. 1978a, b; Larson et al. 1966; Martin and Bloedel 1973; Martin et al. 1980; Nakanishi et al. 1974, 1978; Perot 1973; Perot and Vera 1982; Rowed et al. 1978; Schramm et al. 1979; Shimoji et al. 1979; Singer et al. 1970; Tsumoto et al. 1972; Young 1982; Young et al. 1982) to supplement neurological examination, to assess spinal function in comatose or anesthetized patients, and to detect sensory changes in animal experiments. Although recent advances allow direct recording of subcortical or spinal SEP components (Anziska and Cracco 1980, 1981a, b; Anziska et al. 1978; Arezzo and Vaughan 1982; Arezzo et al. 1981; Beall et al. 1977; Burke et al. 1981; Chiappa et al. 1980; Cracco 1972a, b, 1973; Cracco et al. 1975, 1982; Cracco and Cracco 1976; Desmedt and Cheron 1980, 1982; Dimitrijevic et al. 1978; Ertekin 1978; Jones 1977, 1982; Jones and Small 1978; Kritchevsky and Wiederholt 1978; Pratt et al. 1979; Rossini et al. 1981; Shimoji et al. 1979; Vas et al. 1981; Vasiri et al. 1981; Wiederholt et al. 1982; Yamada et al. 1980), the ease with which cortical SEP can be recorded has made this the most popular spinal cord monitoring technique today.

Over the past 7 years at the NYU-Bellevue Spinal Injury Center, we have collected a large database of cortical SEP in more than 300 SCI patients. SEP were correlated with neurological examinations during both the acute and chronic phases of SCI. Analysis of the database revealed 3 unexpected findings. First, although cortical SEPs obtained early after SCI correlate well with sensory recovery, the incidence of false correlation increased on follow-up studies. Second, 17% of the patients with severe SCI have large long latency negative waves accounting for most of the discrepancies between SEP and sensory function. Third, median nerve SEP amplitudes increase transiently for 2–3 weeks after SCI in paraplegic patients. The implications of these findings will be discussed.

[1] Department of Neurosurgery, New York University Medical Center, 550 First Avenue, New York, NY 10016, USA

Spinal Cord Monitoring
Ed. by J. Schramm and S. J. Jones

Materials and Methods

Patients admitted to the NYU Spinal Trauma Center with a diagnosis of acute SCI and with definite neurological deficits between 1979 and 1984 were studied. On the day of admission, weekly for a month, monthly for 3 months, and then at 6 months and up to a year after SCI, both SEP and neurological examinations were carried out (Berenstein et al. 1983; Young 1982; Young et al. 1982; Young and Berenstein 1985).

SEP responses were recorded from the scalp at Cz-Fz for posterior tibialis (PT) and C3/4-Fz for median (MN) SEP. Scalp potentials were recorded with platinum-iridium needle electrodes and averaged with a Tracor 3000 Evoked Potential System with floppy disk storage. Input filters were set at 10–250 Hz bandpass with −3 dB roll-off. Amplication was 50 microvolt full scale sensitivity. The averager rejected responses exceeding 80% of full scale with a typical rejection rate of 1–3%.

Peripheral nerves were stimulated at 2.3 Hz with bipolar electrodes placed epicutaneously on the MN nerve at the wrist or the PT nerve at the ankles. Constant current (10–20 mA, 0.1 ms) pulses were set to 2× twitch threshold for supramaximal nerve activation at 2.3 Hz rate. In each limb, 3 sets of 200 responses from MN or PT nerve stimulation were averaged, alongside a no-stimulus control. In each test, 12 SEP averages and 4 no-stimulus averages were stored on disk. The procedure typically took 40–60 min to complete. SEP averages are displayed with negative polarity up. The traces are respectively 100 and 200 ms full sweep for MN and PT SEP.

This protocol was developed from trials determining which instrument settings yielded the largest and unambiguous cortical responses in more than 100 spinal-injured patients in our particular recording environment. Parts of this protocol may not be suitable for routine SEP tests. For example, we use needle electrodes rather than surface electrodes because of the greater speed with which needles can be inserted, the ease of maintaining a sterile field during neurosurgery, and our need to compare intraoperative results with SEP obtained before and after surgery. Due to the greater impedance of needle electrodes, typically 5–7 Kilohms, a narrower band-pass filter had to be used. The higher stimulus rate kept the testing time to an hour, i.e. 16 averages of 200 responses. The odd stimulus frequency of 2.3 Hz reduced 60 Hz noise by ensuring there was no consistent phase relation between the noise and the stimulus.

SEP were scored by 3 independent observers who did not know the clinical findings in the patients. The traces were coded and examined separately for MN and PT SEPs. Each SEP was scored on a scale of 0–3 where 0 indicates no response, 1 is a greatly diminished but detectable response, 2 is a response which is relatively large but abnormal in either latency or amplitude, and 3 is normal. In addition to scoring the SEPs in this fashion, the peak-to-peak amplitude of SEP traces within the first 50 ms was measured.

Each patient was examined by a staff neurologist on admission, 1 week, 6 weeks, 6 months, and 1 year after injury. Strength of 14 muscle groups, deep tendon reflexes, pin-prick and touch sensitivity of dermatomes below C1, deep pressure sensibility, joint proprioception, and rectal sphincter tone were documented. Touch and pin sensations were scored on a scale of 0–2 (0 = no, 1 = partial, 2 = normal). Motor

Table 1. Neurological scoring system

Dermatome	Sensory scores				Group	Motor scores		
	Touch	Pin	Maximum	Normalized		Score	Maximum	Normalized
C6	0–2	0–2			Hand flex.	0–5		
C7	0–2	0–2	12	0 → 0	Hand ext.	0–5	15	0 → 0
C8	0–2	0–2		1– 4 → 1	Opp. poll.	0–5		1– 5 → 1
				5– 8 → 2				6–10 → 2
L4	0–2	0–2		9–12 → 3	Foot flex.	0–5		11–15 → 3
L5	0–2	0–2	12		Foot ext.	0–5	15	
S1	0–2	0–2			Opp. poll.	0–5		

Explanation: Scaling and normalization of sensory and motor scores used for comparison with SEP. Each sensory dermatome was graded on a scale of 0–2 where 0 indicated no sensation, 1 is abnormal, and 2 is normal. For motor scores, 0 is paralyzed, 1 indicates barely detectable movement, 2 is ability to move but not against gravity, 3 is movement against gravity but not active resistance, 4 is weak but able to overcome gravity and resistance, 5 means normal strength

function was graded on a scale of 0–5 (0 = paralyzed, 1 = palpable movement, 2 = cannot overcome gravity, 3 = cannot overcome resistance, 4 = overcomes resistance, 5 = normal).

For comparison with MN SEP scores, the sensory scores of C6-8 dermatomes were added for pin and touch separately. Likewise, scores of L4-S1 dermatomes were summed for comparison with PT SEP. The motor scores were likewise summed and normalized on a scale of 0–3. For comparison with MN SEP, hand flexors, hand extensors, and opponens pollicis scores were summed and normalized according to the scale listed in Table 1. For comparison with PT SEP, scores for foot flexors, foot extensors, and extensor hallucis longus were added together.

As part of an ongoing study, 40% of the patients in the database were treated with one of 3 regimes of methylprednisolone: (1) 100 mg initial dose (ID) and 100 mg in 4 divided doses daily for 10 days, (2) 1000 mg ID and 1000 mg daily for 10 days, or (3) 30 mg/kg ID and 3 mg/kg every hour for 23 h. Ten percent of patients received 1–10 mg/kg of naloxone ID and 75% of the loading dose per hour for 23 h. This study has not been completed and will be reported separately.

Results

Of the 300 patients studied, 68% suffered cervical, 28% thoracic, and 4% lumbar injuries. Fifty-eight percent were admitted with no neurological function below the level of the lesion. In this group of so-called "complete" SCI, 93% had no PT SEPs on admission. In patients with residual neurological function, 95% had some PT responses. About 4% of the SEP were deemed uninterpretable due to noise or electrical artifacts present on no-stimulus averages. Excluding studies where SEP were uninterpretable, patients with no SEP from either leg on admission had a 95% probability of no clinically detectable sensory function in the legs. Conversely, SEP in either leg on admission indicated a 97% probability of having some sensory preservation.

In no case were SEPs of normal amplitude and latency associated with severe or complete loss of neurological function. Likewise, we have yet to see a patient with intact sensory function in whom no SEP response could be recorded. In patients with sacral sparing of sensation who were anesthetic in dermatomes supplied by the PT nerves, the majority did not show consistent SEP responses on admission although many proceeded to recover abnormal SEP responses several weeks later. About 7% (20 cases) of patients with no clinically detected sensory function below the lesion on admission had detectable PT SEP responses. Review of these cases indicated that 10 had definite SEP on admission; 6 recovered patches of sensation and 2 had return of some motor function. In the rest, SEPs disappeared days later.

Time Course of SEP Changes

In the acute phase, less severely injured patients sometimes showed rapid recovery of SEP amplitudes during the first two weeks. Figure 1 gives an example of an incomplete T12 injury with SEP recovery in the 2nd week after SCI. Note the depressed LM (left median) SEP on day 0, the day of injury. This is not a common occurrence. Note also the marked decrease of both RP and LP (right and left PT) SEP between days 1 and 5. This may be related to ongoing pathological processes in the spinal cord, such as edema.

In more severe injuries, the SEP often required months to show any permanent recovery, intermittently waxing and waning in amplitude. Figure 2 gives an example of a 42-year-old man with a C5 "incomplete" lesion with only sacral sparing. His

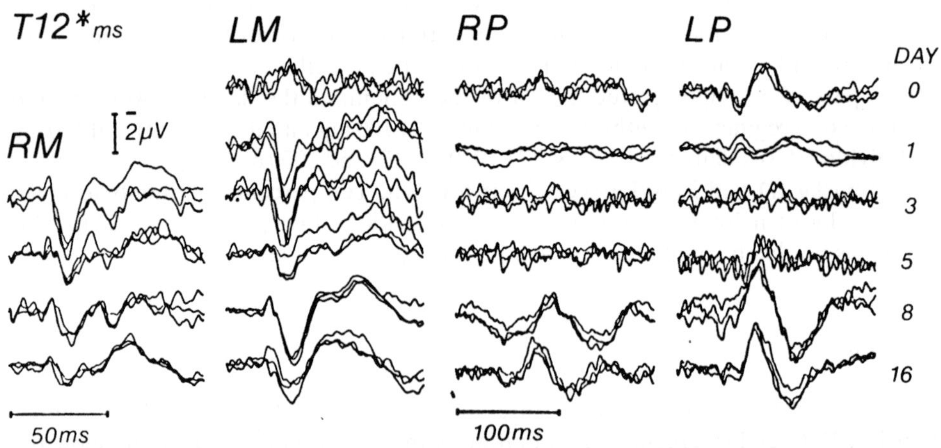

Fig. 1. T12 SCI with unilateral sparing of pinprick sensation on left. Note small response in RP *(right PT)* and larger waveform in LP *(left PT)* SEP. By one week after injury, the SEP recovered in both legs, corresponding to return of touch and pinprick sensation in both legs. Between days 1 and 5, LP and RP responses decreased in amplitude; RM *(right MN)* and LM *(left MN)* SEPs increased in amplitude during the time when lower limb SEPs were depressed. In this and following figures, the full sweep time was 100 ms for MN and 200 ms for PT recordings and negative is up, "Day" refers to the number of days after injury

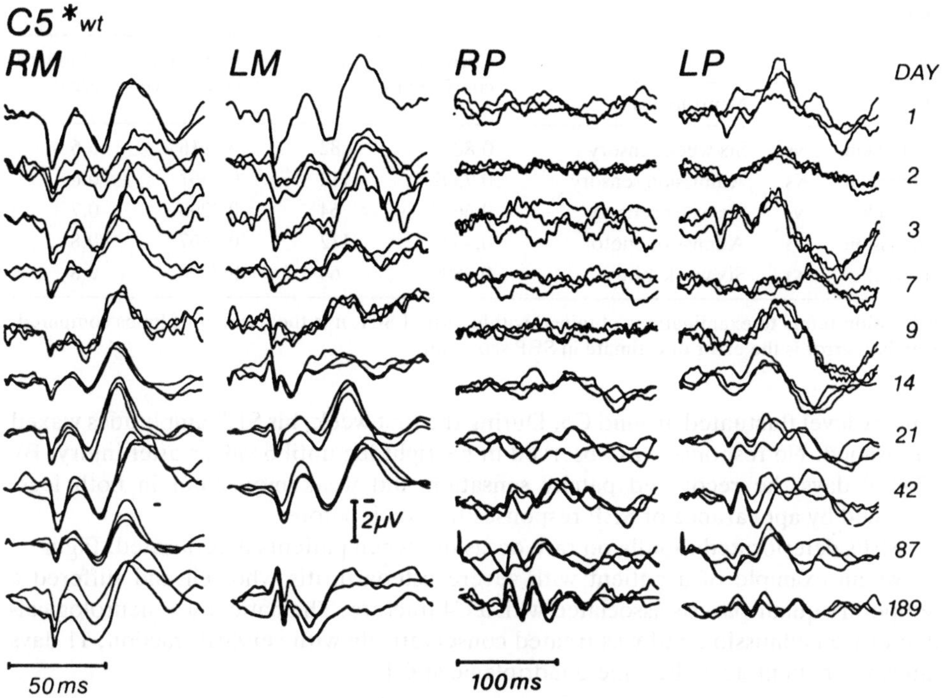

Fig. 2. C5 SCI with only sacral sparing. Note waning of LP *(left PT)*, RM *(right MN)*, and LM *(left MN)* in the first week after injury and the very prolonged recovery time of RP *(right PT)*

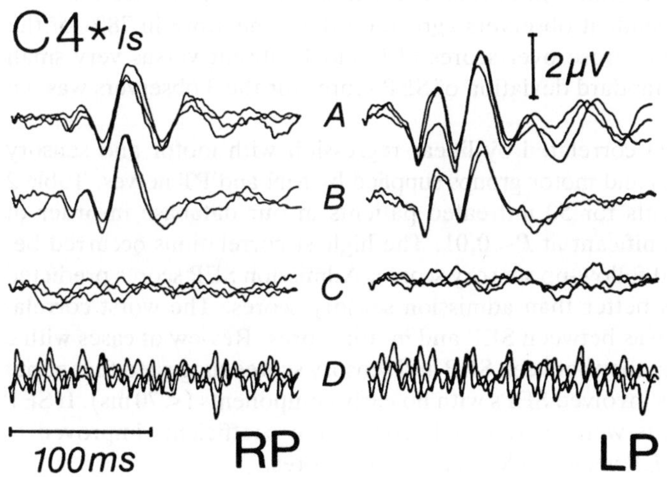

Fig. 3. SEP deterioration in a patient with C4 fracture. This patient was admitted after a transient episode of quadriparesis associated with a fall. X-rays revealed a C4 fracture, severe osteoarthritis, and a narrow spinal canal. Treated conservatively with cervical traction, the patient suffered a quadriplegia at C4. SEPs indicated by A were obtained on day of admission, B at 7 days, C shortly after quadriplegia developed 11 days after admission, and D at 22 days

Table 2

Variables			Correlation coefficient	n	99% limit confidence	Standard error
SEP	vs	Neurological				
Admission	vs	Six week sensory	0.8053	82	0.2916	0.6
Admission	vs	Admission sensory	0.7302	112	0.2467	0.6
Six week	vs	Six week sensory	0.6657	64	0.3298	0.7
Admission	vs	Admission motor	0.4337	112	0.2467	0.8
Six week	vs	Six week motor	0.3899	64	0.3298	0.8

Admission refers to examinations obtained within 48 h of SCI; n is the number of limbs compared. Standard error is the error of estimate in SEP score units

sensory level fluctuated around C5. During the first week, his SEP amplitudes waxed and waned. No response was observed in his right leg until 6 weeks after injury. By 87–189 days, he recovered patchy sensation and weak movement in both legs, preceded by appearance of SEP responses the week before.

SEPs deteriorated rapidly on rare occasions when patients deteriorated. Figure 3 shows an example of a patient with severe osteoarthritis who fell and suffered a temporary quadriparesis associated with a C4 fracture. The patient was neurologically intact on admission and was treated conservatively with cervical traction; 11 days later, the patient again became quadriplegic at C4.

SEP Correlation with Neurological Findings

In most patients, the SEP were so abnormal we could not confidently identify specific SEP components. Only 36% of SEP had recognizable short latency waveforms attributable to components found in normal cortical SEPs. Using the scale of 0–3 described above, 3 independent observers agreed on the same score in 78% of the limbs. Most disagreements arose over scores of 0 and 1, absent versus very small responses. The average standard deviation of SEP scores for the 3 observers was 0.6 units.

SEP scores were cross-correlated by linear regression with motor and sensory scores for the dermatomes and motor groups supplied by MN and PT nerves. Table 2 lists correlation coefficients for 50 untreated patients in our database in order of significance. All were significant at $P<0.01$. The highest correlations occurred between admission SEP and follow-up sensory scores. Admission SEP scores predicted follow-up sensory scores better than admission sensory scores. The worst correlation, though significant, was between SEP and motor scores. Review of cases with 2 or more score unit discrepancy between SEP and sensory scores suggested that about half of such discrepancies involved SEPs with no early components (<70 ms). If SEP with no early components were scored as 1, correlation coefficients improved to >0.9500 for admission SEP versus follow-up sensory scores.

Effect of Stimulus Rate

In selected patients showing SEP and sensory score discrepancies, we varied stimulus rates to ascertain whether this parameter accounted for the difference in function and electrical responses. Decreasing stimulus frequency from 2.3 Hz to 0.11 Hz often resulted in greater amplitudes and shorter latencies of SEP components. In rare patients, the slower stimulus rates resulted in the appearance of responses that were not present at the higher frequencies. Unfortunately, such slow stimulus rates lengthened the testing protocol to several hours, beyond the tolerance of most patients. We could not routinely carry out these tests.

Some patients appeared to show a definite component during the first 50–100 responses but by the time 200 responses were acquired, waveforms were smaller and less definite. We consequently tested several patients on 4 consecutive days, recording groups of 50 responses and then averaging the first, second, third, and fourth sets of 50 responses of each day. In two patients who had no definite SEP responses with our routine protocol, short latency components appeared when we averaged the first 50 responses obtained on consecutive days. Averages of the 2nd, 3rd, and 4th sets of 50 responses on each day showed progressively smaller responses. Averages of 200 responses from each day did not reveal definite SEP responses.

Figure 4 shows an example from a patient with no detectable clinical function below C4-5 until 3 months after injury when he developed an isolated patch of sensibility to touch on the right hip. At 9 months after injury we obtained the SEP recordings shown in Fig. 4. When the right PT nerve was stimulated at 2.3 Hz, the average of 200 responses showed a large long latency wave at >100 ms latency onset with no earlier components. When we averaged the first 50 responses of his SEP obtained on 4 consecutive days, definite early components appeared. The early components were not present in the 2nd set of 50 responses averaged from daily tests. Note that we used a slower stimulus rate of 0.5 Hz, to maximize the size of the early components. The long latency waveform appears less synchronized and of longer latency with the slow stimulus rate.

Large Long Latency Waves

Case by case assessments of false correlations in our database revealed that in virtually every case where SEP scores did not match sensory function, SEP responses had latencies at least twice normal. Furthermore, in such patients, these late waves were not usually associated with shorter latency components. These late negative waves have a stereotypical configuration. In Fig. 4, for example, note the relatively sharp onset of the waveforms and the long tapering tail. The latencies of these waves varied, depending on stimulus rate and also the sensory status of patients (see Fig. 5).

About 17% of the patients in our database showed large long latency waveforms at some point during their recovery. The majority of these patients were severely injured and showed minimal neurological function below the level of the lesion. The long latency waveforms were only occasionally associated with early components. There was reciprocal relationship between the presence of the long latency responses and the early components. As early components recover in the patients, the long latency components get smaller and shorter in latency.

Changes in Median Nerve SEP in Paraplegic Patients

Recently, we observed that paraplegic patients had changes in both sensory function and SEP in their arms. In the first 1–4 weeks after injury, many paraplegic patients

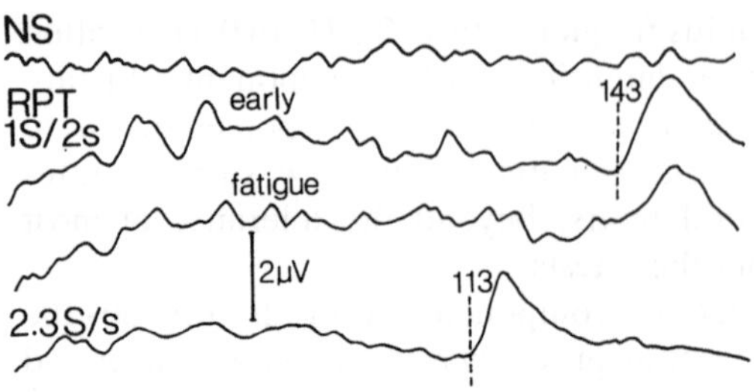

Fig. 4. Fatigue in SEP of a patient with C4-5 SCI. The top trace shows a no-stimulus (NS) control recorded from Cz-Fz. Standard 2.3 stimuli per second (2.3 S/s) to the right PT (RPT) nerve resulted in a long latency negative wave at 113 ms latency *(bottom trace)*. Averaging of the first set of 50 responses on 4 consecutive days resulted in the appearance of early components and a delay of the negative wave to 143 ms latency *(second trace)*. However, average of the second set of 50 responses from 4 consecutive days did not show any early responses. This patient had no evidence of sensory or motor function below C4-5 for at least 3 months after injury. After 3 months, however, he did recover a patch of sensation on the right hip

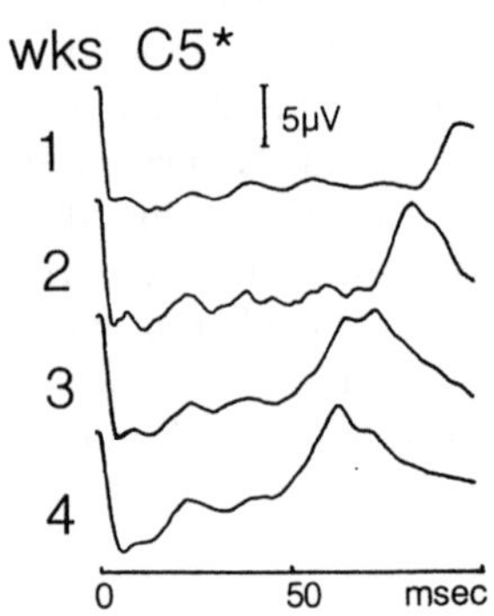

Fig. 5. Decreasing latency of negative wave in C5 SCI. At 3 months after injury, this patient with C5 lesion with minimal sensory sparing in the legs showed some improvement in function associated with progressive decrease in the latency of the SEP over a 4-week period. These SEP were obtained from the right PT and recorded from Cz-Fz. Negative is up

showed intolerance to intense MN nerve stimulation. These patients complained of discomfort with currents of >6–8 mA. We sometimes could not increase stimulus intensity to twice twitch threshold. The sensitivity to MN stimulation often lasted several weeks. The MN SEP amplitude often increased during this period. An example can be seen in Fig. 1. Occasionally, we have seen similar phenomena in patients with low cervical lesions (see Fig. 6).

About 65% of severely injured paraplegic patients show transient increases in median nerve SEP during the first 3 weeks after injury and most of them are under 40 years of age. The mean MN SEP amplitude at 1–2 weeks was larger than control by 164% and larger than 6 month to 1 year SEP amplitudes in the same patients by 182%. Because our control group is small and not age-matched, we cannot attribute definite statistical significance to this comparison with controls. However, the comparison between acute and chronic SEP within individuals is statistically significant ($P<0.001$).

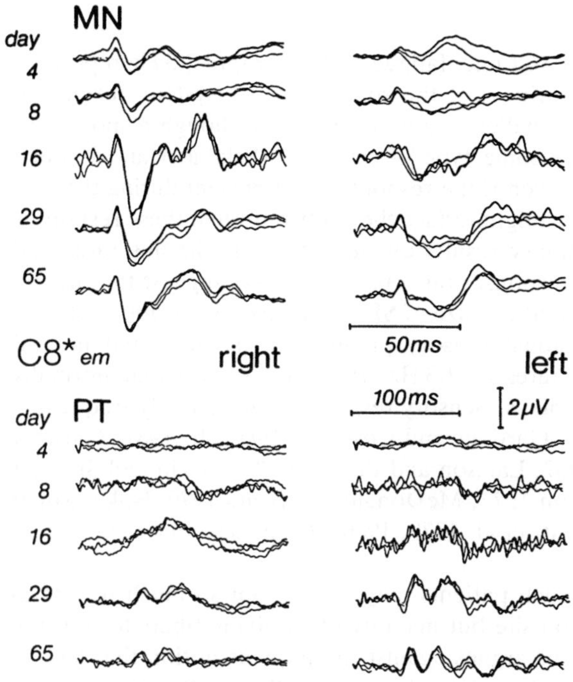

Fig. 6. C8-T1 SCI with minimal sensory sparing on left. MN SEP are shown on the top panels and PT SEPs on the bottom panels, right and left. The days after admission are indicated on the left. Note the transient enlargement of the right MN SEP on day 16. The right MN SEP on days 29–65 is approximately the mean amplitude and latency of our control population. The left MN SEPs on days 29–65 are both later and smaller than control values

Discussion

Our results suggest that cortical SEPs reliably reflect sensory function of SCI patients during the acute period. The absence or presence of SEP correlated in 95% of the cases with whether the patient had clinically detected sensory function below the level of the lesion. The admission SEP correlated particularly well with touch and pin-prick recovery. In more than 97% of the cases, the presence of SEP in a limb predicted recovery of touch or pin-prick sensation in at least a leg dermatome 6–12 weeks after injury.

Although admission SEPs accurately predicted follow-up sensory recovery, our analysis suggested that admission SEPs predicted motor recovery in only 40–50% of the cases. Follow-up SEPs matched sensory scores in only 60–70% of the cases. Note that we define a discrepant correlation as a difference between neurological and SEP scores by more than 4 units out of a total possible sensory score of 12, or 5 units out of a total possible motor score of 15.

The lower rate of correlation between SEP and motor function in chronic spinal injury is not surprising since SEPs are supposed to reflect sensory function. The poorer correlation of SEPs obtained during the chronic phase requires further explanation and investigation. We suspect 3 causes for the poorer correlation of SEP with clinically defined sensory function in the chronic phase of spinal injury: neuronal fatigue, demyelination, and reorganization of brain and spinal cord proximal to the lesion site. As these phenomena are not often considered in interpretation of SEP in spinal cord injury, we will discuss each in sequence below.

Neural Fatigue

Sensory responses of patients with SCI may fatigue with the rapid 2.3 Hz repetitive stimuli applied. Increasing refractoriness of the somatosensory system during the averaging process should result in smaller or absent SEP even though responses to slower or individual stimuli evoke some type of response (Sclabassi and Noreen 1981; Sclabassi et al. 1982). Thus, even if the response were present during the first 20–50 traces of a given SEP, fatigue might reduce the contribution of these responses to the final average. Our preliminary results on the effect of stimulus rates and averaging of 4 sets of the first 50 responses on consecutive days suggest that neural fatigue influences the appearance and the size of SEP responses in chronic SCI.

Our findings suggest that residual sensory pathways in some spinal injured patients may fatigue at stimulus rates of 2.3 Hz. It is well known that normally sensory pathways carry bursts of spikes. Sensory responses are normally frequency-encoded bursts (Beall et al. 1977; Gardner and Costanzo 1980a, b, 1981; Gardner and Tast 1981; Kjellman et al. 1967; Larsson and Prevec 1970; Laskin and Spencer 1979a, b; McDonald and Robertson 1972; McDonald and Sears 1970; Nakanishi et al. 1974; Namerow et al. 1974; Pratt et al. 1979, 1980; Rasminsky 1980; Starr et al. 1981, 1982).

The sensory loss suffered by such patients may be due not so much to loss of axons transmitting across the lesion site but inability of surviving fibers to transmit properly encoded information. This argues against the popular notion that lack of function implies absence of axons. Our data also suggest that SEP testing may be able to detect residual pathways that cannot be discerned clinically.

Demyelination

Demyelination may have two effects on SEP waveforms. First, partial demyelination of surviving tracts may desynchronize action potential volleys activated by peripheral nerve stimulation, causing some axons to transmit at much slower velocities than others. Instead of arriving in synchrony, the components spread out temporally and, upon being averaged, lose amplitude. Second, extensive demyelination may paradoxically produce larger potentials by markedly delaying conduction across the lesion such that incoming action potentials are synchronized into a broad and large long latency wave.

About 17% of the patients in the data base showed negative waves with long (>100 ms) latencies. The presence of these long latency waves correlated negatively with sensory scores in the limb from which the waves were activated. Most of these patients had minimal or no sensory perception in the dermatomes supplied from the nerve being stimulated. Some could not feel the stimulus. SEP with these waves account for a large proportion of discrepancies between SEP and neurological scores. Our observers typically scored these waves as a 2 on a scale of 0 to 3. These waves are sensitive to conditioning stimuli (Young 1982).

In our laboratory, Dr. Blight has systematically recorded in vitro from axons in cats' spinal cords isolated days, weeks, and months after injury, correlating the conduction characteristics with detailed morphometric analyses of the cords. These experiments show that demyelination is not only present but grossly affects axonal conduction across the lesion site. The degree of axonal dysfunction, rather than the number of axons remaining, was often the main differentiating factor between presence or absence of locomotor ability in the cats. Surviving axons in injured spinal cords fatigue with rapid stimulation, conduct slower, cannot support action potential bursts, and block at normal body temperatures. These features are characteristic of demyelinated axons (Eisen and Nudleman 1979; Rasminsky 1980; Rasmusson 1982).

The finding of SEP responses, albeit abnormal, in SCI patients with minimal sensory function suggests that even when signals are transmitted across the lesion site and reach the cortex, the spatio-temporal coding may be so deranged that the signals do not result in sensory perception. Electrical stimulation of peripheral nerve produces an unnatural pattern of sensory signals. Sensory modalities are mixed in a way that seldom occurs naturally (Gardner and Costanzo 1980a, b; Gardner et al. 1984; Gardner and Tast 1981). For example, tactile, pain, and proprioceptive sensations are activated together. The temporal pattern of receptor-activated responses is not duplicated with electrical stimulation of peripheral nerves. Thus, electrically activated cortical SEP is not equivalent to sensory perception.

Reorganization of the Proximal Neural Structures

Reorganization and altered excitability of spinal cord distal to the injury site is well known. Injury-induced disinhibition of spinal reflexes is widely believed to be the cause of spasticity. However, the possibility that there is reorganization of brain and spinal structures proximal to the lesion is seldom considered in the SCI literature. Our finding of transient MN SEP amplitude increases in paraplegic patients supports this possibility.

The prevalence, significance, and origins of the increased MN SEP are uncertain. Many of the patients with thoracic spinal injury could not tolerate stimulus intensities of more than 10 mA. Therefore, uniform stimulus strength was not used in all these patients. But despite lower stimulus levels, MN SEP were enlarged in many of the patients. Increase in cortical evoked potentials to MN stimulation in paraplegic patients may result from many mechanisms, e.g. disinhibition (Gardner et al. 1984; Gardner and Costanzo 1980a, b, 1981; Katz et al. 1978a, b; Laskin and Spencer 1979a, b; Martin et al. 1980) or cortical reorganization similar to that caused by peripheral nerve loss (Abramson and Feibel 1981; Carlen et al. 1978; Franck 1980; Kaas 1983; Kaas et al. 1983; Kalaska and Pomeranz 1979; Merzenich et al. 1983; Metzler and Marks 1979; Paul et al. 1972; Rasmusson 1982; Sur et al. 1980; Wall and Kaas 1984; Wall et al. 1980, 1985). If peripheral nerve lesions can cause cortical reorganization, it would not be surprising for SCI to be associated with reorganization of somatosensory pathways. To our knowledge, no systematic study has been carried out on chronic changes in the proximal somatosensory system in SCI.

Conclusions

(1) SEP are reliable and sensitive indicators of residual neurological function in acute spinal injury,
(2) a simple SEP scoring system accurately predicts sensory recovery but does not correlate as well with motor changes in chronic SCI,

(3) the incidence of discrepancies between SEP and neurological function increases in chronic SCI,
(4) the majority of discrepancies involve the presence of large negative waves with >70 ms onset latency,
(5) response fatigue during SEP averaging may account for some discrepancies between SEP and sensory function,
(6) median nerve SEP may transiently increase in patients with thoracic SCI.

Acknowledgements. Without the SEP technicians, Dan Mollins, Lisa Szambel, Peter Meyer, Henry Merkin, this work would not have been possible. I thank the residents and staff of the Neurosurgery Department for their cooperation and help in conducting this study. Dr. B. Fischer tirelessly performed the neurological examinations on the patients. This work is supported NIH NINCDS NS10164.

References

Abramson AS, Feibel A (1981) The phantom phenomenon: its use and disuse. Bull NY Acad Medicine 57:99–112

Allison T (1962) Recovery function of somatosensory evoked responses in man. Electroencephalogr Clin Neurophysiol 14:331–341

Allison T (1982) Scalp and cortical recordings of initial somatosensory cortex activity to median nerve stimulation in man. Ann NY Acad Sci 388:671–677

Allison R, Hume AL (1981) A comparative analysis of short-latency somatosensory evoked potentials in man, monkey, cat, and rat. Exp Neurol 72:592–611

Anziska B, Cracco RQ (1980) Short latency somatosensory evoked potentials in brain dead patients. Arch Neurol 37:222–225

Anziska B, Cracco RQ (1981a) Short latency SEPs to median nerve stimulation: Comparison of recording methods and origin of components. Electroencephalogr Clin Neurophysiol 52: 531–539

Anziska B, Cracco RQ (1981b) Short latency somatosensory evoked potentials: Studies in patients with focal neurological disease. Electroencephalogr Clin Neurophysiol 49:227–239

Anziska B, Cracco RQ, Cook AW, Feld EW (1978) Somatosensory far field potentials: studies in normal subjects and patients with multiple sclerosis. Electroencephalogr Clin Neurophysiol 45:602–610

Arezzo JC, Schaumburg HH, Spencer PS, Vaughan HR (1980) A novel approach for the detection of toxic axonopathy: The brainstem somatosensory evoked potential. Ann Neurol 8:96

Arezzo JC, Legatt AD, Vaughan HG (1979) Topography and intracranial sources of somatosensory evoked potentials in the monkey. I. Early components. Electroencephalogr Clin Neurophysiol 46:155–172

Arezzo JC, Vaughan HG (1982) The contribution of afferent fiber tracts to the somatosensory evoked potential. Ann NY Acad Sci 388:679–682

Arezzo JC, Vaughan HG, Legatt AD (1981) Topography and intracranial sources of somatosensory evoked potentials in monkey. II. Cortical components. Electroencephalogr Clin Neurophysiol 51:1–18

Beall JE, Applebaum AE, Foreman RD, Willis WD (1977) Spinal cord potentials evoked by cutaneous afferents in the monkey. J Neurophysiol 40:199–211

Berenstein A, Young W, Ransohoff J, Benjamin V, Merkin H (1983) Somatosensory evoked potentials (SEP) during spinal angiography and therapeutic transvascular embolization. J Neurosurgery 60:777–785

Burke D, Skuse NF, Lethlean AK (1981) Cutaneous and muscle afferent components of the cerebral potential evoked by electrical stimulation of human peripheral nerves. Electroencephalogr Clin Neurophysiol 51:579–588

Carlen PL, Wall PD, Nadvorna H, Steinback T (1978) Phantom limbs and related phenomena in recent traumatic amputations. Neurology 28:211–217

Chiappa KH, Choi Sk, Young BR (1980) Short latency somatosensory evoked potentials following median nerve stimulation in patients with neurological lesions. In: Desmedt JE (ed) Progress in clinical neurophysiology, vol 7. Karger, Basel, pp 264–281

Cracco RQ (1972a) The initial positive potential of the human scalp recorded somatosensory evoked response. Electroencephalogr Clin Neurophysiol 32:623–629

Cracco RQ (1972b) Travelling waves of the human scalp recorded somatosensory evoked response. Electroencephalogr Clin Neurophysiol 38:557–566

Cracco RQ (1973) Spinal evoked response: peripheral nerve stimulation in man. Electroencephalogr Clin Neurophysiol 35:379–386

Cracco RQ, Anziska BJ, Cracco JB, Vas GA, Rossini PM, Maccabee PJ (1982) Short-latency somatosensory evoked potentials to median and peroneal nerve stimulation: studies in normal subjects and patients with neurological disease. Ann NY Acad Sci 388:412–425

Cracco RQ, Cracco JB (1976) Somatosensory evoked potential in man: Far field potentials. Electroencephalogr Clin Neurophysiol 41:460–466

Cracco JB, Cracco RQ, Graziani LJ (1975) The spinal evoked responses in infants and children. Neurology 25:31–36

Croft TJ, Brodkey JS, Nulsen FE (1972) Reversible spinal cord trauma: a model for electrical monitoring of spinal cord function. J Neurosurg 36:402–406

Cusick JF, Mycklebust JB, Larson SJ, et al. (1979) Spinal cord evaluation with cortical evoked responses. Arch Neurol 36:140–143

D'Angelo CM, Vangilder JC, Taub A (1973) Evoked clinical potentials in experimental spinal cord trauma. J Neurosurg 38:332–336

Dawson GD (1947) Cerebral responses to electrical stimulation of peripheral nerve in man. J Neurol Neurosurg Psychiat 10:134–140

Desmedt JE, Cheron G (1980) Central somatosensory conduction in man: Neural generators and interpeak latencies of the far field components recorded from neck and right or left scalp and earlobes. Electroencephalogr Clin Neurophysiol 50:382–403

Desmedt JE, Cheron G (1982) Somatosensory evoked potentials in man: subcortical and cortical components and their neural bases. Ann NY Acad Sci 388:388–411

Dimitrijevic MR, Larsson LE, Lehmkuhl D, et al. (1978) Evoked spinal cord and nerve root potentials in humans using a non-invasive recording technique. Electroencephalogr Clin Neurophysiol 45:331–340

Dorfman LJ, Bosley TM, Cummins KL (1978) Electrophysiological localization of central somatosensory lesions in patients with multiple sclerosis. Electroencephalogr Clin Neurophysiol 44: 742–753

Dorfman LJ, Perkash I, Bosley TM, Cummins KL (1980) Use of cerebral evoked potentials to evaluate spinal somatosensory function in patients with traumatic and surgical myelopathies. J Neurosurg 52:654–660

Dorfman LJ (1977) Indirect estimation of spinal cord conduction velocity in man. Electroencephalogr Clin Neurophysiol 42:26–34

Donaghy RMP, Numoto M (1969) Prognostic significance of sensory evoked potential in spinal cord injury. In: Proceedings of the 17th Spinal Cord Injury Conference, Washington DC VAH, pp 9–26

Ducker RB, Salcman M, Lucas JT, Garrison WB, Perot PL Jr (1978) Experimental spinal cord trauma. II. Blood flow, tissue oxygen, evoked potentials in both paretic and plegic monkeys. Surg Neurol 10:64–70

Eisen A, Nudleman K (1979) Cord to cortex conduction in multiple sclerosis. Neurology 29:189–193

Engler LL, Spielholz NI, Bernhard WN, Danziger F, Merkin H, Wolff T (1978) Somatosensory evoked potentials during Harrington instrumentation for scoliosis. J Bone Joint Surg 60: 528–532

Ertekin C (1978) Comparison of the human evoked electrospinogram recorded from the intrathecal, epidural, and cutaneous levels. Electroencephalogr Clin Neurophysiol 44:683–690

Franck JI (1980) Functional reorganization of cat somatic sensory motor cortex (SmI) after selective dorsal root rhizotomies. Brain Research 186:458–462

Gardner EP, Hamalainen HA, Warren S, Davis J, Young W (1984) Somatosensory evoked potentials (SEP) and cortical single unit responses elicited by mechanical tactile stimuli in awake monkeys. Electroencephalogr Clin Neurophysiol 58:537–552

Gardner EP, Costanzo RM (1980a) Spatial integration of multiple-point stimuli in primary somatosensory cortical receptive fields of alert monkeys. J Neurophysiol 43:420–443
Gardner EP, Costanzo RM (1980b) Temporal integration of multiple-point stimuli in primary somatosensory cortical receptive fields of alert monkeys. J Neurophysiol 43:444–468
Gardner EP, Costanzo RM (1981) Properties of kinesthetic neurons in somatosensory cortex of awake monkeys. Brain Research 214:301–319
Gardner EP, Tast JM (1981) Psychophysical measurements of perceived intensity of single-point and multiple-point cutaneous stimuli in human and subhuman primates. J Neurophysiol 46:476–495
Jones SJ (1977) Short latency potentials recorded from the neck and scalp following median nerve stimulation in man. Electroencephalogr Clin Neurophysiol 43:853–863
Jones SJ, Small DG (1978) Spinal and sub-cortical evoked potentials following stimulation of the posterior tibial nerve in man. Electroencephalogr Clin Neurophysiol 44:299–306
Jones SJ (1982) Clinical applications of short-latency somatosensory evoked potentials. Ann NY Acad Sci 388:369–387
Kaas JH (1983) What, if anything, is S-I? Organization of first somatosensory area of cortex. Physiological Reviews 63:206–231
Kaas JH, Merzenich MM, Killackey HP (1983) The reorganization of somatosensory cortex following peripheral nerve damage in adult and developing mammals. Annual Review of Neuroscience 6:325–356
Kalaska J, Pomeranz B (1979) Chronic paw denervation causes an age dependent appearance of novel response from forearm in "paw cortex" of kittens and adult cats. J Neurophysiol 42: 618–633
Katz S, Martin HG, Blackburn JG (1978a) The effects of interaction between large and small diameter fiber systems on the somatosensory evoked potential. Electroencephalogr Clin Neurophysiol 45:45–52
Katz S, Blackburn JG, Perot PL, Lam CF (1978b) The effects of low spinal injury on somatosensory evoked potentials from forelimb stimulation. Electroencephalogr Clin Neurophysiol 44:236–238
Kjellman A, Larsson EL, Prevec TS (1967) Potentials evoked by tapping recorded from human scalp over the cortical somatosensory region. Electroencephalogr Clin Neurophysiol 23:396
Kritchevsky M, Wiederholt WC (1978) Short latency somatosensory evoked responses in man. Arch Neurol 35:706–711
Larson SJ, Sances A Jr, Christenson PC (1966) Evoked somatosensory potentials in man. Arch Neurol 15:88–93
Larsson EL, Prevec TS (1970) Somatosensory response to mechanical stimulation as recorded in human EEG. Electroencephalogr Clin Neurophysiol 29:162–172
Laskin SE, Spencer WA (1979a) Cutaneous masking. I. Psychophysical observations on interactions of multipoint stimuli in man. J Neurophysiol 42:1048–1060
Laskin SE, Spencer WA (1979b) Cutaneous masking. II. Geometry of excitatory and inhibitory receptive fields of single units in the somatosensory cortex of the cat. J Neurophysiol 42:1061–1082
Martin SH, Bloedel JR (1973) Evaluation of experimental spinal cord injury using cortical evoked potentials. J Neurosurg 39:75–81
Martin HF, Katz S, Blackburn JG (1980) Effects of spinal cord lesions on somatic evoked potentials altered by interactions between afferent inputs. Electroencephalogr Clin Neurophysiol 50: 186–195
McDonald WI, Robertson MAH (1972) Changes in conduction during nerve fibre degeneration in the spinal cord. Brain 95:151–162
McDonald WI, Sears TA (1970) The effects of experimental demyelination on conduction in the central nervous system. Brain 93:583–598
Merzenich MM, Kaas JH, Wall J, Nelson RJ, Sur M, Felleman DJ (1983) Topographic reorganization of somatosensory cortical areas 3b and 1 in adult monkeys following restricted deafferentiation. Neuroscience 8:33–55
Metzler J, Marks PS (1979) Functional changes in cat somatic sensory motor cortex during short-term reversible epidural blocks. Brain Research 177:379–383
Nakanishi T, Shimada Y, Sakuta M, Toyokura Y (1978) The initial positive components of the scalp recorded somatosensory evoked potentials in normal subjects and in patients with neurological disorders. Electroencephalogr Clin Neurophysiol 45:26–34

Nakanishi T, Shimada Y, Toyokura Y (1974) Somatosensory evoked responses to mechanical stimulation in normal subjects and in patients with neurological disorders. J Neurol Sci 21: 289–298

Namerow NS, Sclabassi RJ, Enns NF (1974) Somatosensory responses to stimulus trains: Normative data. Electroencephalogr Clin Neurophysiol 37: 11–21

Nash CL, Lorig RA, Schatzinger LA, Brown RH (1977) Spinal cord monitoring during operative treatment of the spine. Clin Orthop 126: 100–105

Paul RL, Goodman H, Merzenich MM (1972) Alterations in mechanoreceptor input to Brodmann's Areas 1 and 3 of the post-central hand area of Macaca mulatta after nerve section and regeneration. Brain Research 39: 1–19

Perot PL (1973) The clinical use of somatosensory evoked potentials in spinal cord injury. Clin Neurosurg 20: 367–382

Perot PL, Vera CL (1982) Scalp-recorded somatosensory evoked potentials to stimulation of nerves in the lower extremities and evaluation of patients with spinal cord trauma. Ann NY Acad Sci 388: 359–368

Pratt HR, Amlie RN, Starr A (1979) Short latency mechanically evoked somatosensory potentials in humans. Electroencephalogr Clin Neurophysiol 47: 524–531

Pratt HR, Amlie RN, Starr A (1979) Somatosensory potentials in humans evoked by both mechanical stimulation of the skin and electrical stimulation of the nerve. In: Kenshalo DR (ed) Sensory functions of the skin of humans. Plenum Press, New York, pp 109–128

Pratt HR, Politoske D, Starr A (1980) Mechanically and electrically evoked somatosensory potentials in humans: Effects of stimulus presentation rate. Electroencephalogr Clin Neurophysiol 51: 138–147

Rasminsky M (1980) Physiology of conduction in demyelinated axons. In: Waxman SG (ed) Physiology and pathobiology of axons. Raven Press, New York, pp 361–376

Rasmusson DD (1982) Reorganization of raccoon somatosensory cortex following removal of the fifth digit. J Comp Neurol 205: 313–326

Rossini PM, Cracco RQ, Cracco JB, House WJ (1981) Short latency somatosensory evoked potentials to peroneal nerve stimulation: Scalp topography and effect of different frequency filters. Electroencephalogr Clin Neurophysiol 52: 540–552

Rowed DW, McLean JAG, Tator CH (1978) Somatosensory evoked potentials in acute spinal cord injury: prognostic value. Surg Neurol 9: 203–210

Schramm J, Hashizume K, Fukushima T, Takahashi H (1979) Experimental spinal cord injury produced by slow, gradual compression. Alterations of cortical and spinal evoked potentials. J Neurosurg 50: 48–57

Sclabassi RJ, Vries JK, Bursick DM (1982) Somatosensory evoked potentials to random stimulus trains. Ann NY Acad Sci 388: 695–701

Sclabassi RJ, Noreen GK (1981) Complex pattern evoked somatosensory responses in the study of multiple sclerosis. Proc IEEE 65: 626–633

Shimoji K, Matsuki M, Shimizu H (1979) Waveform characteristics and spatial distribution of evoked spinal electrogram in man. J Neurosurg 46: 304–312

Singer JM, Russel GV, Coe JE (1970) Changes in evoked potentials after experimental cervical spinal cord injury in the monkey. Exp Neurol 29: 449–461

Starr A, McKeon B, Skuse N, Burke D (1981) Cerebral evoked potentials to muscle stretch in man. Brain 104: 149–166

Starr A, Pratt H, Burke D (1982) Natural stimuli evoking somatosensory evoked potentials. Ann NY Acad Sci 388: 702–705

Sur M, Merzenich MM, Kaas JH (1980) Magnification, receptive field area, and "hypercolumn" size in Areas 3b and 1 of somatosensory cortex in owl monkeys. J Neurophysiol 44: 295–311

Tsumoto T, Hirose N, Nonaka S, et al. (1972) Analysis of somatosensory evoked potentials to lateral popliteal nerve stimulation in man. Electroencephalogr Clin Neurophysiol 33: 379–388

Vas GA, Cracco JB, Cracco RQ (1981) Scalp recorded short latency cortical and subcortical somatosensory evoked potentials to peroneal nerve stimulation. Electroencephalogr Clin Neurophysiol 52: 1–8

Vasiri D, Pratt H, Saiki JK, Starr A (1981) Evaluation of somatosensory pathway by short latency evoked potentials in patients with end-stage renal disease maintained on hemodialysis. Int J Artificial Organs 4: 17–22

Wall JT, Kaas JH (1984) Cortical reorganization and sensory recovery following nerve damage and regeneration. (To be published)
Wall JT, Cusick CG, Kaas JH (1985) Cutaneous responsiveness of S-I hindpaw cortex before and after partial hindpaw deafferentation in adult rats. (To be published)
Wall JT, Merzenich MM, Sur M, Nelson RJ, Felleman DJ, Kaas JH (1980) Organization of the representations of the hand in Areas 3b and 1 of post-central somatosensory cortex of monkeys after section and regeneration of the median nerve. Society of Neuroscience Abstracts 6:651
Wiederholt WC, Meyer-Hardting E, Budnick B, McKeown KL (1982) Stimulating and recording methods used in obtaining short-latency somatosensory evoked potentials in patients with central and peripheral neurologic disorders. Ann NY Acad Sci 388:349–358
Yamada TM, Kimura J, Nitz DM (1980) Short latency somatosensory evoked potentials following median nerve stimulation in man. Electroencephalogr Clin Neurophysiol 48:367–376
Young W (1982) Correlation of somatosensory evoked potentials and neurological findings in spinal cord. In: Tator CH (ed) Early management of acute spinal cord injury. Raven Press, New York, pp 153–165
Young W, Cohen A, Merkin H, Fisher B, Berenstein A, Ransohoff J (1982) Somatosensory evoked potential changes in spinal injury and during intraoperative manipulation. J American Paraplegia Soc 5:44–48
Young W, Berenstein A (1985) Somatosensory evoked potential monitoring of intraoperative procedures. In: Schramm J, Jones SJ (eds) Spinal cord monitoring. Springer, Berlin Heidelberg New York Tokyo, pp 197–203

Therapeutic Trial of Hypertension and Hyperoxia Combined with Hypercarbia on Experimental Acute Spinal Cord Injury in the Dog

S. Hukuda[1], T. Mochizuki, M. Ogata, S. Fujii, and K. Shichikawa

Summary

Hypertension and hyperoxia combined with hypercarbia, which in previous studies had been found to improve post-traumatic spinal cord hypoxia but to have no therapeutic effects when applied 3 h after injury, were used for 3 h beginning immediately after trauma in dogs subjected to experimental cord injury, separately and in combination. For comparison of the effects of different anesthetic agents 2 control groups with barbiturate anesthesia and halothane anesthesia were also investigated. The experimental animals were examined neurologically and electrophysiologically for one week. It was concluded that hypertension as well as hyperoxia accompanied by hypercarbia, combined or alone, had no aborting effect against the post-traumatic progress of intramedullary damage. Rather, hypertension was presumed to work adversely by worsening intramedullary hemorrhage. Neither was any beneficial effect of barbiturate on acute spinal cord injury verified.

Introduction

In our previous studies (Hukuda et al. 1980a, b), combined hypertension, hyperoxia and hypercarbia was found to improve post-traumatic spinal cord hypoxia and was used in dogs subjected to experimental spinal cord injury with the aim of preventing the development of an irreversible intramedullary lesion. However it was shown to have no therapeutic effect against acute spinal cord trauma after eight week observation when it was started three hours after injury. Our previous tissue oxygen study revealed the appearance of spinal cord hypoxia as early as one hour after injury, suggesting that the three hour delay in the previous therapeutic trial might have been responsible for the negative outcome. In the present study we examined experimentally the effects of the same therapeutic method when applied immediately after injury.

Material and Methods

Fifty female mongrel dogs weighing around 10 kg were used. Anesthetic procedures, using secobarbiturate intravenously with the dog intubated and ventilated with a Manley Pulmovent volume respirator, and surgical preparations, in which a femoral artery was catheterized for monitoring the arterial blood pressure and collecting blood gas samples, were exactly the same as those described in the previous report

[1]Dept. of Orthopaedic Surgery, Shiga University of Medical Science, Otsu, Shiga 520-21, Japan

Spinal Cord Monitoring
Ed. by J. Schramm and S. J. Jones

(Hukuda et al. 1980b). After the dogs were fixed in prone position on a Takahashi stereotactic instrument the spinal cord was traumatized at the 5th thoracic segment by suddenly inflating to 400 mm Hg a 7 mm wide tourniquet that encircled the spinal cord epidurally; this inflation pressure was maintained for 5 min. Somatosensory evoked potentials (SEPs) were monitored with a disc electrode attached to the vertex of the scalp and a reference electrode on the forehead using Medelec MS6 apparatus before and immediately after the trauma. The sciatic nerve was stimulated percutaneously at the ischial level 256 times at a rate of 2 Hz with square waves of 0.2 ms duration; the analysis time was 200 ms with a frequency band of 100 to 1000 Hz. In one control group anesthesia was induced by intravenous injection of ketamin (10 mg/kg) and maintained with halothane (see below).

Out of the above mentioned 50 dogs 30 were subjected to the following experimental procedures, each group including 10 dogs: Combined hypertension, hyperoxia and hypercarbia which was started immediately after trauma and maintained for 3 h (Experiment 1); hypertension alone of 3 h duration beginning immediately after trauma (Experiment 2); hyperoxia combined with hypercarbia of 3 h duration beginning immediately after trauma (Experiment 3). Hypertension, with the mean arterial blood pressure between 170 mm Hg and 200 mm Hg, was induced and continued for 3 h by intravenous infusion of 0.002% epinephrine. Hyperoxia combined with hypercarbia was produced intermittently by alternating 95% O_2 : 5% CO_2 gas ventilation for 15 min with air ventilation for 10 min. The remaining 20 dogs which served as controls were divided into 2 groups, each of which included 10 dogs. One group was anesthetized in the same way as the experimental groups using barbiturate, and the anesthesia was maintained for 3 h after cord injury (Control 1). Ketamin and halothane were used in the other control group, in which anesthesia was discontinued after trauma (Control 2).

Neurological examinations were repeated every day for one week, after which final SEPs were recorded before sacrifice. The thoracic spinal cords were studied pathologically thereafter. Those dogs which did not survive for a week were replaced with new ones.

Results

Influence of Experimental Procedures on Arterial Pressure

The maneuver of encircling the dural tube with a 7 mm wide tourniquet sometimes caused transient systemic hypertension in the dogs anesthetized with barbiturate, but this was not observed in the dogs receiving halothane anesthesia (Control 2). Sudden inflation of the tourniquet to 400 mm Hg almost always provoked an abrupt increase in systemic arterial pressure up to 300 mm Hg or more, which lasted for approximately one minute. This injury-induced hypertension was also far less remarkable under halothane anesthesia (Table 1).

Arterial Blood Gas Analysis

Arterial blood gas was monitored in Experimental groups 1 and 3 in which air ventilation was alternated with 95% O_2 : 5% CO_2 gas ventilation. Blood sampling was

Table 1

Exp. group	Dog #	Arterial pressure (mm Hg)		Neurol.[a] Grade (lw.)	SEP latency (ms)		Histology	
		Tourniquet	Injury		Bef. inj.	1 week	Lesion (%)	Hemorrhage
Exp. 1	89	150	150	1	19	—	100	Moderate
	90	125	270	0	16	—	78	Slight
	91	190	330	1	17	22	39	Moderate
	92	150	290	3	15	20	21	Slight
	93	140	270	0	15	—	36	Slight
	94	145	240	3	19	21	10	Moderate
	95	200	250	3	18	20	11	Slight
	96	240	250	0	17	—	89	Moderate
	98	270	170	0	19	—	93	Marked
	99	270	175	1	15	—	63	Marked
		188.0 ±55.1	239.5 ±57.6	1.3 ±1.5	17.0 ±1.7		53.9 ±34.9	
Exp. 2	100	230	240	1	18	—	50	Marked
	101	250	330	3	17	21	27	Marked
	102	300	370	0	18	30	71	Marked
	103	250	225	0	18	—	100	Marked
	104	210	300	0	17	—	97	Marked
	106	150	200	0	18	—	83	Slight
	108	200	135	1	18	—	98	Slight
	109	125	250	1	18	—	44	Moderate
	110	150	270	1	20	25	28	Slight
	111	170	320	1	19	—	37	Marked
		203.5 ±55.2	264.0 ±69.2	1.1 ±1.2	18.1 ±0.9		63.4 ±29.7	
Exp. 3	112	130	225	3	17	20	39	Slight
	113	225	220	0	15	—	97	Slight
	114	250	300	3	17	20	5	Slight
	115	180	300	3	16	23	25	Moderate
	116	240	180	0	17	—	100	Marked
	117	180	250	2	17	21	62	Slight
	118	90	320	3	15	20	22	Slight
	119	230	300	1	17	—	94	Moderate
	120	250	220	0	16	—	61	Moderate
	121	160	330	3	19	26	30	Moderate
		193.5 ±54.9	262.5 ±53.9	1.9 ±1.4	16.6 ±1.2		53.5 ±34.5	

Table 1 (continued)

Exp. group	Dog #	Arterial pressure (mm Hg)		Neurol.[a] Grade (lw.)	SEP latency (ms)		Histology	
		Tourniquet	Injury		Bef. inj.	1 week	Lesion (%)	Hemorrhage
Control 1	124	115	300	0	16	—	100	Slight
	125	165	225	0	18	—	100	Marked
	126	155	250	3	17	20	33	Slight
	128	120	280	0	17	—	69	Marked
	129	200	290	3	14	18	35	Slight
	130	220	300	3	16	19	27	Moderate
	131	240	350	3	16	21	8	Moderate
	132	340	300	3	16	—	53	Slight
	133	150	300	3	20	25	35	Moderate
	134	250	350	0	17	Not examined	92	Moderate
		195.5 ±69.3	294.5 ±38.5	1.9 ±1.6	16.7 ±1.6	—	55.2 ±33.1	
Control 2	135	140	210	0	14	—	29	Slight
	136	75	130	3	17	20	9	Slight
	137	110	100	1	15	25	29	Slight
	139	120	75	3	18	22	26	Slight
	140	100	120	0	17	—	63	Slight
	141	100	140	3	15	21	39	Moderate
	142	110	130	0	16	—	97	Marked
	143	80	165	3	14	23	39	Marked
	144	75	75	0	16	—	98	Moderate
	145	90	125	3	17	23	25	Moderate
		100.0 ±20.9	127.0 ±40.4	1.6 ±1.5	15.9 ±1.3		45.4 ±30.7	

[a] Grade 0: complete paraplegia; Grade 1: flicker movement in the hindlegs; Grade 2: good movements of joints, but no weight bearing; Grade 3: use of legs for weight bearing; Grade 4: normal

repeatedly performed at every alternation of the air and the mixed gas ventilation throughout the 3 h experimental procedure. Arterial PO_2 and PCO_2 were around 25 mm Hg and 130 mm Hg respectively during air ventilation, but went up to around 50 mm Hg and 430 mm Hg respectively after the 95% O_2 : 5% CO_2 mixture was administered for 15 min (Table 2).

Neurological Outcome

Neurological grades were recorded according to the following criteria: Grade 0, complete paraplegia; Grade 1, flicker movement in the hindlegs; Grade 2, good movements of joints, but no weight bearing capability; Grade 3, use of legs for

Table 2. Blood gas analysis

Exp. group	Dog #	Air ventilation				$95\%\ O_2 \cdot 5\%\ CO_2$ ventilation			
		Hb	PH	PCO_2	PO_2	Hb	PH	PCO_2	PO_2
Exp. 1	89	7.3±1.0	7.3±0.1	30.5±3.1	138± 5	8.4±0.2	7.1±0.0	55.5±4.8	471±52
	90	14.0±0.5	7.3±0.1	32.4±2.7	131± 5	15.1±0.4	7.1±0.0	58.7±4.7	416±30
	91	10.3±0.8	7.4±0.0	22.1±1.1	126± 8	11.3±0.6	7.1±0.0	47.7±1.7	426±26
	92	8.1±0.3	7.4±0.0	23.9±1.0	127± 7	8.9±0.2	7.1±0.0	47.8±3.1	430±19
	93	7.0±0.6	7.3±0.0	24.7±1.4	114± 6	7.4±0.2	7.1±0.0	46.7±2.0	404±33
	94	11.3±0.8	7.4±0.1	23.1±2.4	139±12	12.2±0.2	7.1±0.0	47.5±1.2	420±26
	95	14.9±0.9	7.4±0.1	23.9±1.8	155± 3	16.7±0.4	7.1±0.0	55.0±1.2	498±18
	96	12.2±1.4	7.4±0.0	20.4±1.3	151± 4	14.7±0.5	7.1±0.0	47.9±2.4	476±20
	98	11.1±1.5	7.3±0.0	29.8±1.4	114± 9	13.1±0.3	7.1±0.0	50.2±1.8	438±38
	99	7.9±0.3	7.3±0.1	30.0±2.8	119± 4	8.2±0.2	7.1±0.0	56.4±4.5	472±36
		10.4±2.8	7.3±0.1	26.1±4.2	131±14	11.6±3.3	7.1±0.0	51.3±4.6	445±31
Exp. 2	112	11.0±0.4	7.3±0.1	24.7±2.9	103±15	11.8±0.2	7.1±0.0	46.5±1.9	366±16
	113	9.6±0.9	7.3±0.0	24.1±3.3	128± 8	10.3±0.3	7.1±0.0	48.2±2.8	417±24
	114	12.7±0.4	7.4±0.0	14.6±0.9	145± 6	13.7±0.5	7.1±0.0	40.3±1.2	419±28
	115	10.2±1.0	7.2±0.0	38.7±4.2	135± 7	11.1±0.3	7.1±0.0	57.9±3.6	443±20
	116	11.0±0.8	7.4±0.1	19.3±1.5	151± 6	12.2±0.4	7.1±0.0	46.2±2.0	413±11
	117	11.9±0.8	7.4±0.1	15.6±1.8	174± 6	13.2±0.4	7.0±0.0	49.0±2.7	502±33
	118	10.1±0.8	7.2±0.1	15.1±1.0	160± 4	10.3±0.4	7.0±0.0	38.1±1.7	462±41
	119	8.9±1.0	7.3±0.0	21.5±1.2	131± 3	10.5±0.5	7.1±0.0	42.8±4.2	396±36
	120	12.6±0.5	7.5±0.0	15.8±0.9	147± 5	13.9±0.3	7.2±0.0	40.5±4.2	442±15
	121	10.9±0.3	7.3±0.0	30.5±1.1	128± 5	11.8±0.2	7.1±0.0	53.8±1.6	413±19
		10.9±1.2	7.3±0.1	22.0±7.8	140±20	11.9±1.4	7.1±0.1	46.3±6.2	427±37

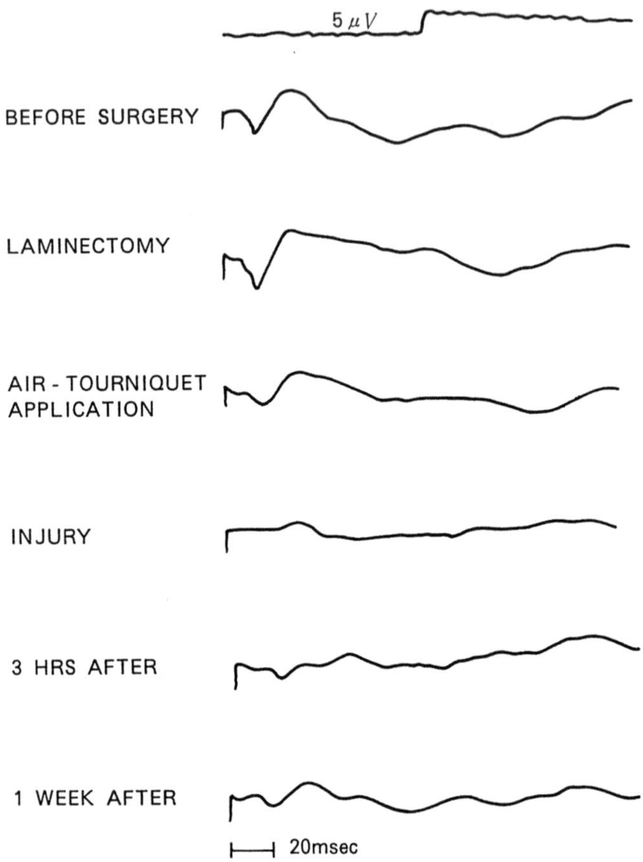

Fig. 1. Somatosensory evoked potentials recorded from the scalp of the dog following sciatic nerve stimulation. Positive polarity at the vertex gives rise to a downward deflection (forehead reference)

weight bearing; Grade 4, normal. No dogs recovered to Grade 4 in one week. Useful recovery of the hindlegs (Grades 2 and 3) was observed in 50% or more of animals in Experiment 3 as well as Control groups 1 and 2, but was observed in only 30% and 10% respectively of animals in Experiments 1 and 2, both of which involved the procedure of induced hypertension (Table 1).

SEPs of the dog consisted of an initial sharp positive peak and a subsequent broad negative wave. The latter was often obscure. The amplitude was around 5 microvolts for both potentials (Fig. 1). SEPs could be obtained in all 50 animals when examined before laminectomy, with an average latency of 16.8 ± 1.5 ms. SEPs survived the experimental spinal cord trauma in 10% (5 out of 50) but the surviving potentials showed aberrant waveforms with low amplitude as well as prolongation of latency by 9 ms on average. After one week SEPs were observed in 50% with a latency delay of 5.1 ms on average compared with the initial recordings. Recovery of SEPs was worse in the animals of Experiment 1 and 2 compared with both control groups (Table 1).

Histology of the spinal cord 1 week after trauma revealed that the central gray and the surrounding white matter had fallen into relatively well demarcated necrosis with interstitial hemorrhage of different severity. The spinal cord segments directly beneath the compression cuff always demonstrated the most severe change which extended, although gradually tapering, 1.5 to 2 cm in both rostral and caudal directions. At the upper and lower end of the lesion the necrotic focus was usually located in the central grey substance dorsal to the central canal (Fig. 2). The size of maximal demarcation, which was selected from around 150 serial transections taken from each spinal cord lesion, was measured using a profile projector. The severity of the hemorrhage in or around the necrotic foci was also evaluated. The size of the maximal lesion was between 45 and 65% of the transverse section of the spinal cord with the standard deviation around 30% in all groups. The animals of Experiment 2 (induced hypertension for 3 h) showed the largest lesion and Control group 2 the smallest, but there were no statistically significant differences. Hemorrhage was also more remarkable in the animals of Experiments 1 and 2 (Table 1).

Discussion

The present study indicated that combined hypertension, hyperoxia and hypercarbia had no concrete effect in preventing the secondary destructive process of the acutely traumatized spinal cord, from the view point of clinical improvement, SEP recovery or histological outcome. Several possible reasons for this conclusion were considered. It was virtually impossible, even by the most careful performance of the experimental procedures, to produce a spinal cord lesion consistently of the same severity, so that subtle differences in the final outcome between each experimental group may have been obscured. In order to proceed further with an experiment of this kind, much larger numbers of experimental animals would be required. Recently Walker et al. (1977, 1979) and Anderson et al. (1980) reported that metabolic changes might precede ischemia and hypoxia in the traumatized spinal cord. If this is true, the cord might be unable to utilize oxygen and nutrient substances even if they were transported to the lesion site by blood flow. Thus, the improvement of the metabolic disturbance might be more imperative than improvement of hypoxia and ischemia in the treatment of acute spinal cord injury.

In the previous study of eight weeks survival (Hukuda et al. 1980b) combined hypertension and hypercarbia was not found to increase the severity of interstitial hemorrhage at the site of the traumatic lesion. The present study revealed, however, that the hemorrhage was more marked when hypertension was induced.

Fresh traumatic lesions of the spinal cord are invariably hemorrhagic, but the hemorrhage subsequently becomes obscured as the lesion is transformed into tissue necrosis and finally into cavity. Therefore, induced hypertension when applied in the acute phase might provoke increased interstitial hemorrhage in the cord lesion. This might account for the finding that neurological and electrophysiological recovery as well as the largest histological lesions were observed in the animals of Experiment 2. In the previous study of 8 weeks observation the lesions were thought to have been complete and the pathology did not reflect precisely the initial grade of the hemorrhage. Rawe et al. (1978) demonstrated experimentally that elevation of systemic

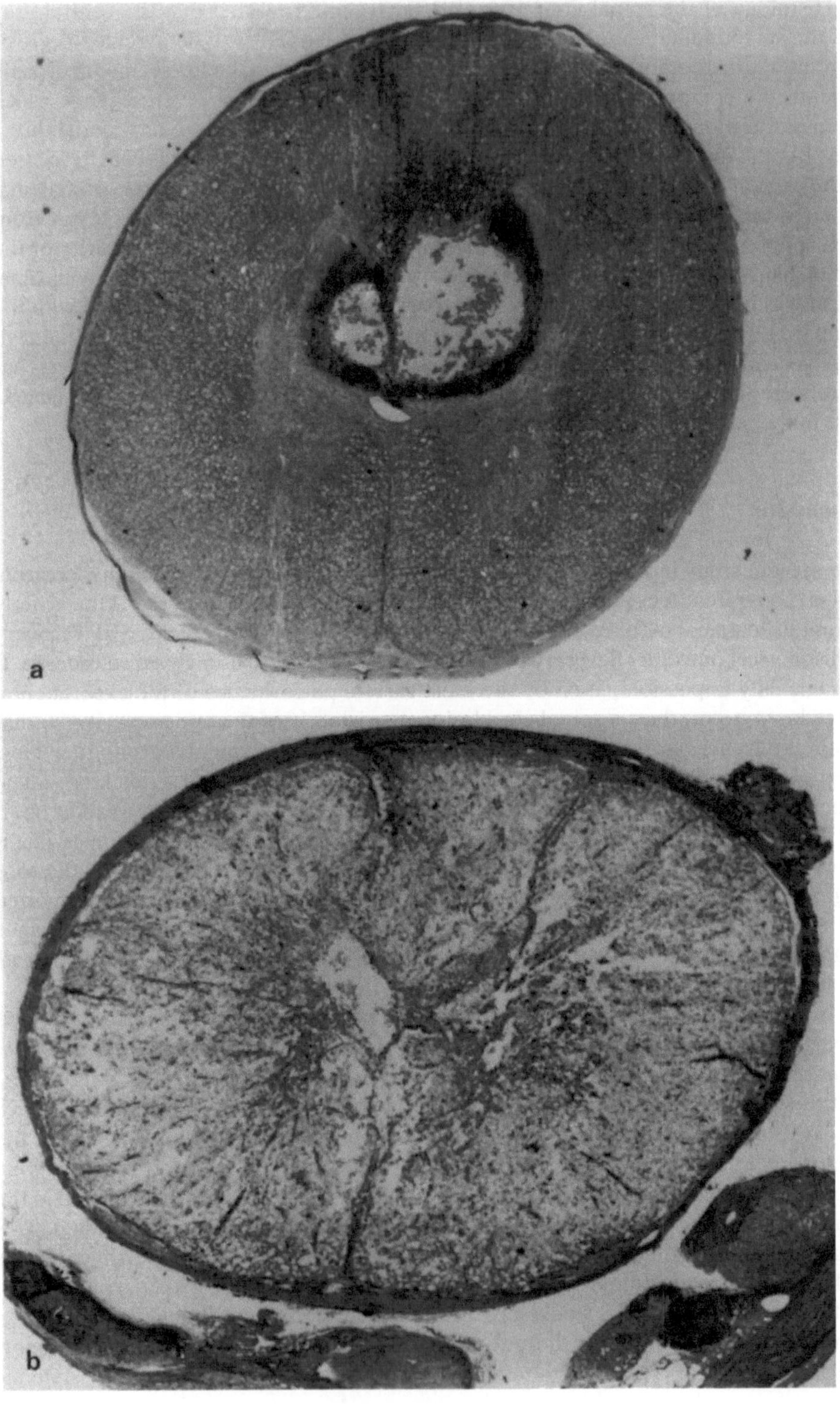
a
b

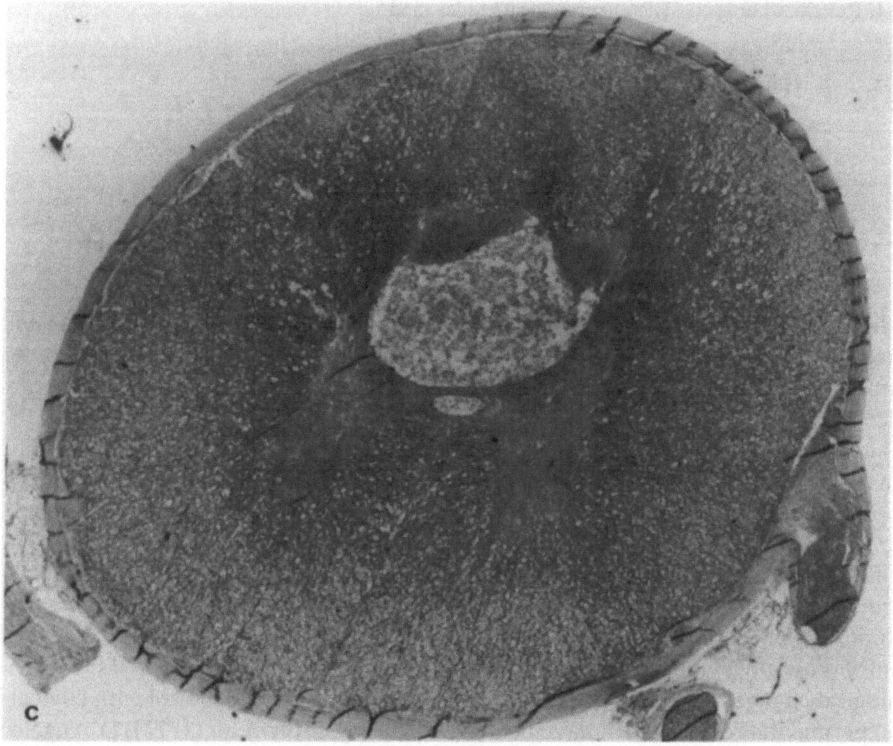

Fig. 2a–c. Transverse sections of the spinal cord lesion in dog No. 142. **a** A well-demarcated necrotic cavity surrounded by a hemorrhagic zone is seen behind the central canal at the rostral end of the lesion. **b** Almost the whole area of the transverse section is fallen into necrosis at the site of direct compression. **c** Section at the caudal end of the lesion shows a well-demarcated necrotic focus again behind the central canal

Table 3. Amount of barbiturate used during experiment

Experimental group	Secobarbital sodium (mg)
Experiment 1	424 ± 173
Experiment 2	400 ± 121
Experiment 3	333 ± 154
Control 1	425 ± 101
Control 2	0

blood pressure in the post-injury period enhanced the hemorrhagic involvement of the spinal cord at the site of injury.

There are experimental studies by Oldfield et al. (1982) and Ducati et al. (1984) which appear to demonstrate that barbiturate effectively minimized the clinical and pathological damage of acute spinal cord ischemia. The total amount of barbiturate used per animal in the present study is shown in Table 3. This was around 40 mg/kg,

Table 4. Relations between SEPs, neurological grades, and histological changes

SEPs at sacrifice	Neurological grade	Size of lesion	No. dogs
Yes	2.6±0.9	28.9±15.6	24
No	0.4±0.7	77.1±24.2	25
	(P<0.001)	(P<0.001)	

and slightly smaller than the doses of 65–90 mg/kg in the cat (Ducati et al. 1984) and 52.5 mg/kg in the rabbit (Oldfield et al. 1982). Since the experimental results of Control groups 1 and 2 were similar in the present study, no prophylactic or therapeutic effect of barbiturate against acute spinal cord trauma was demonstrated.

Somatosensory evoked potentials were revealed to be very sensitive to the trauma applied to the spinal cord. They were sometimes abolished, albeit transiently, by the procedure of applying the epidural air tourniquet, and even by laminectomy alone. Even if SEPs did not disappear after such surgical procedures, their latency was always increased. However, the transient disappearance of SEPs and prolongation of latency did not have any correlation with the final neurological and histological outcome. Their persistence after trauma or reappearance in the post-traumatic period was usually evidence of meaningful functional recovery and also of less severe histological changes as shown in Table 4, and this finding coincided well with clinical and experimental results which have been described by Croft et al. (1972), D'Angelo et al. (1973), Martin and Bloedel (1973), Rowed et al. (1978), Spielholz et al. (1979) and Dorfman et al. (1980).

Conclusions

1. Hypertension and hyperoxia accompanied by hypercarbia, combined or alone, were not effective in aborting the expansion of irreversible histological damage in the acutely traumatized spinal cord. Induced hypertension seemed to have an adverse effect by aggravating the intramedullary hemorrhage.
2. The persistence or reappearance of somatosensory evoked potentials after trauma was a good indicator of eventual functional recovery.
3. Barbiturate was not shown to be effective in reducing spinal cord damage after acute spinal cord trauma.

Acknowledgements. This study was partially supported by the Japan Education Ministry Grant-in-Aid for Scientific Research. Projects No. 557342 and No. 57570541. The authors wish to thank Ms. Ishiko Yamato, R.N. for her assistance in the animal experiments and Ms. Aiko Tsuruoka for her efforts in the preparation of this manuscript.

References

Anderson DK, Means ED, Waters TR, Spears CJ (1980) Spinal cord energy metabolism following compression trauma to the feline spinal cord. J Neurosurg 53:375–380

Croft TJ, Brodkey JS, Nulsen FE (1972) Reversible spinal cord trauma; a model for electrical monitoring of spinal cord function. J Neurosurg 36:402–406
D'Angelo CM, Van Gilder JC, Taub A (1973) Evoked cortical potentials in experimental spinal cord trauma. J Neurosurg 38:332–336
Dorfman LJ, Perkash I, Bosley TM, Cummins KL (1980) Use of cerebral evoked potentials to evaluate spinal somatosensory function in patients with traumatic and surgical myelopathy. J Neurosurg 52:654–660
Ducati A, Schieppati M, Giovanelli MA (1984) Effects of deep barbiturate coma on acute spinal cord injury in the cat. Surg Neurol 21:405–413
Hukuda S, Mochizuki T, Ogata M (1980a) Effects of hypertension and hypercarbia on spinal cord tissue oxygen in acute experimental spinal cord injury. Neurosurgery 6:639–643
Hukuda S, Mochizuki T, Ogata M (1980b) Therapeutic trial of combined hypertension and hypercarbia on experimental acute spinal cord injury. Neurosurgery 6:644–648
Martin SH, Bloedel JR (1973) Evaluation of experimental spinal cord injury using cortical evoked potentials. J Neurosurg 39:75–81
Oldfield EH, Plunkett RJ, Nylander WA, Meacham WF (1982) Barbiturate protection in acute experimental spinal cord ischemia. J Neurosurg 56:511–516
Rawe SE, Lee WA, Perot PL Jr (1978) The histopathology of experimental spinal cord trauma. J Neurosurg 48:1002–1007
Rowed DW, McLean JAG, Tator CH (1978) Somatosensory evoked potentials in acute spinal cord injury; prognostic value. Surg Neurol 9:203–210
Spielholz NI, Benjamin MV, Engler C, Ransohoff J (1979) Somatosensory evoked potentials and clinical outcome in spinal cord injury. In: Tator CH (ed) Neural trauma. Raven Press, New York, pp 217–222
Walker JG, Yates RR, O'Neill JJ, et al. (1977) Canine spinal cord energy state after experimental trauma. J Neurochem 29:929–932
Walker JG, Yates RR, Yashon D (1979) Regional canine spinal cord energy state after experimental trauma. J Neurochem 33:397–401

Cortical Evoked Potentials in Experimental Scoliosis

K. MITZKAT[1] and M. BARTELS

Summary

Since the reliability of information provided by cortical evoked potentials during distraction operations for scoliosis has not yet been established, this electrophysiologic method of measurement cannot replace the forensically valid wake-up test (Vauzelle et al. 1973). A corrective scoliosis operation was simulated in 20 rabbits by experimental induction of progressive dorsolumbar scolioses with convexity to the right, in some cases with pronounced kyphotic components produced via a compression system. The subsequent scolioses ranged between 45° and 60°, according to Cobb. At a second session, the scoliosis was subjected to maximal distraction. Cortical somatosensory evoked potentials were continuously recorded during distraction. The crucial variable proved to be the change of potential amplitude during maximal distraction, i.e., radiologically demonstrated total correction of the scoliotic curve. The potential amplitude was obliterated at distraction distances exceeding 21 mm.

Introduction

In 1975, MacEwen et al. (1975) reported a neurologic complication rate of up to 1% for corrective interventions on the spine. It is understandable, therefore, that electrophysiologic techniques for intraoperative monitoring of spinal cord function during distraction, such as somatosensory evoked potentials (SEP), are being used in orthopaedic surgery. Since intraoperatively measured SEPs are influenced by many variables, it was necessary to identify the crucial measurement variable during distraction. The two variables that came into question were peak latency and potential amplitude.

Materials and Method

We induced progressive scoliosis at the dorsolumbar junction of rabbits using a stapling technique (Fig. 1). The tests were performed on 20 rabbits in a two-stage procedure. In the first session, staples were inserted in 6- to 8-week-old rabbits. SEPs were measured during distraction in a second session, 16 to 20 weeks later. Almost total correction of the scoliotic spines was achieved with an outrigger system. The tests were performed under sodium pentobarbital anesthesia (30 mg/kg, ip).

[1] Orthopädische Universitäts- und Poliklinik, Calwer Str. 7, D-7400 Tübingen

Spinal Cord Monitoring
Ed. by J. Schramm and S. J. Jones

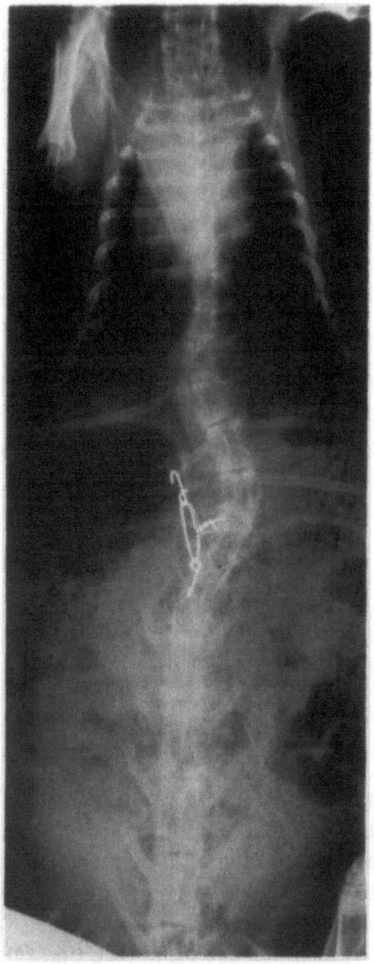

Fig. 1. Experimentally induced dorsolumbar scoliosis with convexity to right in an 18-week-old rabbit

Tibial nerve SEPs during distraction were recorded with a single-channel standard myograph set at 2 stimuli per second. Stimulus intensity ranged between 40 and 50 volts. The recording electrode was placed in the dorsal third of the marginal fissure which, in the rabbit, is located on the right at the paramesial calvaria and at the level of the dorsal orbital margin. It was firmly anchored in the periosteum, contralateral to the stimulus electrode. The reference electrode was located at the root of the nose. The left tibial nerve was stimulated with surface electrodes (felt electrodes). The animal was connected to ground at the right hind paw. In addition, intra-arterial blood pressure was monitored (Statham system) plus the continuous EEG and body temperature.

Results

Two parameters (i.e., potential amplitude, peak latency) were evaluated during step-by-step (i.e., millimeter-by-millimeter) distraction of the scoliotic spine (Fig. 2).

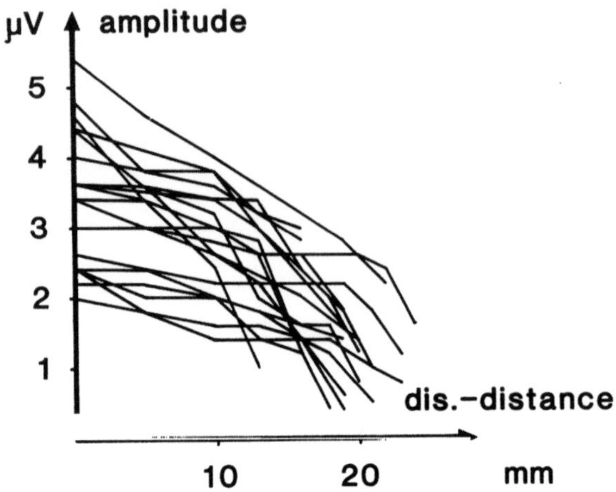

Fig. 2. Comparison between amplitude reduction curves to potential extinction and distraction distance

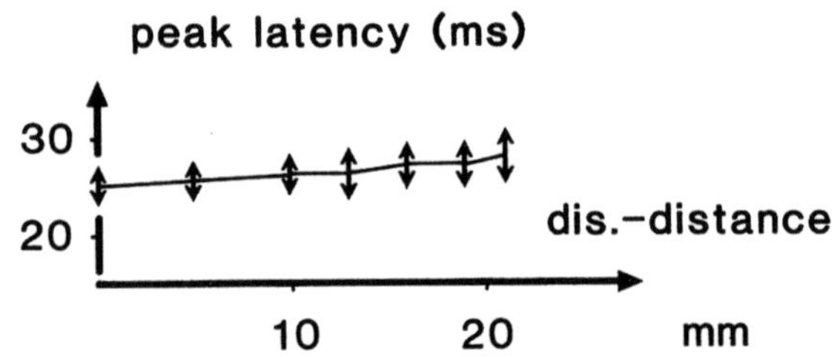

Fig. 3. Mean peak latencies in relation to distraction distance, showing only a slight increase plus an increase in standard deviation

The scoliotic curves initially measured from 45° to 60°, according to Cobb. Potential amplitude was found to respond sensitively to distraction, and sometimes declined so drastically after distractions of more than 16 mm that it could no longer be measured. This drastic reduction occurred particularly when the scoliotic component was accompanied by pronounced kyphosis. Nevertheless, scoliotic spines, regardless of the degree of curvature, could usually be distracted by as much as 21 mm (see Fig. 2). Once tissue resistance is overcome, the force of distraction is transmitted directly to the spinal cord. Potential amplitude was reduced in all distractions of 10 or more millimeters (Fig. 2).

With the exception of a few interindividual differences, i.e., slight latency delays, no essential changes of peak latency were detected during distraction (Fig. 3). The changes, in any case, were not distinctive enough for them to serve as intraoperative control variables.

In all 20 distraction tests potential amplitude was reduced to total extinction. The obliteration of potential amplitude during maximal distraction that occurred in all 20 rabbits was irreversible.

Discussion

Our series of animal experiments showed that, similar to the distracting scoliosis operation, the extent of SEP amplitude reduction is a sensitive variable during distraction of experimentally induced scolioses. This reduction of potential amplitude directly after step-by-step distraction represents an immediate response on the part of the afferent pathways to the distraction. The primary reasons for amplitude reduction of tibial nerve SEPs could be stretching of axons in the afferent pathways of the posterior funiculus of the spinal cord.

One important aspect of this series is that the entire experimentally induced scoliosis was distracted and not just one or two segments (Nordwall et al. 1979; Larson et al. 1980). Similar to the scoliosis operation, all scoliotic segments (i.e., 6–7) were distracted under preservation of their anatomic relations. Our study showed that the crucial control variable during spinal cord distraction is the change of potential amplitude.

References

Larson SJ, Walsh PR, Sances A, Cusick JF, Hemmy DC, Mahlers H (1980) Evoked potentials in experimental myelopathy. Spine 5:299–302

MacEwen GD, Bunnell WP, Sriram K (1975) Acute neurological complications in the treatment of scoliosis: a report of the Scoliosis Research Society. J Bone Joint Surg (Am) 57-A:404–408

Nordwall A, Axelgaard J, Harada Y, Valencia P, McNeal DR, Brown JC (1979) Spinal cord monitoring using evoked potentials recorded from feline vertebral bone. Spine 4:486–494

Vauzelle C, Stagnara P, Jouvinseux P (1973) Functional monitoring of spinal cord activity during spinal surgery. Clin Orthop 93:173–178

Monitoring —
Cortical Evoked Potentials

Intraoperative Deterioration of Evoked Potentials During Spinal Surgery

A. KOHT[1], T. SLOAN, A. RONAI, and J. R. TOLEIKIS

Summary

Successful intraoperative monitoring using cortical SSEP was achieved in 381 out of 395 patients undergoing corrective surgery for pathology of the spinal cord or spinal column. SSEP deterioration due to physical, anesthetic and technical factors in 36 cases was not associated with neurological deterioration. Likewise in 14 cases reversible SSEP changes due to surgical factors had no neurological sequelae, but 8 patients in whom SSEP deterioration could not be reversed had postoperative neurological deficits.

Introduction

Recently, there has been an increase in the number of operations for stabilizing the spinal column utilizing evoked potential monitoring (Nash and Brown 1979). These operations include Luque and Harrington rod instrumentation with anterior, posterior, and combined anterior/posterior spinal fusion. These, as well as operations for spinal cord tumors, have been associated with injury and altered spinal function postoperatively (MacEwen et al. 1978). Intraoperative somatosensory evoked potential (SSEP) monitoring may reduce the number of these complications as understanding of factors involved in its recording is improved (Grundy 1982, 1983). At Northwestern University-McGaw Medical Center, 395 such patients have had intraoperative somatosensory evoked potential monitoring over the 20-month period ending November, 1984. This report summarizes the incidence and type of events associated with acute intraoperative SSEP deteriorations in these patients.

The intraoperative environment offers many personal, technical, physiological, pharmacological and surgical challenges not present in the diagnostic laboratory. Since the foundation of successful monitoring of evoked potentials is experience in dealing with the myriad of problems and compromises that face the monitoring team, recognition of these events will assist in understanding intraoperative changes and allow improved anesthetic and technical management.

Methods

From April, 1983 to November, 1984, 395 patients with pathology of the spinal cord or spinal column underwent corrective surgery with SSEP monitoring. All were tested preoperatively and had reproducible somatosensory evoked potentials from

[1]Northwestern University Medical School, 303 East Chicago Ave., Chicago, IL 60611, USA

Spinal Cord Monitoring
Ed. by J. Schramm and S. J. Jones

stimulation distal to the planned operative site (typically posterior tibial nerve or common peroneal nerve).

The operations included 111 spinal fusions to stabilize the spinal column after trauma, 105 spinal instrumentations (83 Luque, 22 Harrington rods), 170 lumbar or thoracic laminectomies (decompression of spinal cord, resection of disc, relief of root entrapment and spinal stenosis or resection of spinal tumors) and combined anterior/posterior spinal fusion.

Preoperative testing of all patients was done to establish a baseline for intraoperative monitoring. Stimulation was done by bar electrode over the major nerve trunk; median nerve (MN) at the wrist, posterior tibial nerve (PTN) at the ankle or the common peroneal nerve at the knee. Recording was achieved by disc electrodes positioned over C_3', C_4' for median nerve or C_z' for lower extremities. Cervical spinal electrodes were placed at SC_1, or SC_7. A non-cephalic ground and reference electrode at F_z were used for all cases. Stimulation was done by constant current square wave of 300 microsecond duration delivered at one to two milliamperes above motor threshold at a rate of 5.4–8.7 Hz. The average of 250–500 repetitions was recorded from the scalp and the cervical spine using bandpass filtration from 5–250 Hz and an analysis window of 50 milliseconds (MN) or 100 milliseconds (PTN). All tracings were done in duplicate to assure the reproducibility of the test. Intraoperative monitoring of SSEP used subdermal needle electrodes for stimulation and recording.

The patients were premedicated with either 0.4 mg atropine or scopolamine injected intramuscularly with either oral diazepam (0.1 mg/kg) or intramuscular morpine (0.1 mg/kg). Patients were also monitored intraoperatively with continuous EKG, blood pressure, arterial line (when indicated), and end tidal CO_2 monitor. Anesthesia was induced with intravenous sodium thiopental (3–5 mg/kg) and incremental doses of fentanyl. Motor threshold for the nerve to be tested was established before full paralysis using pancuronium or d-tubocurarine. All patients had endotracheal intubation with controlled ventilation. Maintenance anesthesia consisted of fentanyl (20–60 mcg/kg), muscle relaxation and thiopental drip (1.5 mg/kg/h) or low concentrations of ethrane, isoflurane or halothane (0.25–0.5 MAC).

Criteria chosen for this study were acute deteriorations from baseline recordings comprising: 1) drop in amplitude of more than 50%; 2) increase in latency of 10% or greater. When such a change occurred in two successive recordings, a search for associated technical, surgical, physiological and anesthetic events was made. A detailed review was made of any changes in the anesthetic concentrations, the addition of other agents or the injection of a bolus dose of an intravenous agent. Changes in blood pressure, temperature, blood gases or other physiological parameters were noted. The raw EEG was examined and technical aspects were assessed including machine settings and recording and stimulating electrodes. The surgeon was notified and asked to exclude any surgical events including irrigation solutions. After the events were identified, an attempt to correct any technical, physiological or anesthetic change was made. In most cases, this was sufficient to re-establish the signals. For example, if the event was associated with a retractor, the release of the traction on the spinal column often resulted in recovery of the SSEP.

If the evoked response could not be re-established, the following anesthetic changes were made to remove depressant anesthetic effects and improve the spinal cord perfusion; inhalation agents or the thiopental infusion were stopped and inspired oxygen concentration was increased to 100%. The surgical field was irrigated with warm saline and surgery temporarily halted. Arterial blood pressure, CVP, and hemoglobin were raised. If after these measures the signals were not restored, antecedant surgical manipulations were reversed where possible (e.g. Luque wires were removed or Harrington distraction was decreased). In most cases, these methods of resuscitation were enough to restore the SSEP signals and surgery was completed. Postoperative evoked response testing was conducted in similar manner to preoperative testing. Postoperatively patients were examined for evidence of new neurological deficits.

Results

Of the 395 patients who had preoperative evoked potential responses, 381 were successfully monitored. Fourteen patients (3.5%) were not monitored because of failure to obtain good SSEP signals after the start of anesthesia, and all had abnormal pre-

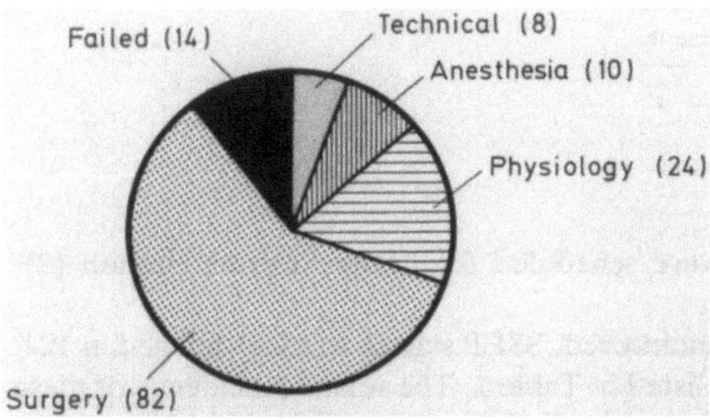

Fig. 1. Relative incidence of surgical, physiological, anesthetic and technical factors causing intraoperative SSEP deterioration

Table 1. Intraoperative experience with SSEP at Northwestern University

Procedure	Total	Cases with SSEP deteriorations	Related events			
			Surgical	Physiological	Anesthetic	Technical
Luque	81	33	29	2	2	0
Harrington	22	7	7	0	0	0
Spinal fusion	105	39	15	15	4	5
Laminectomy	164	42	29	6	4	3
A/P fusion	9	3	2	1	0	0
Total	381	124	82	24	10	8

This table does not include the 14 patients where monitoring failed to establish a usable response

Table 2. Reversibility of SSEP deteriorations[a]

Procedure	Non-surgical events	Surgical events
Luque	3/3	27/29
Harrington	0/0	6/7
Spinal fusion	22/22	14/15
Laminectomy	10/10	25/29
Ant/pos fusion	1/1	2/2
Total	36/36	74/82

[a] Presented as the number of reversible cases/total cases with deteriorations

Table 3. Association of postoperative neurological deficit with intraoperative deteriorations

Type of event	Postoperative deficit	
	Absent	Present
Reversible	110	0
Non-reversible	0	8

operative signals. These patients were scheduled for Luque instrumentation (2), spinal fusion (6) or laminectomy (6).

Of the 381 patients successfully monitored, SSEP signals acutely changed in 124. The events and types of surgery are listed in Table 1. The relative incidence of these changes is depicted in Fig. 1.

Technical events associated with deterioration included electrical noise (3), dislodged electrodes (4) and locking on alpha rhythm (1). Physiological events included leg tourniquet ischemia (3), cold fluid infusion in the stimulated arm (2), decreased BP or acute blood loss (3), cold irrigation of the spinal cord (2), positioning (head rotation (1), shoulder retraction (1)), carotid retraction (2) and hypothermia (2).

Anesthetic events included loss of SSEP due to inhalational agents (5), bolus injection of intravenous agents (3) and too light anesthesia (2). Surgical events which were reversible included sublaminar wire passage (21), root or cord manipulation (22), graft insertion or surgery on the bony lamina (13), distraction (11), cord retractors (4) and blunt cord trauma (2). Surgical events causing SSEP changes which were not reversible included surgery of the bony lamina (4), dissection of cord tumor (1), spine straightening with Luque wiring (1) and loss of cottonoid into canal with subsequent removal (2).

The reversibility of these deteriorations is tabulated in Table 2. In all cases of technical, anesthetic and physiological deteriorations, the SSEP recovered and no new neurological deficit occurred. In surgically related reversible changes the outcome was generally similar, but surgically related deteriorations which were not reversible were always associated with new neurological deficits (Table 3).

Discussion

Intraoperatively, acute changes in previously stable responses were found to be associated with various surgical and non-surgical events. One hundred and twenty-four patients (32.5%) showed intraoperative changes in SSEP. Of these changes, 60% were related to surgical intervention while 40% were associated with anesthetic manipulation, physiological alterations or technical problems. The identification of non-surgical causes of changes in SSEP would avoid unnecessary surgical intervention, such as removal of wires, rods, retractors or bone grafts. Past experience testifies to the poor tolerance of the spinal cord to prolonged adverse surgical manipulation.

Technical causes of evoked potential changes are more difficult to identify in the operating room than in the diagnostic laboratory. The reason for this is related to limited access to the patient and electrical interference from other equipment. In addition, long hours of intense concentration can fatigue the monitoring technician.

Physiologic and anesthetic changes can affect the responses abruptly, such that a quick check of arterial blood gases, temperature, the hemodynamic parameters and the anesthetic record are helpful in identifying possible events related to the SSEP changes. Further, recording evoked potentials from more than one site helps differentiate between local events and anesthetic effects.

These results demonstrate that SSEP deteriorations occurred in each type of operation monitored. Patients with scoliosis undergoing surgical instrumentation had a higher incidence of SSEP deteriorations than patients undergoing other types of surgery. The overall incidence of SSEP changes in scoliosis patients was 40.7% of the total number of patients monitored, compared with 25.6% in laminectomy patients. The frequency of surgically associated changes was 35.8% in scoliosis and 17.7% during laminectomy. Changes of SSEP in the other surgical procedures fell in between.

Intraoperatively, satisfactory monitoring was conducted in 381 patients. Recordings were unsatisfactory for monitoring in 14 patients. The success rate (96.5%) is higher than reported by Raudzens (1982). Monitoring success in these patients can be attributed to several factors, including superb technical support and identification of optimal stimulus and recording parameters for each patient by preoperative study. In several patients, reproducible SSEP responses were of good quality at the cervical sites but poor in the cortical leads. This allowed monitoring in cases where the cortical responses would have been of marginal quality for monitoring. Further, an anesthetic technique designed to avoid agents known to depress the evoked response has been used.

Anesthetic management was planned according to the size of the preoperative signals. Patients with poor responses could be easily rendered unmonitorable by inhalational agents. In contrast, patients with robust responses tolerated larger doses of depressant anesthetics. The choice of anesthetics made by an anesthesiologist knowledgeable about the effect of different anesthetic agents on these signals clearly assisted in monitoring. This knowledge and good cooperation with the monitoring team is essential for successful monitoring.

Somatosensory evoked potentials would be a very important tool if they could predict the outcome and decrease the incidence of morbidity following spinal surgery. Unfortunately, such information is not available at this time and a double-blind study to demonstrate their value is unethical. At present anecdotal reports must be utilized. In these cases, three patients had SSEP responses which were totally lost and never recovered; all three patients awoke with permanent neurological damage. Three additional patients in whom SSEPs were lost exhibited a simultaneous motor deficit on "wake-up" tests. The responses recovered and the patients had recovery of their motor function on awakening at the conclusion of surgery. In all cases, where the loss of SSEP was reversible the postoperative outcome was good, but where the loss interoperatively was permanent the patient awoke with a sensory deficit. All non-surgical events were reversible and were associated with good outcome.

In summary, there was a high rate of success in monitoring SSEP intraoperatively. SSEP changes related to several events were seen in different types of surgical procedures. By understanding how these events relate to evoked response deteriorations and thereby avoiding unnecessary changes in surgical procedure, monitoring can focus on the surgically related deteriorations that appear to prognosticate adverse neurological outcome.

References

Grundy B (1982) Monitoring of sensory-evoked potentials during neurosurgical operations: Methods and applications. Neurosurgery 11:556–575

Grundy B (1983) Intraoperative monitoring of sensory-evoked potentials. Anesthesiology 58:72–87

MacEwen GD, Bunnel WP, Sriram K (1978) Acute neurological complications in the treatment of scoliosis: A report of the Scoliosis Research Society. J Bone Joint Surg 57-A:404–408

Nash CL, Brown RH (1979) The intraoperative monitoring of spinal cord function: Its growth and current status. Orthopedic Clinics of North America 10:919–926

Raudzens PA (1982) Intraoperative monitoring of evoked potentials. In: Bodis-Wollner I (ed) Evoked potentials. Ann NY Acad Sci 388:308–325

Cortical Somatosensory Evoked Potentials During Surgery for Scoliosis and Coarctation of the Aorta

R. C. McWilliam[1], A. N. Conner, and J. C. S. Pollock

Summary

Three particular aspects of spinal cord monitoring have been studied. Firstly observations have been made on the effects of varying degrees of interruption to the blood supply to the spinal cord during aortic occlusion. Loss of signal for periods of up to 28 min are compatible with complete recovery of function. Secondly we have studied the effects of halothane on recordings during spinal surgery. Concentrations of up to 0.5 vol% have no significant effect on amplitude or latency of the response. Above this there is a gradual deterioration of signal and at concentrations above 1.0 vol% it is lost completely. Finally we have recorded responses in patients with neurological disorders. Stimulus and recording parameters may require modification in these patients in order to obtain a useful record and abnormalities of central sensory conduction may be revealed.

Introduction

The value of recording evoked potentials for the intraoperative assessment of spinal cord function is well recognised. In spite of this many centres continue to perform operations which put the spinal cord at risk without monitoring its function. There is a need to simplify equipment, to understand better the factors which influence responses, and to share our findings in unusual cases so that those who regard the technique as mysterious may be encouraged to take up the challenge.

We report our experience in recording cortical somatosensory evoked potentials (SEPs) during aortic and spinal surgery with particular comment on three aspects of our work viz a) the effect of interruption of the blood supply to the spinal cord by aortic occlusion, b) the effects of halothane, and c) the response in children with neuromuscular disease.

Patients and Methods

During the past two years 35 patients have had intraoperative spinal cord monitoring using cortical SEPs. In thirteen (aged 2–50 years) recordings were made during

[1] Department of Paediatric Neurology, Royal Hospital for Sick Children, Yorkhill, Glasgow, Great Britain

Spinal Cord Monitoring
Ed. by J. Schramm and S. J. Jones

surgery for coarctation of the aorta; in 22 (aged 4–16 years) during Harrington or Luque instrumentation. Excluding four children with neuromuscular scoliosis (Friedreich's ataxia, Duchenne muscular dystrophy, neurofibromatosis, Larsen's syndrome), preoperative neurological examination was normal.

We have used a Medelec Sensor, comprising an ST10 stimulator and an ER94 amplifier, averager, analyser and display unit (Medelec Limited, Surrey, England). All recordings were made after the induction of anaesthesia except in the patient with Friedreich's ataxia in whom a preoperative record was obtained. The stimulus was applied through a pair of 6 mm silver/silver chloride cup electrodes filled with conductive gel and taped over the posterior tibial nerve behind both knees. The inter-electrode distance was 2 cm, the cathode being placed proximally. The signal was recorded using similar electrodes fixed to the scalp with collodion. Contact impedance was reduced to less than 2 kΩ by gently abrading the skin with a blunted needle. The active electrode was positioned at Pz (International 10-20 system) and a reference electrode at either Cz or Fz. In individual patients the reference which gave the best signal/noise ratio was selected. An electrode at FPz served as a ground.

A square wave, constant current stimulus was used. The duration was 0.3 ms except in the patient with Friedreich's ataxia where a stimulus duration of 1 ms was required to elicit a response. Stimulus strength was increased until no further increase was seen in the amplitude of the evoked response (usually 25 mamps). A repetition rate of 2/s was generally employed although in a few patients an adequate response could be obtained more rapidly using a faster rate. In the patient with Friedreich's ataxia the rate was reduced to 1/s in order to produce a satisfactory record.

Initially a 200 ms period following each stimulus was analysed but this was reduced to 100 ms if the response was complete within that time. Exceptionally the analysis time was increased to 500 ms in order to record a very late potential. The amplifier gain was adjusted to give the maximum displayed signal with the minimum artefact rejection (the averager automatically rejects any signal producing a screen deflection of more than 5 cm except during a period following the stimulus equal to one eighth of the analysis time), usually 20 μv/cm. The bandpass was 10 Hz–1.5 kHz. Averaging was continued until a satisfactory waveform was obtained (100–200 sweeps) and this averaged response was then digitally amplified to give a final gain of 0.5–2 μv/cm.

Recordings were made from each leg in turn during spinal surgery and from both legs together during cardiac surgery in the following situations:

1. After induction of anaesthesia.
2. Following any change in level of anaesthesia, in particular after changes in halothane concentration (see below).
3. Immediately before spinal distraction.
4. Ten and 20 min after spinal distraction.
5. Continuously following aortic cross clamping or intercostal artery occlusion, until 20 min after reperfusion.

In 12 patients we observed the effects of varying concentrations of halothane on the evoked response during scoliosis surgery. The halothane concentration was measured using a Normac Anaesthetic Agent Monitor (Datex, Helsinki) continuously sampling the gas via a T-piece attached to the endotracheal tube. When the inspired and expired gas concentrations were equal we assumed that a steady state representing alveolar gas concentration had been reached.

Results

A reproducable record was obtained in all patients, although considerable interindividual variations were seen (Fig. 1–3). Minor degrees of asymmetry, both in latency and amplitude, were observed between the left and right leg in most tracings from neurologically normal patients. Except in one patient there were no differences between traces before and after spinal instrumentation. In this patient the response from the right leg was lost after distraction but an immediate wake-up test showed normal motor function and the response had returned 24 h later. No additional postoperative neurological abnormality was recognised in any of our patients.

Aortic Occlusion

During interruption of distal aortic perfusion three types of response were recognized:

1. (6 patients) There was no change in the SEP during the period of aortic cross clamping (up to 23 min).
2. (6 patients) Progressive reduction in amplitude and increase in latency were observed to occur between 15 and 20 min after aortic cross clamping. Following reperfusion the signal reappeared and within five minutes had returned to normal (Fig. 1). (Maximum aortic cross clamp time—28 min)
3. (1 patient) This five-year-old boy had unusually large intercostal collateral vessels and the waveform began to deteriorate shortly after several of these arteries were

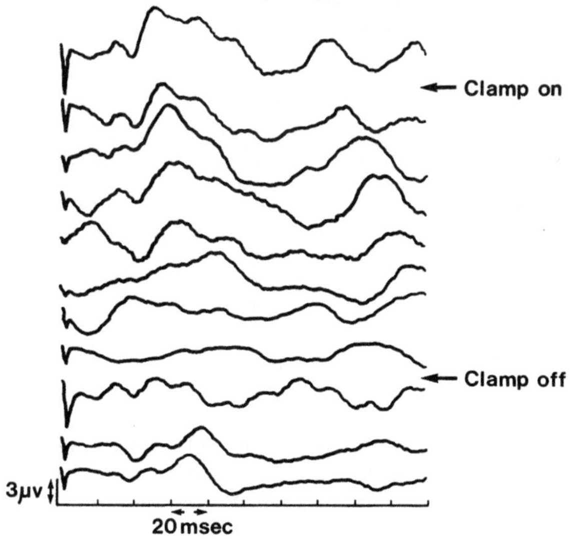

Fig. 1. Serial recordings during repair of coarctation of the aorta showing a 'type 2' response. The *arrows* indicate application and release of the aortic clamp

occluded. Within 10 min, and before the aorta was cross clamped, the signal could not be detected and remained absent for a total of 28 min. The signal returned within three minutes of release of the clamps (cross clamp time 21 min).

Halothane

We found little effect on the amplitude or latency of the evoked potentials provided the halothane concentration did not exceed 0.5 vol%. Higher concentrations of halothane were associated with a progressive deterioration in signal and above 1.0 vol% satisfactory records could not be obtained. A typical study is illustrated in Fig. 2.

Neurological Disorders

In the patient with muscular dystrophy a stimulus of 25 mamps for 0.3 ms produced a normal response. The patient (female—15 years) with Friedreich's ataxia was examined preoperatively because of the well recognised difficulty in recording SEPs in this disorder. A normal intensity stimulus (0.3 ms, 19 mamp) produced a well defined SEP at greatly increased latency, an analysis time of 500 ms being required to display it adequately. After induction of anaesthesia a similar response could be

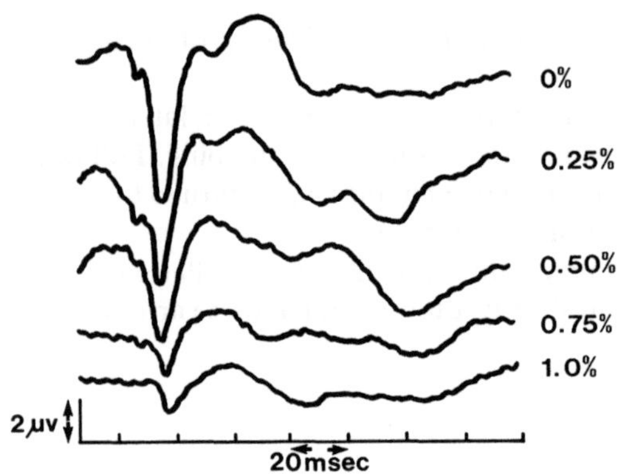

Fig. 2. Serial recordings during spinal surgery showing effect of halothane (right leg only stimulated)

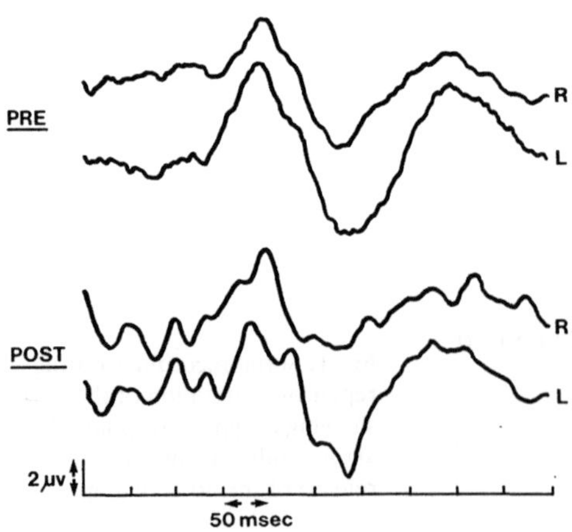

Fig. 3. Cortical SEP in Friedreich's ataxia before and after distraction. The latency of the first negative peak is 186 ms

obtained but only with an unusually large stimulus (1 ms, 100 mamp) (Fig. 3). The remaining two patients had asymmetrical neurological (motor) signs due to neurofibromatosis and Larsen's syndrome. In both patients there was a delayed response from the more severely affected leg.

Discussion

The instability of cortical SEPs during anaesthesia (Nuwer and Dawson 1984) and specifically the adverse influence of halothane and enflurane (Engler et al. 1978; Nash et al. 1977) has prompted some workers to adopt spinal monitoring techniques (Jones et al. 1983; Lamont et al. 1983). We favour cortical recording because it is suitable for cardiac surgery and is divorced from the operative field during spinal work. It has proved compatible with our standard anaesthetic for scoliosis surgery. After induction with thiopentone (5 mg/kg) we maintain anaesthesia with nitrous oxide, oxygen, muscle relaxant (pancuronium 0.1 mg/kg as necessary) and morphine infusion (0.1 mg/kg/h). The addition of halothane helps maintain a dry operative field by reducing the autonomic responses to surgery and, provided the concentration does not exceed 0.5 vol%, does not interfere with monitoring. Instability which does not appear to be caused by anaesthetic agents may be solved by alternative electrode sites. Nuwer and Dawson (1984) examined different montages and found a bipolar channel between Pz and Cz gave the best signal/noise ratio. In our experience this was often so but in some patients a reference at Fz gave the clearer signal. We have no experience with other montages but a channel between Cz and the upper cervical spine might enhance detection of the earlier (and more stable) peaks. Digital filtering (Eisen et al. 1984) may also improve recordings.

Cortical SEP measurement appears to be a sensitive indicator of spinal cord ischaemia as evidenced by the loss of signal in seven patients following interruption to distal aortic perfusion. That all our patients recovered without neurological deficit suggests that the method may be too sensitive. We were encouraged by the early return of the SEP in these patients but this does not always predict escape from permanent ischaemic damage. Mizrahi and Crawford (1984) described a patient in whom the SEP was lost for 59 min with a cross clamp time of 87 min. This patient showed prompt recovery of signal following reperfusion but on recovery from anaesthesia had a typical anterior spinal artery syndrome.

It seems unlikely that there could be significant spinal cord ischaemia without at least a temporary change in SEP and we are not aware of a case of permanent cord damage where the recording has remained stable. In aortic surgery we would advise ignoring any deterioration in signal unless the period of aortic occlusion exceeds 30 min. In that event a temporary bypass graft should be established as a means of rapidly restoring distal aortic perfusion. In scoliosis surgery a wake-up test should be performed if there is an obvious alteration in response following distraction although just what one means by 'obvious' must remain at present subjective. However, changes in latency are probably more important than changes in amplitude (Maccabee et al. 1982). In the future, tests of anterior function such as F wave, crossed extensor response and H reflex may prove more valuable indicators of cord damage.

In patients with neurological scoliosis responses are more variable and different stimulus and recording parameters may be required. There have been reports of difficulty in eliciting responses in patients with Friedreich's ataxia (LaMont et al. 1983; Jones et al. 1983). There are two possible reasons. Firstly, the peripheral sensory nerve fibres are reduced in number and the remainder may be relatively insensitive to electrical stimulation, in keeping with the well established sensory preponderance of the peripheral neuropathy in this disorder (Ackroyd et al. 1984). Secondly, the occurrence of a very late cortical SEP may not be detected unless a sufficiently prolonged analysis time is used. The finding of a potential at this latency, while not reported previously with leg stimulation, is consistent with the greatly increased latency of the later (cerebral) component of long loop 'reflexes' seen in Friedreich's ataxia (Diener et al. 1984). The asymmetry of response in patients with neurofibromatosis and Larsen's syndrome corresponding with their asymmetry of motor signs but in the presence of normal sensation suggests a sensitivity greater than that of clinical examination. Cord monitoring is worthwhile in patients with muscular dystrophy because of the risk to sphincter function—one of the few remaining areas of independence for patients in the advanced stages of the disease.

Evoked potential measurement is time consuming and since, at present, interpretation requires the presence of an experienced neurophysiologist or technician it is demanding of manpower. We are currently investigating the feasibility of computer controlled operation utilising exponential averaging in an attempt to overcome some of these difficulties.

Acknowledgements. We acknowledge Dr. C. S. Cairns, consultant anaesthetist, for his help and advice with the halothane studies. We are grateful to the Department of Medical Illustration, Royal Hospital for Sick Children for preparation of the figures.

References

Ackroyd RS, Finnegan JA, Green SH (1984) Friedreich's ataxia. Arch Dis Child 59:217–221

Diener HC, Dichgans J, Bacher M, Guschlbauer B (1984) Characteristic alterations of long-loop 'reflexes' in patients with Friedreich's disease and late atrophy of the cerebellar anterior lobe. J Neurol Neurosurg Psychiat 47:679–685

Eisen A, Roberts K, Low M, Hoirch M, Lawrence P (1984) Questions regarding the sequential neural generator theory of the somatosensory evoked potential raised by digital filtering. Electroenceph Clin Neurophysiol 59:388–395

Engler GL, Spielholz NI, Bernhard WN, Danziger F, Merkin H, Wolff T (1978) Somatosensory evoked potentials during Harrington instrumentation for scoliosis. J Bone Joint Surg (Am) 60-A:528–532

Jones SJ, Edgar MA, Ransford AO, Thomas NP (1983) A system for the electrophysiological monitoring of the spinal cord during operations for scoliosis. J Bone Joint Surg (Br) 65-B: 134–139

LaMont RL, Wasson SL, Green MA (1983) Spinal cord monitoring during spinal surgery using somatosensory spinal evoked potentials. J Pediatr Orthop 3:31–36

Maccabee P, Levine DB, Kahanovita N, Pinkhasov E (1982) Monitoring of spinal and subcortical somatosensory evoked potentials during Harrington rod instrumentation. Orthop Trans 6

Mizrahi EM, Crawford ES (1984) Somatosensory evoked potentials during reversible spinal cord ischaemia in man. Electroenceph Clin Neurophysiol 58:120–126

Nash CL, Lorig RA, Schatzinger LA, Brown RH (1977) Spinal cord monitoring during operative treatment of the spine. Clin Orthop 126:100–105

Nuwer MR, Dawson E (1984) Intraoperative evoked potential monitoring of the spinal cord: enhanced stability of cortical recordings. Electroenceph Clin Neurophysiol 59:318–327

Scalp Recorded Somatosensory Evoked Potentials During Spinal Surgery

S. BREITNER[1] and K. A. MATZEN

Summary

The experiences in 124 intraoperative examinations with scalp SEP in spinal surgery including 38 scoliosis operations and 35 traumatic cases are reported. Technically reliable results were obtained in 92% of cases. Among the changes observed where a continuously decreasing amplitude in 26%, constant amplitudes in 40% and unclassifiable fluctuation in the remainder cases. Neurologically-induced changes were seen in two cases, changes induced by shock and acidosis were found in another two cases. The value of a reproducable P40/N50-complex is stressed and the problems of a grey zone of responses with difficult interpretation is discussed.

Introduction

We have been using somatosensory evoked potentials in our clinic on a routine basis for the past 2 years to monitor spinal cord function on nearly all spinal surgery in which the spinal cord is vulnerable to injury. In this paper we describe the technique we have been using and present its results.

Methods

In routine clinical practice we generally record somatosensory cortical evoked potentials using a 2-channel-electro-myograph with integrated averager and X-Y plotter which is coupled to the system. The posterior tibial nerve behind the medial malleolus is stimulated by means of silver electrodes attached to the leg; the potential is recorded bipolarly between Cz and a frontal reference, using platinum needle-electrodes. By placing the leads on the median line, we can take measurements from the left and right limbs separately even when we only have a 1-channel-EMG at our disposal. The tibial nerve is stimulated by an impulse lasting 0.5 ms at a frequency of 3 Hz and a range of 150–250 volts; just sufficient to obtain a motor response. The band pass of the amplifiers is 32–1600 Hz. An average of 200 sweeps with a duration of 200 ms each is averaged and the potential is then plotted out on paper. The first measurement is usually taken after anaesthesia has been induced but before the patient is positioned, and subsequent recordings are made approximately every 20 min. During critical phases of surgery, measurements are taken frequently. The so-called "wake-up-test" is only performed when the potentials are not reliable enough to entirely rule out sensory pathway involvement at the time. The wake-up-test has

[1] Orthopädische Klinik und Poliklinik, Klinikum Großhadern, Ludwig-Maximilians-Universität München, Marchioninistr. 15, D-8000 München 70

Spinal Cord Monitoring
Ed. by J. Schramm and S. J. Jones

been performed a total of 6 times in 124 operations. Generally, neuroleptic anaesthesia is given in the form of such agents as Fentanyl, Pancuronium, Dehydrobenzperidol, as well as nitrous oxide. Patient positioning differs according to the surgical procedure. Apart from prone, supine and lateral positions, recordings have also been made from patients who were operated on while in a sitting position.

Material

124 intraoperative examinations on 119 patients have been carried out over a period of less than 2 years. Five patients were operated on in 2 sessions during which data were recorded. The total comprised 38 scoliosis operations (Harrington instrumentation and VDS ≙ ventral derotation spondylodesis), 2 patients who underwent ventral redressement following traumatic kyphosis, 35 unstable fractures and subluxations of the spine at levels C_3–L_2 and 49 other spinal operations. The age of the patients ranged from 8–80 with an average age of 32.2 years. Thirty-five patients had preoperative neurologic deficits, 26 of these slight with mostly transient, radicular or one-sided symptoms; 9 patients had severe conditions, some extending to complete paralysis. As expected, the last 9 patients had either absent or unutilizable potentials, which is why they were not included in the results being presented. In one case of spinal metastases, considerable postoperative deterioration in the neurologic state was observed. In a total of 3 cases, slow recovery or disappearance of the preoperative neurologic deficits was observed. In all other cases, no change in the neurologic state was seen.

Results

In 84 examinations, a constant $P_{40}N_{50}P_{60}N_{70}$ formation was evident (Table 1), indicating that the sensory pathway was intact. In 21 examinations, an almost constant $P_{40}N_{50}$ formation was present, although there was considerable variability of P_{60} and N_{70}.

The last 6 examinations performed in this group showed changes in the overall waveform of the potentials, which made it impossible to obtain a reliable estimate of the neurologic state. In 4 cases responses could not be recorded for partially inexplicable technical reasons.

A continuous increase in latencies of between 2 and 5 ms was observed during surgical procedures in 49 of the 105 utilizable examinations. A continuous decrease

Table 1. Results

I. Constant "W" configuration	84	(73%)
II. Constant P40-N50 complex, inconstant P60-N70 complex	21	(18%)
III. Inconstant "W" configuration	6	(4%)
IV. Failure for technical reasons	4	(4%)
	$n = 115$	

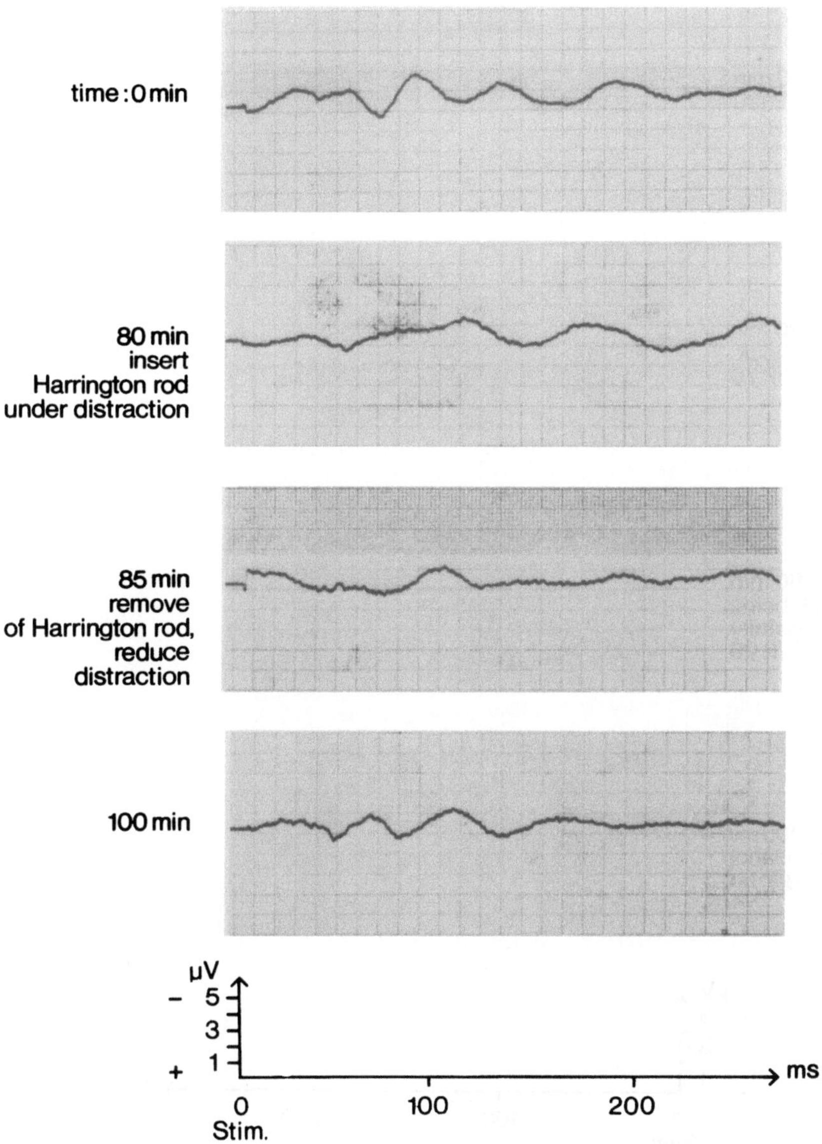

Fig. 1. Transient degradation and recovery of potentials during Harrington instrumentation, with no neurologic deficit

in amplitude was observed in 27 (26%) of the examinations, whereas in 42 (40%) examinations amplitude remained constant. These figures are both in reference to P_{40}–N_{50}. The remaining examinations showed unclassifiable fluctuations, whereby discrepancies of up to 50% were observed in individual cases, having no effect upon the neurologic state.

In 2 cases we observed positive neurologically-induced changes (Fig. 1): In one case the potentials almost completely disappeared immediately after implantation of

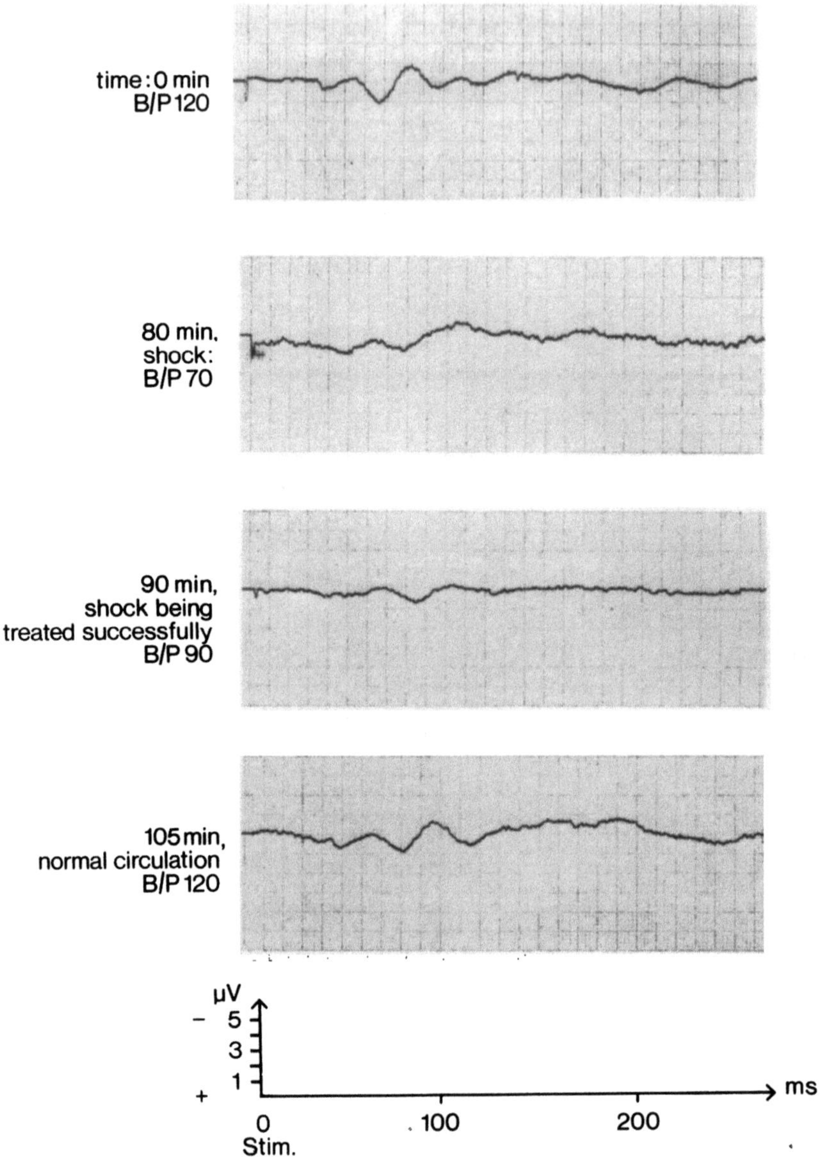

Fig. 2. Marked change in potentials following haemodynamic shock and acidosis, with subsequent recovery

the Harrington instrumentation during scoliosis surgery. The potentials returned to their initial form some minutes after the distraction force was reduced and the patient suffered no neurologic deficit. In another case the potentials abruptly disappeared during removal of an infiltrated axial spinal vertebra and remained undetectable for the remainder of the procedure. The patient suffered almost complete paralysis, being left with only minimal residual sensory function. It was several weeks before the neurologic deficits began to recede.

We observed marked changes in potentials in each of 2 cases following haemodynamic shock and acidosis (Fig. 2).

In both cases, after positive counter-regulatory action was taken, the potentials returned to their initial form. Neurologically-induced changes in the potentials were ruled out in both cases.

Discussion

The analysis of our results shows that, owing to inter- and intra-individual variations in amplitude and latency no absolute values can be used as criteria for indication of a possible onset of spinal cord injury or of damage which has already occurred. In our opinion, these variations are most likely caused by anaesthesia. In any case, circulatory factors particularly blood pressure, respiration, temperature and metabolism do not appear to influence the potentials to a great extent, unless they pose a threat to the patient's physical condition. As far as the various anaesthetic agents are concerned, we found that Halothane always caused a change in potentials whereas Enflurane proved to be more suitable. However, a larger patient population needs to be studied before significant statements can be made.

We feel that the safest criterion in helping to form an opinion with regard to the patient's neurologic status is a stable $P_{40}N_{50}$ complex, which we found in all examinations where no severe neurologic deficits or circulatory or metabolic dysfunctions ensued. Obviously the $P_{40}N_{50}$ complex is likely to be less influenced by anaesthetic and circulatory factors than the $P_{60}N_{70}$ complex.

If one uses the stable reproducible $P_{40}N_{50}$ complex as a criterion for certifying an intact neurologic state, then our clinical series has been able to document reliable or at least adequate results in nearly 92% of all cases examined. One disadvantage we see in the use of somatosensory cortical evoked potentials is the problem of inexplicable variation in amplitude, making it extremely difficult to obtain results in some cases. On the other hand, the use of somatosensory cortical evoked potentials does have a few distinct advantages: the system can be used at any time, and on either conscious or anaesthesized patients, permitting preoperative evaluations. The stimulus electrodes are safely fastened in place, making it easier for the examiner to monitor them intraoperatively. The monitoring apparatus is technically easy to operate so that monitoring does not interfere with the surgical procedure. The responses could be recorded from all sections of the spine including the cervical spine, independent of the patient's position.

Without doubt, an improvement in results can be expected when multi-channel systems are employed, which allow for data to be recorded at various levels, both above and below the operation site.

Lastly, a few remarks should be made concerning the principle use of cortical evoked potentials:

In patients with neurologic deficits somatosensory cortical evoked potentials should only be examined when good or satisfactorily reproducible responses can be recorded. Patients with distinct neurologic deficits (incomplete paralysis) are generally unsuitable for evaluation of somatosensory cortical evoked potentials. Also unsuitable for evaluation of somatosensory cortical evoked potentials are patients with cardiac pace-makers and halo. We also feel that close co-operation should be established with attending anaesthesiologists, since only when a continuous flow system of anaesthesia is used avoiding bolus injection, and early detection of metabolic dysfunction can be improved, is there a chance of being able to record reliable cortical evoked potentials.

The usefulness of SEPs has also been proven by others in animal experiments. Bennett (1983) described changes after anterior and posterior compression of the spinal cord. Ischaemia also produces SEP changes (Kobrine et al. 1979).

For routine clinical intraoperative purposes most authors use the cortical SEP (Eulert et al. 1979; Greenberg and Ducker 1982; Nash et al. 1977). They also found alterations of the potentials not associated with neurologic changes. Worth et al. (1982) described changes of amplitude, Engler et al. (1978) also latency shifts. Effects from anesthesia have been discussed but are still not fully understood (Grundy 1983). For that reason some authors prefer epidural recordings with good results (Jones et al. 1983). In our opinion the main disadvantages of these techniques are the insecure needle positioning and the limited possibilities of application (e.g. in cervical spinal surgery). Cord stimulating systems, as they are used mainly in Japan (Tsuyama et al. 1978), seem to be too invasive for routine purposes. Our results, presented here, show that there is a "grey zone" of responses with difficult interpretation. Further research is necessary, to find and understand the reason of these variations, and so to enhance the reliability of the procedure.

In summary, our results show a high rate of reliability even with employment of this relatively simple technique. Despite the limitations previously mentioned, we will continue to use cortical evoked potentials in our clinic on a routine basis on all spinal cord endangering surgery, so that in time we can totally avoid having to use the, in our opinion, very unreliable wake-up-test.

References

Bennett MH (1983) Effects of compression and ischemia on spinal cord evoked potentials. Exp Neurol 80: 508–519

Engler LG, Spielholz NJ, Bernhard WN, Danzinger F, Markin H, Wolff T (1978) Somatosensory evoked potentials during Harrington instrumentation of scoliosis. J Bone Joint Surg 60-A: 528–532

Eulert J, Mitzkat K, Stöhr M, Gekeler J (1979) Zur Anwendung evozierter Potentiale bei Skolioseoperationen. Presented at 66th Meeting of Deutsche Gesellschaft Orthopädie und Traumatologie, Basel

Greenberg RP, Ducker TB (1982) Evoked potentials in the clinical neurosciences. J Neurosurg 56: 1–18

Grundy BL (1983) Intraoperative monitoring of sensory evoked potentials. Anaesthesiology 58: 72–87

Jones SJ, Edgar MA, Ransford AO, Thomas NP (1983) A system for the electro-physiological monitoring of the spinal cord during operations for scoliosis. J Bone Joint Surg 65-B: 134–139

Kobrine AJ, Evans DE, Rizzoli HV (1979) Experimental acute balloon compression of the spinal cord: factors affecting disappearance and return of the spinal evoked response. J Neurosurg 51: 841–845

Nash CL, Lorig RA, Schatzinger LA, Brown RH (1977) Spinal cord monitoring during operative treatment of the spine. Clin Orthop 126: 100–105

Tsuyama N, Tsuzuki N, Kurokawa T, Imai T (1978) Clinical application of spinal cord action potential measurement. Int Orthop 2: 39–46

Worth RM, Markand ON, de Rosa GP, Warren CH (1982) Intraoperative somatosensory evoked response monitoring during spinal cord injury. In: Courjon J, Maugière F, Revol M (eds) Advances in neurology, vol 32: Clinical applications of evoked potentials in neurology. Raven Press, New York, pp 367–374

The "Grey Zone" in Intra-operative S.C.E.P. Monitoring

R. H. Brown[1] and C. L. Nash Jr.

Introduction

Our intra-operative somatosensory cortical evoked potential (SCEP) monitoring experience of over 600 cases from four separate surgeons over the last six years has shown that, given a reliable monitoring system and primary complexes which are maintained with reasonable amplitude and latency values, no significant intra-operative neurologic complications can be expected. This experience has also shown that under the same conditions an abrupt change in the records obtained from stimulation caudal to the operative site of more than 50% reduction in amplitude and/or more than 3 ms increase in latency, which does not tend to resolve back toward the baseline intra-operative control record within 15 min, indicates the cord to be significantly "at risk".

However, in our series there were a significant number of cases in which the monitoring records did not meet either of these criteria but rather fell into a "Grey Zone" which we have defined as "a perceivable change in the SCM records which cannot be interpreted in a predictive manner". That is, the records neither underwent an abrupt change, nor remained totally stable, nor resolved totally back to the intra-operative control records by the end of the case following an intra-operative perturbation.

Methods

In this study all monitoring was conducted by the first author using the same monitoring system. This system was an in-house design based on a Data General Computer. The method used was that of somatosensory cortical evoked potential monitoring. The median and posterior tibial nerves were stimulated using percutaneous needle electrodes. Sixty-four averaged reponses were recorded from corresponding contralateral scalp locations using differential amplification with a bandwidth of 0.3 Hz to 125 Hz. The EEG was also continuously monitored. Pre-operative medication was standardized at 2 mg/kg of Secobarbital and intra-operative anesthesia was the balanced nitrous oxide/narcotic/muscle relaxant technique utilizing continuous infusion of narcotic.

In conducting this review, pre-induction, immediate post-induction, pre- and post-instrumentation and start-of-closure records were evaluated for each patient.

[1]Division of Surgical Research, Saint Luke's Hospital, Shaker Boulevard 11311, Cleveland, OH 44104, USA

Spinal Cord Monitoring
Ed. by J. Schramm and S. J. Jones

Table 1. Patient population

Group	Surgery	No. of patients
I	HRI + SSI	49
II	SLI + SSI	20
III	HRI only	68

HRI, Harrington rod instrumentation; SSI, sublaminar wiring; SLI, standard Luque instrumentation

However, since records are routinely taken every 15 min under our protocol, additional records were available from each patient for supplemental evaluation as required.

The existence of a "grey zone" was confirmed by this review of 137 consecutive cases of posterior spinal fusion where monitoring was employed. This population was broken down into 3 groups (Table 1). Group I (HRI and SSI) consisted of 49 patients in whom Harrington instrumentation together with sublaminar wiring was employed. Group II (SLI and SSI) consisted of 20 patients in whom standard Luque instrumentation with sublaminar wiring was employed and Group III (HRI only) consisted of 68 patients in whom only Harrington instrumentation was used.

Results

Cases

The following two cases are representative of the type of data used in this review: In all cases the records are 512 ms in length.

Case 1 was a 40-year-old female with a 63° idiopathic scoliosis which was corrected with Harrington instrumentation and sublaminar wiring from T4 to L4. Her monitoring records were stable until the final distraction of one notch after tightening of all sublaminar wires. At that time her lower limb responses lost in excess of 50% amplitude and showed a distinct increase in latency of greater than three milliseconds. Since these responses did not resolve within our 15 min rule, a wake-up test was conducted, showing only a slight motion of the right leg and no motion of the left. Distraction was let down one notch, which resulted in no improvement in the records, and a second wake-up test showed no significant change in motor function. The distraction was then taken down completely, but the wires and rod were left intact. The subsequent records improved and the third wake-up test showed return of motor function.

Figure 1 shows her operative records. The pre-distraction record at the top shows a normal complex. Second is the record obtained immediately following the final distraction. Note the almost total loss of the complex. The third one down is after the distraction was released one notch and the bottom one shows the record obtained after distraction was taken off completely. Her motor response on the third wake-up graded approximately 3 out of 5. Note that although the bottom complex has not returned back to baseline values it is beginning to show a distinct primary complex in

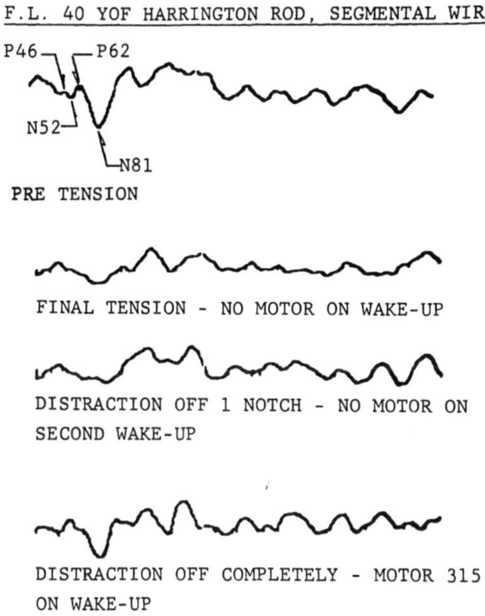

Fig. 1. Intra-operative records from adult scoliosis patient with intra-operatively documented neurologic compromise. Responses from stimulation of left posterior tibial nerve. Analysis time 512 ms

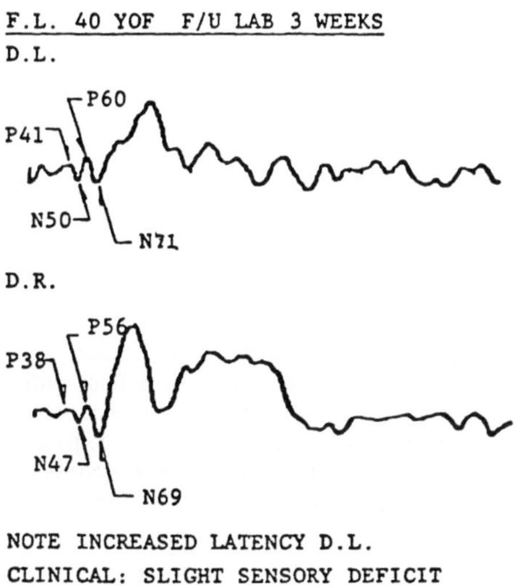

Fig. 2. Three week follow-up of adult scoliosis patient following intra-operative neurologic episode—DL and DR refer to left and right lower extremity responses respectively. Analysis time 512 ms

the early part of the record. Based on this, the decision was made to leave the instrumentation in, closure was effected and the patient sent to recovery room. Immediately post-operatively she demonstrated a distinct partial neurologic deficit in her left leg.

Three weeks post-operatively she still demonstrated a minimal sensory deficit in her left leg. While she had strong SCEP complexes from both lower extremities, she demonstrated a 2 ms latency delay in the left primary complex as compared with the right (Fig. 2). This case is an example from that group of patients who demonstrated

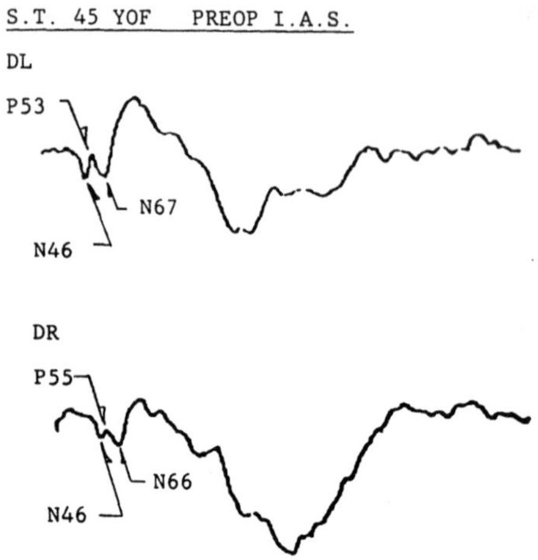

Fig. 3. Pre-operative records from adult scoliosis patient showing normal responses. Analysis time 512 ms

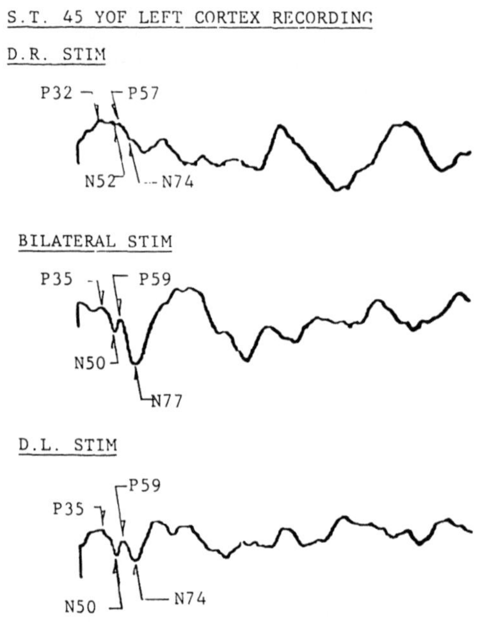

Fig. 4. Intra-operative posterior tibial responses from adult scoliosis patient with compromised right posterior tibial nerve (DR) response. Post-operatively the patient exhibited transient right lower extremity paresthesia. Analysis time 512 ms

persistent neurologic changes as a result of the surgery. All cases in this group were identified intra-operatively with monitoring. They do not fall into the Grey Zone.

Case 2 demonstrates more clearly the ambiguous zone in monitoring. This was a 45-year-old female with scoliosis undergoing a posterior fusion with Harrington rod and segmental wiring. Intra-operatively she demonstrated decreased amplitudes, by slightly less than 50%, from her right lower extremity. Based on this, a wake-up test was conducted during which she demonstrated a good motor response from both lower extremities. However, post-operatively, she did have a slight paresthesia in the right lower extremity which resolved prior to discharge.

Table 2. Summary of monitoring changes

Group	No change	Transient	Permanent
I (49)	32 (65%)	15 (31%)	2 (4%)
II (20)	9 (45%)	10 (50%)	1 (5%)
III (68)	38 (56%)	29 (43%)	1 (1%)
All (137)	79 (58%)	54 (39%)	4 (3%)

Table 3. Post-operative neurologic status

Group	No change	Transient	Permanent
I (49)	41 (84%)	6 (12%)	2 (4%)
II (20)	16 (80%)	3 (15%)	1 (5%)
III (68)	67 (98%)	0 (0%)	1 (2%)
All (137)	124 (90%)	9 (7%)	4 (3%)

Her pre-operative records as shown in Fig. 3 demonstrate good conduction from both lower extremities. Intra-operatively, as you can see from the top record in Fig. 4, the right lower extremity fundamentally suffered a loss of its primary complex. Bilateral stimulation and left lower limb stimulation demonstrated that this was indeed a neurologic problem as both these responses were within normal limits. The second case represents a group in which the monitoring showed itself to be somewhat more sensitive than the wake-up test, in that there was a transient neurologic change post-operatively which was implied by monitoring but not confirmed on wake-up.

In addition, there was another distinct group of patients in this series who, although their records were essentially normal throughout the entire course of their procedure, did wake-up post-operatively with a transient neurologic problem.

Records

Review of the SCEP records from this study showed that in Group I, 32 (65%) had totally normal intra-operative results while 15 (31%) demonstrated some transient latency and/or amplitude change intra-operatively and two (4%) had permanent monitoring changes intra-operatively. In Group II, the standard Luque group, nine (45%) had totally normal records intra-operatively, 10 (50%) demonstrated some transient change and one (5%) had a permanent monitoring change. In Group III, 38 (56%) demonstrated totally normal records, 29 (43%) demonstrated a transient change and one (1%) had permanent change in the monitoring record. These data are summarized in Table 2. Taken as a whole, these data show that only 58% of the patients in this series demonstrated totally normal intra-operative records while 39% demonstrated distinct transient monitoring changes generally related to the procedure and 3% demonstrated significant SCEP changes which did not resolve by the end of the procedure.

Table 4. SCEP vs post-op. status

SCEP changes intra-op.		No post-op. problems	Transient post-op. problems	Major post-op. problems
None	79	75 (95%)	4 (5%)	0
Transient amplitude	40	37 (93%)	3 (7%)	0
Transient latency	14	12 (86%)	2 (14%)	0
Permanent	4	0	0	4 (100%)

The post-operative neurologic status of each of the patients in this study was independently assessed. As can be seen from Table 3, in Group I, 12% demonstrated transient sensory deficits post-operatively while 4% had a major deficit. In Group II, 15% demonstrated some transient sensory deficit post-operatively while 5% had a major motor/sensory complication. In Group III, the standard Harrington group, no transient sensory deficits were noted post-operatively while there was a 2% major neurologic complication rate.

Summary

These data were then combined to assess the predictiveness of the monitoring and are tabulated in Table 4. Of the 79 patients who had totally normal intra-operative monitoring, 75 (95%) had no post-operative neurologic sequelae. However, four (5%) of these patients did suffer a transient sensory deficit post-operatively which resolved within two weeks or less.

Forty patients demonstrated transient intra-operative amplitude changes. Fifteen (37%) resolved completely and 25 (63%) improved to near baseline values prior to closure. Thirty-seven (93%) had no post-operative problems while three (7%) had a transient sensory problem post-operatively which resolved prior to discharge. Of the 14 who demonstrated transient latency shifts, 12 (86%) had no post-operative deficits while two (14%) did have a transient sensory problem post-op. Of the 4 patients whose monitoring changed by more than 50% in amplitude and/or 3 ms in latency intra-operatively, and did not resolve by the time of closure, all (100%) had a major neurologic complication documented post-operatively.

Conclusion

From the data the following statistics emerge:

First, all of the four patients who had permanent monitoring changes at the time of closure had a major neurologic complication post-operatively, for a 100% correlation in this group.

Second, having removed the four patients with permanent changes from the study and examining the remaining 133 patients, 108 (81%) had final intra-operative records which were essentially unchanged from the intra-operative control records. The other 25 patients (19%) had records which demonstrated transient intra-operative changes. These were resolving back to baseline controls but at the time of final closure were not considered to be completely resolved. Of the 108 with good records at the time of closure, four had a transient neurologic deficit post-operatively for a false negative rate of 3.7%. Of the 25 patients whose records were resolving toward baseline at the time of closure, 5 (20%) had transient post-operative neurologic deficits and 20 (80%) were normal. We feel that this group of 25 patients represents the "Grey Zone". In these cases, the records had not stabilized by the time of closure, and therefore no definitive prediction regarding the post-operative neurologic status could be made. Therefore, they represent neither false positives nor false negatives. It is important to note, however, that no serious neurologic complications resulted in this group of patients. Also, given that their records had not totally resolved by the time of final closure, monitoring identified this group as "potentially at a higher risk" for a post-operative neurologic problem. This was reflected in the fact that five (20%) of this group did have a transient post-operative neurologic deficit. However, the monitoring was not able to distinguish patients in this group with respect to which ones would and which ones would not have a post-operative problem. Given that 20 (80%) had no post-operative neurologic problems, improvement of the records toward baseline is considered to be generally predictive of a good outcome.

Finally, in considering these data, it is important to remember that with the exception of the four major neurologic complications—all of which were predicted by monitoring—all other post-operative neurologic complications were slight, sacral or lumbo-sacral in nature, were associated with sublaminar wiring procedures and all resolved prior to discharge. The possibility that "late" complications may have affected these data must also be considered. Even so, these data do indicate that we still have much to do with regard to monitoring the lower portion of the cord, and to understand better those perceivable changes in SCEP records which cannot be interpreted in a predictive manner—the "Grey Zone".

Variance of Latency and Amplitude in SEPs Monitored During Spinal Operations with and Without Cord Manipulation

J. SCHRAMM[1], J. ROMSTÖCK, F. THURNER, and R. FAHLBUSCH

Summary

Variability in amplitude and latency after median nerve, peroneal nerve and cauda equina stimulation has been evaluated in both healthy and diseased and manipulated and non-manipulated cords using the paired *t*-test. Three periods during operation were compared in individual cases. It could be demonstrated that amplitudes are more variable than latencies, that variability increases towards the end of the procedure and that variability is larger in diseased and manipulated cords than in healthy non-diseased cords. Cauda equina stimulation yielded relatively little variability in patients with manipulated spinal cords, similar to the variability after median nerve stimulation in healthy persons.

Introduction

A certain degree of variance occurring during continuous recording of SEPs is natural. One of the major issues in monitoring is the discrimination between changes reflecting normal variance and changes indicating spinal cord damage (Nordwall et al. 1979). Few data have been published concerning typical potential changes during monitoring (Gonzalez et al. 1984; Brown and Nash 1985; Worth et al. 1982). The purpose of this paper is to try to quantify the degree of variance occurring naturally when SEP-recordings are being performed in the hostile environment of the operating theatre.

Patients and Methods

Four patient groups were formed:

Group 1: 10 patients without spinal cord manipulation and without spinal cord lesion. Peroneal nerve stimulation.

Group 2: 10 patients with spinal cord lesion and spinal cord manipulation. Peroneal nerve stimulation.

Group 3: 8 patients with spinal cord manipulation and spinal cord lesion. Cauda equina stimulation.

Group 4: 10 patients without cord lesion and without cord manipulation. Median nerve stimulation.

The group with spinal cord manipulation and spinal cord pathology consisted of patients with intra- and extramedullary space-occupying lesions (tumors and thoracic

[1]Neurochirurgische Klinik, Schwabachanlage 6, D-8520 Erlangen

Spinal Cord Monitoring
Ed. by J. Schramm and S. J. Jones

discs); the group without spinal cord pathology consisted of patients operated on because of a lumbar disc herniation who gave informed consent for peroneal nerve or median nerve stimulation and scalp recording. The major difference between the groups was manipulation of the cord on the one hand and the existence of the pathologic process compressing the spinal cord on the other.

Stimulus Parameters

Frequency 5.1 Hz, stimulus duration 100 to 300 μs, 300 repetitions, stimulus strength 10 to 20 mA, effecting maximal responses.

Recording Parameters

Bandpass 30–3000 Hz (in 4 cases 5–1500 Hz), cortical electrode position C_z'-F_z for peroneal nerve and cauda equina stimulation. C_3',C_4'-Fz for median nerve stimulation. Analysis time 50 or 100 ms for peroneal nerve and cauda equina stimulation, 30 to 100 ms for median nerve stimulation. Intraoperative recordings were performed continuously through the whole duration of the operation. The total number of recordings being variable because of different length of surgery, 15 to 90 were made per operation.

From the patients with spinal cord pathology, clinically none demonstrated a complete transsectional syndrome and none showed a complete transsectional lesion on the preoperative SEP evaluation. A protocol was maintained including surgical commentary as given by the surgeon, remarks concerning the EP data as made by the monitoring person, and anesthesiologic data concerning blood pressure, blood gas analyses and anesthetic drugs used.

Data Analysis

In no case did the intraoperative recordings become flat or demonstrate a Zero-recording or so-called "block". The quality of recordings obtained intraoperatively was rather variable, however, when comparing recordings from patient to patient. In peroneal nerve stimulation a negative deflection at around 40 ms ($\overline{\text{N40}}$) was most constant and therefore the $\overline{\text{N40}}$ latency and the $\overline{\text{P30}}/\overline{\text{N40}}$ amplitude were evaluated. With cauda equina stimulation a positive deflection at 30 ms was most constant and therefore a $\overline{\text{N25}}/\overline{\text{P30}}$ amplitude and a $\overline{\text{P30}}$ latency were used. With median nerve stimulation the $\overline{\text{P25}}$ peak was most constant and therefore the $\overline{\text{N20}}/\overline{\text{P25}}$ amplitude and the $\overline{\text{P25}}$ latency were evaluated.

As the duration of surgery varied considerably and thus the number of recordings per surgical procedure varied considerably the following step was undertaken to achieve comparable data. From each patient 5 consecutive tracings (consisting of 300 averages) from the same 3 periods of the surgical procedure were chosen: From the beginning (setting up the patient—wound opening), the middle (laminectomy, dura opening and tumor preparation) and the end (wound closure). From the patient group without cord manipulation (i.e. the lumbar disc patients) similar episodes were chosen. Thus, from each patient 3 blocks of 5 consecutive tracings were obtained. From each 5-tracing block 3 values for the component to be evaluated were calculated for every patient: the mean, the standard deviation and the coefficient of

Table 1. Original measurements for N40 latency and P30/N40 amplitude in two patients and statistical data calculated therefrom

Case	Period	Measurement					Mean ($\bar{x}$)	Standard deviation (SD)	Coefficient of variability (%)	Value pair difference (d_i)
$\overline{\text{N40}}$ latency	Beginning	29.2	29.2	29.2	30.8	30.4	29.8	0.78	2.62	1.07
Patient H.H.	Middle	29.6	30.0	29.2	29.2	28.8	29.4	0.46	1.55	0.32
Group 1 (ms)	End	28.8	28.8	29.2	29.6	28.8	29.0	0.36	1.23	
$\overline{\text{P30}}/\overline{\text{N40}}$ amplitude	Beginning	1.20	1.40	1.45	1.35	1.73	1.43	0.19	13.60	9.77
Patient H.H.	Middle	1.70	1.60	1.60	1.55	1.55	1.60	0.06	3.83	12.54
Group 1 (μV)	End	2.18	1.63	1.48	1.78	1.53	1.72	0.28	16.37	
$\overline{\text{N40}}$ latency	Beginning	46.2	43.6	44.8	43.6	44.8	44.6	1.08	2.41	2.19
Patient K.L.	Middle	47.2	43.2	42.4	42.4	43.2	43.7	2.01	4.60	0.13
Group 2 (ms)	End	40.8	45.6	44.4	44.4	45.6	44.2	1.97	4.47	
$\overline{\text{P30}}/\overline{\text{N40}}$ amplitude	Beginning	0.98	0.93	0.75	1.11	0.86	0.93	0.13	14.50	8.67
Patient K.L.	Middle	1.00	0.79	0.79	0.66	0.53	0.75	0.17	23.17	57.52
Group 2 (μV)	End	0.93	0.73	0.20	0.10	0.28	0.45	0.36	80.69	

Table 2. Statistical data from mean of amplitudes in groups 2 and 3

Patient	Mean $\bar{x}$ (μV)			SD			Coefficient of variability			Value pair differences	
	Beginning	Middle	End	Beginning	Middle	End	Beginning	Middle	End	$d_i = v_m - v_b$	$d_i = v_e - v_m$
Group 2 (peroneal nerve stimulation, with cord manipulation)											
1	1.30	2.04	2.31	0.57	0.45	1.10	44.11	21.90	47.65	−22.21	25.75
2	0.93	0.75	0.34	0.13	0.18	0.24	14.44	23.30	71.20	8.86	47.90
3	0.07	0.03	0.09	0.03	0.02	0.04	48.63	58.33	44.42	9.70	−13.91
4	0.37	1.02	0.39	0.30	0.51	0.12	82.19	50.30	31.11	−31.89	−19.19
5	1.03	2.26	1.21	0.43	1.37	1.20	41.76	60.73	98.79	18.97	38.06
6	0.15	0.12	0.14	0.05	0.05	0.11	29.97	37.55	80.72	7.58	43.17
7	0.62	0.88	0.70	0.36	0.43	0.63	57.34	49.33	89.74	−8.01	40.41
8	0.42	1.11	0.70	0.11	0.48	0.35	25.29	43.27	49.23	17.98	5.96
9	0.19	0.16	0.11	0.06	0.05	0.08	29.91	30.81	67.85	0.90	37.04
10	0.24	0.16	0.79	0.11	0.12	0.35	46.45	73.50	44.59	27.05	−28.91
t-values for whole group										$t=0.49$	$t=1.92$
Group 3 (cauda equina stimulation, with cord manipulation)											
1	0.54	0.52	0.28	0.27	0.37	0.08	49.06	70.96	29.73	21.90	−41.23
2	0.36	0.42	0.54	0.10	0.10	0.16	27.05	24.60	29.56	−2.45	4.96
3	0.95	0.68	0.71	0.28	0.14	0.51	29.01	20.33	71.06	−8.68	50.73
4	1.47	1.57	1.10	0.19	0.58	0.23	13.22	36.91	21.21	23.69	−15.70
5	0.41	0.48	0.46	0.08	0.10	0.11	20.44	20.00	22.73	−0.44	2.73
6	0.29	0.46	0.38	0.19	0.35	0.16	66.66	76.24	43.24	9.64	−33.00
7	0.62	0.20	0.41	0.30	0.13	0.41	47.67	63.38	98.85	15.71	35.47
8	0.55	0.18	0.15	0.09	0.18	0.05	16.99	96.69	37.08	79.70	−59.61
t-values for whole group										$t=1.77$	$t=-0.52$

variability (Tables 2, 3). In order to statistically evaluate a difference in the variability occurring at the end, the middle and the beginning of the procedure the value pair difference d_i of the two coefficients of variability (v) for two periods was calculated for each patient. The sum $\sum d_i$ for one particular group was formed and the *t*-value calculated for that group of *n* patients:

$$t = \frac{\frac{\sum d_i}{n}}{\sqrt{\frac{\sum (d_i)^2 - \frac{(\sum d_i)^2}{n}}{n(n-1)}}}$$

Table 3. Statistical data from mean latencies in groups 2 and 3

Patient	Mean $\bar{x}$ (ms)			SD			Coefficient of variability			Value pair differences	
	Beginning	Middle	End	Beginning	Middle	End	Beginning	Middle	End	$d_i = v_m - v_b$	$d_i = v_e - v_m$
Group 2 (peroneal nerve stimulation, with cord manipulation)											
1	42.6	38.8	41.3	2.34	1.63	2.43	5.50	4.19	5.89	−1.31	1.70
2	44.6	43.8	44.2	1.15	1.91	1.99	2.58	4.36	4.51	1.78	0.15
3	42.8	43.7	43.6	1.02	2.31	3.86	2.38	5.27	8.85	2.89	3.58
4	42.5	43.0	41.3	0.72	0.88	0.52	1.68	2.04	1.26	0.36	−0.78
5	41.2	40.7	37.2	1.02	1.53	1.74	2.48	3.77	4.69	1.29	0.92
6	41.0	41.7	40.2	0.22	0.52	1.22	0.53	1.25	3.03	0.72	1.78
7	53.6	50.5	51.3	1.60	2.18	4.34	2.99	4.32	8.46	1.33	4.14
8	38.0	44.2	41.0	1.02	2.85	3.74	2.68	6.45	9.12	3.77	2.67
9	34.6	34.8	34.2	0.48	0.80	0.92	1.38	2.30	2.69	0.92	0.39
10	42.6	44.6	48.9	2.17	3.97	3.89	5.08	8.89	8.00	3.81	−0.89
t-values for whole group										$t=3.10$	$t=2.50$
Group 3 (cauda equina stimulation, with cord manipulation)											
1	27.4	28.1	30.3	1.97	1.59	1.40	7.17	5.66	4.61	−1.51	−1.05
2	26.0	25.4	28.6	0.75	0.85	1.88	2.88	3.36	6.60	0.48	3.24
3	29.9	29.9	28.2	1.45	1.04	1.26	4.84	3.47	4.48	−1.37	1.01
4	28.2	28.1	28.9	0.46	1.27	1.03	1.62	4.52	3.55	2.90	−0.97
5	28.2	28.0	27.8	0.36	0.54	0.54	1.27	1.92	1.93	0.65	0.01
6	34.5	43.5	41.3	3.10	3.36	1.79	8.99	7.74	4.34	−1.25	−3.40
7	42.3	40.1	42.4	7.71	2.29	5.97	18.22	5.70	14.07	−12.52	8.37
8	42.3	42.6	39.2	1.04	1.19	1.60	2.45	2.79	4.08	0.34	1.29
t-values for whole group										$t=-0.93$	$t=0.85$

Using appropriate tables for *t*-values the significance of changes in variability comparing two periods of surgery could be determined. Examples of original measurements are given in Table 1.

Results

Peroneal Nerve Stimulation

The 2 groups with peroneal nerve stimulation differed in two factors—the spinal cord was either diseased and manipulated intraoperatively, or it was healthy and not manipulated. Comparing both groups, a prominent finding is the different variability of latency and amplitude during the operative procedure. Although a certain degree of variability of *amplitudes* is found in both groups, in the group with cord manipula-

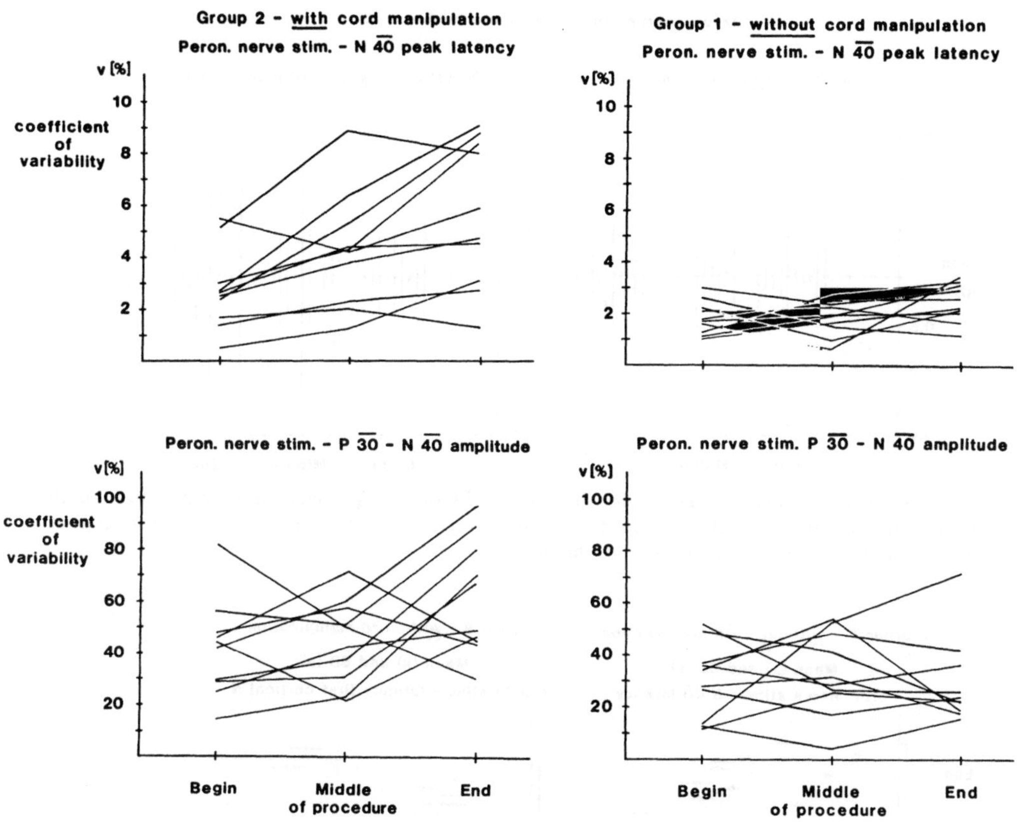

Fig. 1. Coefficient of variability for N40 latency and amplitude shown for the beginning, the middle and the end of the procedure in patients with normal spinal cord without cord manipulation (group 1) and patients with diseased spinal cord with cord manipulation (group 2). Values are shown for each patient, calculated from five consecutive tracings at the beginning, the middle and the end of the procedure

tion the variability is even larger (Fig. 1, lower left). The mean values and standard deviations of the $\overline{P30}/\overline{N40}$ amplitude of all patients demonstrate considerable differences, both with and without cord manipulation (Fig. 2). The mean and standard deviation of the N40 latency, comparing the middle, the end and the beginning of the procedure, show very small fluctuations (Fig. 3, left).

Considering the coefficient of variability, the $\overline{P30}/\overline{N40}$ amplitude remains at about the same level throughout the procedure only in the group without cord manipulation. In the group with cord manipulation the variability increases towards the end of the procedure, for both latency and amplitude (Fig. 1).

The changes in variability of the N40 latency and P30/N40 amplitude were significant in group 2, although there was no neurological deterioration postoperatively (Table 4).

Therefore changes found to be significant on purely mathematical grounds are not necessarily associated with clinically significant changes.

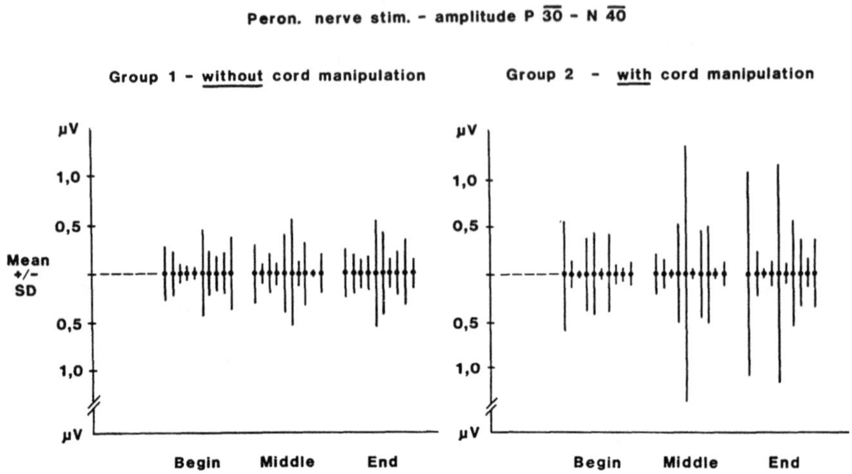

Fig. 2. Mean and standard deviations for the N40 amplitude after peroneal nerve stimulation for all 10 patients with and without spinal cord manipulation, calculated from five consecutive tracings at the beginning, the middle and the end of the procedure

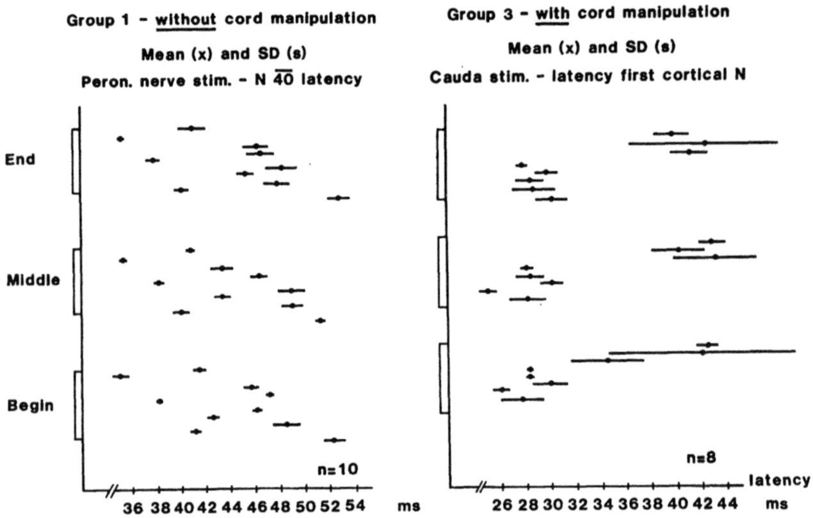

Fig. 3. Mean and standard deviation calculated from five consecutive tracings taken from the end, the middle and the beginning of the procedure for all patients of group 1 (no cord disease, no cord manipulation, peroneal nerve stimulation) and group 3 (cord disease, cord manipulation, cauda equina stimulation)

Cauda Equina Stimulation

The 8 patients with cauda equina stimulation (group 3) all had a space-occupying lesion and their cords were manipulated during surgery. The degree of latency fluctuation, given as mean and standard deviation for the blocks of 5 tracings from every patient, is demonstrated again comparing the three periods of the surgical

Table 4. Significance of differences in the coefficients of variability. The differences were determined by Student's *t*-test for paired values. Paired values were obtained by comparing the coefficients from the middle and the beginning, and the middle and the end of the procedure

	Variability of N40 latency: Beginning → Middle	Variability of N40 latency: Middle → End	Variability of P30/N40 amplitude: Beginning → Middle	Variability of P30/N40 amplitude: Middle → End
Group 1 ($n=10$) Without manipulation	N.S.	Significant $P=0.05$	N.S.	N.S.
Group 2 ($n=10$) With manipulation	Significant $P=0.01$	Significant $P=0.025$	N.S.	Significant $P\leq 0.05$

Table 5. Different variability of EP-parameters in group 3 (cauda equina stimulation in spinal cord pathology and spinal cord manipulation). The difference between the coefficients of variability at the beginning and the middle, or the middle and the end of the procedure is statistically not significant

Group 3 With manipulation	Variability of first cortical peak, Cauda equina stimulation—cortical recording: Beginning → Middle	Middle → End
Latency 1st N	N.S.	N.S.
Amplitude 1st P/N	N.S.	N.S.

procedure (Fig. 3, right). There is not much difference from group 1 (no spinal cord manipulation, peroneal nerve stimulation), except in some cases where the standard deviation is considerably larger. This finding seems to indicate that, although a spinal cord lesion was present and the spinal cord was manipulated, the variability in latency was much lower than in cases with cord lesion when peroneal nerve was stimulated (Table 3, Fig. 3). Comparing different stages of the operative procedure in the cauda equina group, it can be shown that the differences in the coefficients of variability are not significant (Table 5). This finding is true for both the latency and the amplitude of the first negative cortical deflection after cauda equina stimulation.

Therefore, with cauda equina stimulation variability tends to be lower throughout the procedure and lower than with peroneal nerve stimulation in patients with cord lesions.

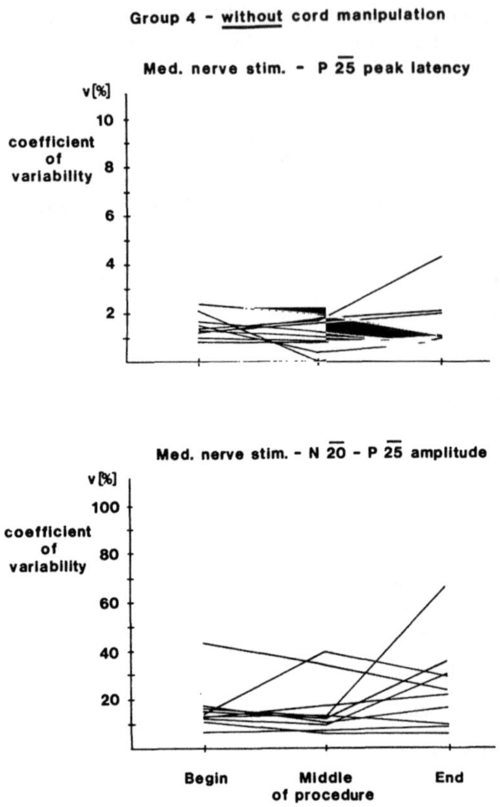

Fig. 4. Coefficients of variability for P25 latency and amplitude after median nerve stimulation, calculated from five consecutive tracings taken from the beginning, the middle and the end of the procedure for each patient of group 4 (no cord pathology, no cord manipulation, median nerve stimulation)

Median Nerve Stimulation

Group 4, consisting of 10 patients with lumbar disc operations, had no spinal cord space-occupying lesion and no spinal cord manipulation. Thus, the absolute values for the P25 peak latency and the N20/P25 amplitude showed little fluctuation. This is reflected in the coefficient of variability (Fig. 4) which remains fairly constant throughout the operative procedure, with some tendency for the amplitude to be more variable towards the end.

Latencies

Latencies were most consistent in the group without cord manipulation, in both median and peroneal nerve stimulation for the P25 and N40 peak respectively. A similar degree of constancy was seen in cauda equina stimulation for the first negative cortical deflection. Latencies varied considerably showing significant differences in the coefficients of variability throughout the procedure with peroneal nerve stimulation in the group with spinal cord manipulation. If latency fluctuations are evaluated under conditions most likely to occur in monitoring, that is with manipulated and diseased spinal cords, cauda equina stimulation tends to produce the smallest variance.

Amplitudes

Generally speaking, amplitude variability was much larger than latency variability in all groups, regardless of whether spinal cord disease was present or not. In addition the amplitude variability was even larger in the group with the spinal cord manipulation when using peroneal nerve stimulation, whereas changes in variability were *not* significant for cauda equina stimulation. Amplitude variability was rather low in the non-diseased, non-manipulated cord group after median nerve stimulation.

Summary of Results

Cortical SEP after peroneal nerve stimulation in diseased and manipulated cords show a large degree of variability, whereas cortical potentials after cauda equina stimulation show a low degree of variability coming close to the very low degree of variability found after median nerve stimulation in healthy non-manipulated spinal cords.

There is an increase in peroneal nerve evoked cortical SEP variability during the operation. Amplitudes show a larger variability in all groups as compared with latencies.

Consequently, when comparing differences in variability for the amplitudes the differences are not so often statistically significant, whereas latency changes are more often *statistically* significant. Statistical correlation does not make sense, however, without keeping the patient in mind. As the patients did not show postoperative impairment in neurological functions these statistically significant changes in latency variability appeared *clinically* not significant. It cannot be excluded, however, that these changes in variability indicated a temporary conduction impairment which was yet too small to be clinically detectable. The point that variance changes per se were of little clinical importance is made considering that there was no overall change in mean latency and amplitude.

Discussion

From our results it seems that cauda equina stimulation yields more reliable cortical potentials than peroneal nerve stimulation when looking at patients with a diseased spinal cord and with spinal cord manipulation.

Both factors are present in the typical neurosurgical monitoring case. Presumably there are two main reasons for this. With cauda equina stimulation many more nerve fibers are stimulated simultaneously and the distance to be travelled is much shorter, so that temporal dispersion is less important than with peroneal nerve stimulation. Even in healthy persons the degree of variability (i.e. the standard deviation) is larger after peroneal nerve stimulation than after median nerve stimulation. In spinal cord pathology with altered peroneal nerve evoked potentials the degree of variability is already large in the awake patient tested preoperatively. When the hostile environment of the operating theatre and the influence of narcotics and anesthetics is added, it does not come as a surprise that the degree of variability increases further.

The poorer the quality of signals preoperatively, the lower the chance to evaluate the significance of intraoperative EP changes.

This means that those very patients who need monitoring most urgently, because they already show poor evoked potentials preoperatively, have the least chance of profiting from intraoperative monitoring.

The larger degree of variability towards the end of the operative procedure, as commonly found in our groups and also reported by Gonzalez et al. (1984) for some components of cortical and spinal SEP, is not easy to explain. As it is found not only in the group with diseased spinal cord and cord manipulation (Fig. 1) but also to a lesser degree with median nerve stimulation in patients with healthy cords (Fig. 4), the most likely explanation would be the operative procedure itself and the combination of anesthetic drugs (Thurner et al. 1985). As the effect was also observed in the group without spinal cord manipulation it seems unlikely that alterations in the nerve fibers at the site of the spinal cord lesion were involved in this variability increase. The intraoperative changes seem to be different in scoliosis cases without cord pathology, where latencies and amplitudes change at the begin of anesthesia, then remain largely stable but change significantly again in the post instrumentation or closure period (Gonzalez et al. 1984; Brown and Nash 1985).

As can be seen from Figs. 2 and 3, a considerable amount of spontaneous fluctuation in amplitude and latency may occur, more for amplitude than for latency. As this is also the case in those patients with spinal cord pathology, where the spinal cord is manipulated quite naturally during the operation without causing any deterioration in the neurological status, it follows that amplitude and latency changes have to be very severe in order to be an indicator of spinal cord damage during intraoperative monitoring.

Brown and Nash have already pointed out that for a change to be significant it is necessary for it to be present for at least ten to fifteen minutes. The development of criteria for intraoperative intervention (Nordwall et al. 1979) seems to be more difficult when taking the results of this study into account.

References

Brown RH, Nash CL (1985) Intraoperative somatosensorisch evozierte kortikale Potentiale bei spinalen Operationen – Rückenmarkmonitoring. In: Schramm J (Hrsg) Evozierte Potentiale in der Praxis. Springer, Berlin Heidelberg NewYork Tokyo, S 153–182

Gonzalez EG, Hajdu M, Keim H, Brand L (1984) Quantification of intraoperative somatosensory evoked potentials. Arch Phys Med Rehabil 65:721–725

Nordwall A, Axelgaard J, Harado Y, Valencia P, McNeil D, Brown J (1979) Spinal cord monitoring using evoked potentials recorded from vertebral bone in cat. Spine 4:486–494

Thurner F, Schramm J, Romstöck J, Fahlbusch R (1985) Effects of Fentanyl and Enflurane on cortical and subcortical SEP during general anesthesia in man. This volume, pp 82–89

Worth RM, Markand ON, DeRosa GP, Warren C (1982) Intraoperative somatosensory evoked response monitoring during spinal cord surgery. In: Courjon J, Mauguière F, Revol M (eds) Clinical applications of evoked potentials in neurology. Raven Press, NewYork, pp 367–373

Somatosensory Evoked Potential Monitoring of Intraoperative Procedures

W. YOUNG[1] and A. BERENSTEIN

Summary

Cortical SEP have been extensively used to monitor operative procedures which put the spinal cord at risk of iatrogenic injury. SEP were highly sensitive to anterior spinal artery occlusion accompanying angiography and embolization, resulting in a reduced incidence of neurological complications. Other procedures in which SEP changes have influenced patient management include anterior cervical fusion, myelotomy and repair of thoracic aorta aneurysm.

Introduction

Procedures which involve surgical manipulation of the spinal cord or which alter blood flow to the spinal cord can pose considerable risks of neurological deficits. Because of its compact longitudinal organization, relatively small lesions in the spinal cord can cause significant functional loss. Brief periods of ischemia in the cord, especially in patients who already have decreased spinal perfusion, can cause serious morbidity ranging from weakness and sensory loss to paraplegia. Likewise, even slight pressure or other mechanical disturbances can injure the spinal cord.

At the NYU Medical Center, we have extensively used cortical somatosensory evoked potentials (SEPs) to monitor operative procedures which put the spinal cord at risk of iatrogenic injury. In our experience, SEP monitoring of the spinal cord can reduce the risks of spinal cord angiography, transvascular embolizations of arterio-venous malformations and tumors, and other procedures which reduce blood flow or mechanically disturb the spinal cord. In addition, we have found SEPs useful for detection of acute insults to the cord during surgical procedures.

Materials and Methods

Our techniques of monitoring cortical SEP have been described elsewhere (Berenstein et al. 1983). Posterior tibialis (PT) nerves were stimulated with epicutaneous bipolar electrodes at the ankle with 10–20 mA constant current pulses of 0.1–0.3 ms duration and at 2.3 Hz frequency. A Tracor Northern TN-3000 clinical signal averager

[1]Department of Neurosurgery, New York University Medical Center, 550 First Avenue, New York, NY 10016, USA

Spinal Cord Monitoring
Ed. by J. Schramm and S. J. Jones

was used to amplify, average, and store the responses. The SEP responses were recorded from needle electrodes inserted into the scalp at the Cz-Fz scalp locations for left and right PT nerve stimulation, as well as combined left and right stimulation. Median nerve SEP and other stimulation sites were sometimes used as controls for procedures involving the thoracic cord.

Results

Over the past year, we have added more than 30 cases of SEP monitoring during spinal angiography and embolization to our original database of 80 cases. We have found no exceptions to our earlier finding (Berenstein et al. 1983) that contrast dye injection into the anterior spinal artery (ASA) produces rapid decreases in SEP amplitudes. An example of a recent case is shown in Fig. 1, this being a 4-year-old girl with a high flow cervical AVM who showed dramatic improvement of SEP amplitudes after embolization of the arteriovenous malformation. Neurological improvements accompanied the SEP change.

We continue to find SEP sensitive, rapid, and reliable indicators of local spinal cord ischemia produced by brief occlusion of the ASA. We were able to use SEP tests to identify dye injection into the anterior spinal artery, to detect potentially injurious steps in the embolization procedure by brief balloon occlusions of the arteries prior to embolization, and to use this information for planning multi-staged procedures. Monitoring of SEP seems to have reduced the morbidity usually associated with transvascular therapy of lesions in the spinal cord. Since we started SEP monitoring in these angiographic and embolization procedures, the incidence of

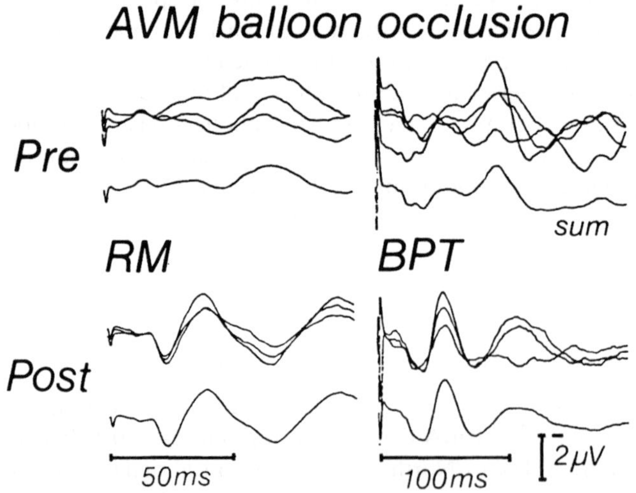

Fig. 1. SEP from a 4-year-old girl with a symptomatic high cervical arteriovenous malformation with neurological deficits greater on left than right. Prior to embolization, her SEP were very diminished, especially on the left. After the procedure, she made rapid recovery. The SEP responses were from bilateral posterior tibialis *(BPT)* and right median *(RM)* nerves. Negative is up in this and all following figures

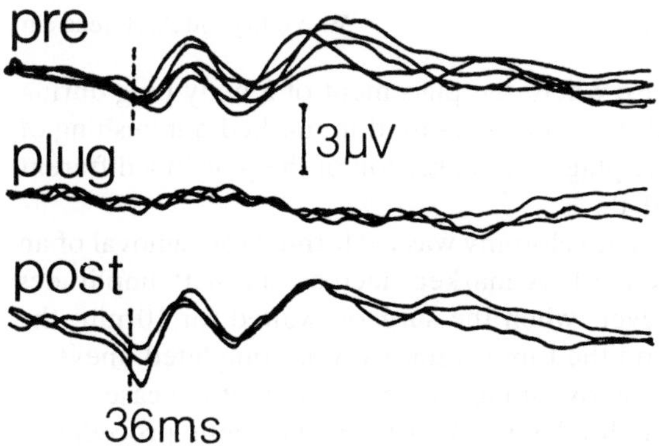

Fig. 2. An example of PT SEP change during placement of a bony plug at C5-6 for an anterior fusion. Note that the SEP were abnormal in both latency and amplitude to begin with. Upon first placement of the plug, the amplitude of the SEP responses diminished. The plug was removed and replaced in a different position. The SEP returned within 10 min

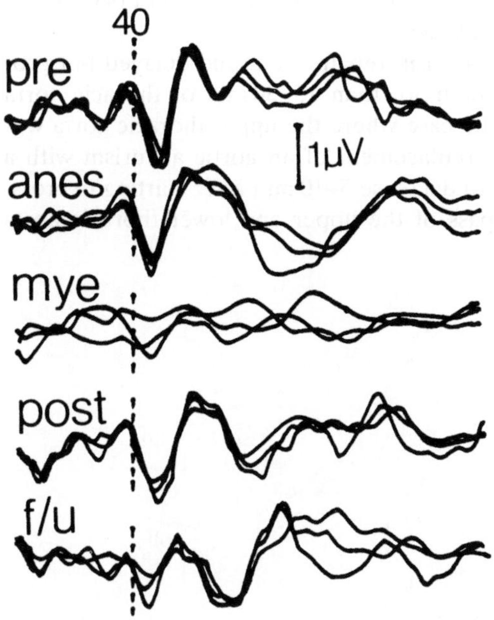

Fig. 3. SEP changes during myelotomy for removal of intramedullary tumor. The amplitude of SEP responses decreased with the myelotomy. The procedure was stopped and SEP recovered. The remainder of the tumor was removed uneventfully

neurological complications in selective angiography and embolization of AVM and spinal tumors has fallen from about 20% to less than 1%. The one isolated complication reported earlier (Berenstein et al. 1983), occurring in a young boy several hours after the embolization procedure and involving a high flow arteriovenous malformation in the anterior spinal artery axis, remains the only one.

About 50 cases of SEP changes (out of more than 200 cases which have been monitored) during surgical manipulation of the spinal cord have been documented. Typically, acute SEP changes involve amplitude shifts with large latency changes occurring only rarely. Because each case represents a different situation, we present representative examples of SEP changes which had an impact on patient care decisions.

Figure 2, for example, shows a case where the placement of a bony plug during anterior cervical fusion (i.e. Cloward procedure) resulted in marked diminishing of SEP amplitudes. Removal of the bone plug and reinsertion of the plug in a different position resulted in recovery of the SEP.

Figure 3 illustrates a case in which a myelotomy was performed for removal of an intramedullary tumor in the thoracic cord. A marked decrease in SEP amplitudes occurred with the myelotomy. However, when the surgeon waited for 10 min, the SEP amplitude recovered partially and the tumor removal was completed uneventfully. The SEP responses were only slightly diminished by the end of the case.

On occasions, changes in SEP amplitudes resulted from anesthetic procedures. Figure 4 shows an example of a case in which the SEP amplitude decreased significantly during a spinal operation when the spinal cord was not being touched. A quick check of the arterial blood gases revealed a PCO_2 of 60 mm Hg. Respiratory correction produced a rapid return of SEP amplitudes.

In about 10 cases of thoracic aorta aneurism repair, we found delayed but profound depression of SEP responses after more than 15–30 min of thoracic aorta occlusion. Figure 5 shows an example of a case where the upper thoracic aorta was clamped for 30 min during excision and replacement of an aortic aneurism with a dacron graft. The SEP responses began to decrease 5–10 min after starting to cross-clamp, disappearing by 20–30 min. Bypass of the upper and lower thoracic aorta

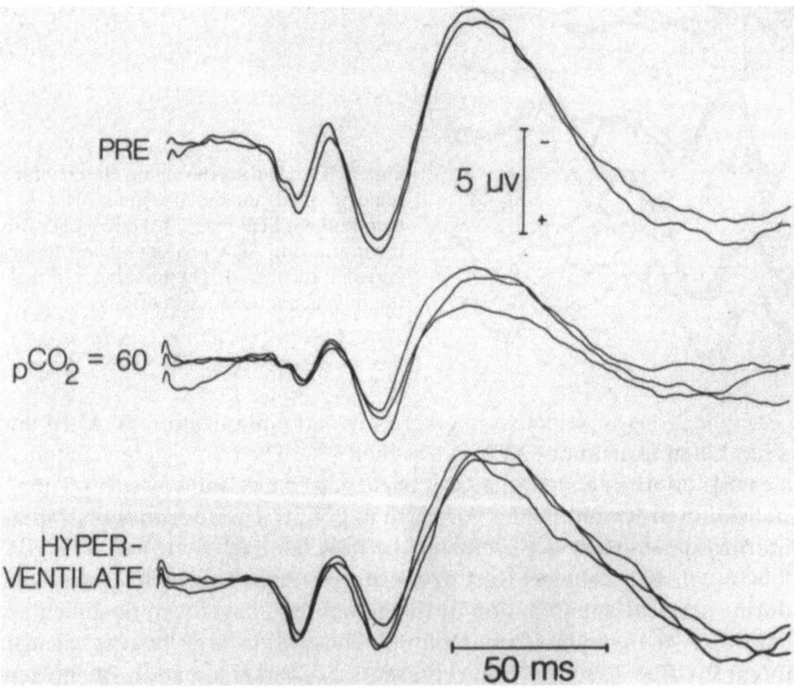

Fig. 4. SEP changes related to PCO_2. During a spinal cord operation, SEP amplitude decreased at a time when the spinal cord was not being touched. Investigation showed that the PCO_2 was 60 mm Hg. Respiratory correction of PCO_2 to 35 mm Hg resulted in restoration of SEP amplitude

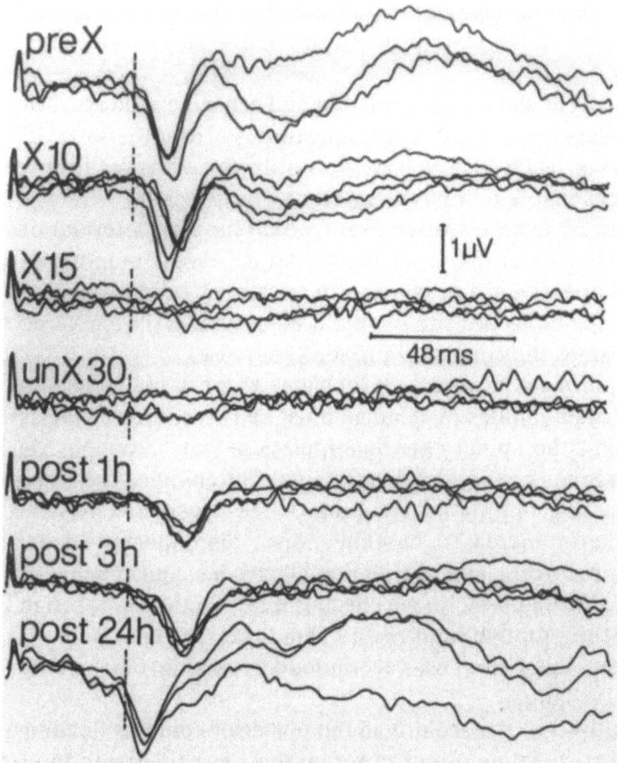

Fig. 5. SEP change due to thoracic aortic occlusion. In a 54-year-old man with a dissecting aneurism of the thoracic aorta, cross-clamping of the upper thoracic aorta during the aneurism repair produced gradual loss of SEP over a 10–15 min period. At 24 h, the SEP recovered almost to preoperative level. Note that the SEP responses were abnormal even prior to cross-clamping, suggesting that the spinal cord may already have been ischemic

during cross-clamping reduced the incidence and severity of SEP changes seen. SEP responses did not recover for many hours postoperatively.

Discussion

We have consistently found a high correlation between SEP changes and certain intraoperative events. Although some SEP changes can be attributed to anesthetic procedures or systemic events, these situations are relatively easy to rule out. Instances of persistent iatrogenically induced losses of SEP fortunately have been rare, but in the few cases (e.g. of tumor removal) where SEP loss was prolonged and present postoperatively, neurological deficits were found.

To date, we have not observed any case in which cortical SEP did not change while the patient developed significant neurological deficits. In over 100 cases of selective spinal angiography monitored by SEP, we have yet to encounter any exceptions to our earlier observation that dye opacification or occlusion of ASA produces a rapid decrease of SEP amplitude. Likewise, occlusion of the thoracic aorta for periods longer than 15 min invariably resulted in loss of SEP responses. In 50 cases of neurosurgical manipulation of the spinal cord, we have documented changes in the SEP associated with either compression or cutting into spinal cord tissues. In the majority of the cases, stopping the compressive action leading to the SEP change or restoring the condition of the cord to the state prior to the SEP change resulted in return of the evoked responses.

We thus believe that cortical SEP changes are reliable and sensitive indicators of intraoperative injury to the spinal cord. The difficult question is at what point SEP changes should be considered significant. Though we do not have an answer to this question, several criteria are reasonable and have served our purposes. If a change in SEP is associated with a specific intraoperative event which has the potential of harming the cord, if the change is greater than baseline variations of SEP amplitude, if spurious causes such as noise interference or changes in anesthetic levels are ruled out, and if the SEP change is specific to the site of spinal manipulation, we assume that the SEP change signifies risk to the cord until proven otherwise.

Our first action upon encountering a persistent SEP change that can be duplicated at least 3 times, which automatically implies the change must be present for 5 min, is to warn the clinician performing the spinal cord manipulation that a change has occurred. Depending on the urgency of the SEP loss (a total flattening of the trace being a matter of greater concern), temporary stopping of the procedure may be recommended. We then check for increased baseline noise, disconnected leads, increased contact impedance of recording and stimulating electrodes, and changes in anesthesia status of the patient. When possible, we check the SEP activated through pathways not passing through the putative injury site. This takes 2–3 min more. If the problem is not resolved by this time, we recommend reversing the activity thought to have caused the SEP change.

SEP is a test of ascending pathways, particularly in the posterior columns, leading from peripheral nerve to the cortex. Some of our observations have suggested to us that conditions which cause SEP changes may reflect the condition of the spinal cord in general. For example, our finding that SEP is exquisitely sensitive to slight compromises of the ASA surprised us. This artery provides critical blood supply to the gray matter and anterior-lateral columns of the spinal cord. The posterior spinal radicular arteries, however, supply the posterior columns. But we have found, not only that SEP are rapidly suppressed by ASA occlusion but also that compromise of the posterior spinal artery does not seem to have a major effect on SEP.

From our experience with spinal ischemia produced by thoracic aorta occlusion in humans and cats, we expected changes in SEP during spinal vascular embolization to be slow and delayed by 10–15 min. But, whenever the ASA was opacified or occluded, the SEP became significantly smaller by the first average obtained within 2–3 min. The SEP responses were typically abolished by the time the second average was obtained at 3–5 min, if the ASA remained occluded.

The unexpected sensitivity of SEP to ASA occlusion led us to investigate the phenomenon in cats. Cats also lost SEP responses rapidly when the ventral spinal artery was occluded. Histological examination of the spinal cords revealed that the spinal cord at the site of the vascular occlusion was not severely damaged by the ventral artery occlusion. This suggests that ischemia in the lower lumbar segments may be responsible.

Identification of the site of conduction blockage produced by ASA occlusion awaits more detailed studies of SEP responses of the spinal cord below and above the level of arterial compromise. Many mechanisms may contribute to a rapid conduction block in the spinal cord, including increase in extracellular potassium levels (Young et al. 1982a, b) at the site of vascular occlusion. We suspect such a mechanism because of the rapidity with which the conduction block occurs. Metabolic compromise due to ischemia usually takes at least 5–6 min, approximately the time course of the SEP changes that we found with thoracic aorta occlusion.

Our data does rule out several possible causes. For example, the SEP changes are unlikely to be due to contrast dye toxicity since we were able to produce SEP changes with balloon occlusion in humans and coagulation of the artery in cats, neither of which involve dye administration to the cord. Furthermore, if dye toxicity is involved, we should see reduction of SEP amplitudes with dye injections into the posterior spinal arteries. This did not occur.

In conclusion, properly carried out, SEP monitoring is a useful means of detecting spinal cord dysfunction during intraoperative procedures that carry some risk of damage to the spinal cord. We do not yet understand the mechanisms by which certain insults to the spinal cord, such as ASA occlusion, produce rapid and dramatic changes in SEP. Nevertheless, our data suggest that attention to SEP changes will reduce the morbidity normally associated with risky spinal procedures.

Acknowledgements. We gratefully acknowledge the generosity of the Tracor Northern Company for their loan of a Tracor 3500 signal averager for some of our studies. We thank the residents and staff of the Neurosurgery department at NYU Medical Center and Bellevue Hospital for their help and cooperation. Our SEP technicians, Peter Mayer, Lisa Szambel and Henry Merkin carried out many of the SEP procedures. Dr. Vincent DeCrescito performed the animal experiments. Fred Holmes, Ben Ayala, James Sinaly, and James Gore provided excellent technical assistance.

References

Berenstein A, Young W, Ransohoff J, Benjamin V, Merkin H (1983) Somatosensory evoked potentials (SEP) during spinal angiography and therapeutic transvascular embolization. J Neurosurg 60: 777–785

Young W, Koreh I, Yen V, Lindsay A (1982a) Effect of sympathectomy on extracellular potassium activity and blood flow in experimental spinal cord injury. Brain Research 253: 115–125

Young W, Yen V, Blight A (1982b) Extracellular calcium activity in experimental spinal cord contusion. Brain Research 235: 105–113

Evoked Potential Monitoring During Interventional Neuroradiology: Experimental Aspects of Special Catheter Techniques

W. HACKE[1], H. BUCHNER, A. FERBERT, H. ZEUMER, and G. BÖCKER

Summary

The use of evoked potential monitoring, both brain-stem acoustic evoked potentials and somatosensory evoked potentials, is reviewed considering our previous experiences and taking into consideration the experiences of other authors. In illustrative cases monitoring is described during occluding and recanalizing techniques of interventional neuroradiology as well as during intraarterial application of pharmacological agents. In discussing the results it is stressed that intraoperative monitoring does not only allow one to supervise interventional neuroradiological procedures but also provides new insights into hemodynamic mechanisms.

Introduction

Interventional neuroradiology (INR) makes use of different diagnostic and therapeutic catheterisation techniques. Some of the methods make it most important to check the patient's neurological state during the procedure. This often being impossible using conventional methods of neurological clinical examination, intraoperative monitoring of evoked potentials offers an opportunity to obtain objective information about changes in the patient's neurological state (Hacke et al. 1983). The use of different catheter techniques provides the opportunity for investigating some additional experimental aspects of intraoperative monitoring during interventional neuroradiology: for example, these techniques allow for repetitive test performance of reversible balloon occlusion in order to determine whether permanent occlusion of a vessel will be tolerated (Hacke et al. 1981, 1982). Occlusion of different vessels singly provides information with regard to collateral blood flow and the identification of the vessel which can be occluded with least risk (Berenstein et al. 1984). Additionally, selective injection of contrast media or short acting barbiturates allows the identification of impending but reversible intraoperative disturbances and the distinction between hemodynamic and toxic disturbances in both brain and spinal cord supplying vessels. A short introduction to the development of intraoperative monitoring in interventional neuroradiology will be followed by examples of these new aspects concerning spinal cord and brain-stem supplying arteries. Some examples of intraoperative monitoring with regard to hemispheric function will also be demonstrated.

[1] Abteilung Neurologie, Klinikum der RWTH Aachen, Pauwelsstraße, D-5100 Aachen

Spinal Cord Monitoring
Ed. by J. Schramm and S. J. Jones

Definition and Methods

Table 1 summarizes the different methods used in INR. Generally, INR consists of recanalizing and occluding techniques and, additionally, the intraarterial application of drugs. Recanalizing techniques can be performed pharmacologically (local intraarterial fibrinolytic therapy (Hacke et al. 1982; Zeumer et al. 1982, 1983a, b) and mechanically (percutaneous transluminal angioplasty) (Ringelstein and Zeumer 1984; Zeumer et al. 1983b). Occluding techniques are either diagnostic (reversible) or therapeutic. The permanent occlusion of a vessel can be performed with detachable balloons or by injection of embolizing agents. Drugs being administered intraarterially are cytostatic agents, short acting barbiturates, fibrinolytic agents or embolizing substances. The following monitoring methods have been used during INR: online EEG analysis, SEP and BAEP monitoring and continuous CW-Doppler-ultrasound of brain-supplying arteries (Berenstein et al. 1984; Hacke et al. 1983; Ringelstein and Zeumer 1984). The same technical problems that are known from the operating theatre have to be taken into account in the X-ray-room also (Grundy 1983; Hacke 1984a, b; Lueders et al. 1982; MacCallum and Bennett 1975; Nash et al. 1977; Worth et al. 1982; Raudzens 1982; Spielholz et al. 1979). Although we have encountered no cauterization artefacts, high voltage artefacts during serial angiography and fluoroscopy are to be expected. Sometimes the decisive steps of interventional procedures are very short and require changes in analysis time, stimulus frequency and the number of sweeps averaged for evoked potentials (Hacke 1984a). In our department evoked potentials are recorded using either the Nicolet Instru-

Table 1. Methods of interventional neuroradiology

A) Occlusion of cerebral or spinal arteries
   - Reversible balloon occlusion
   - Permanent balloon occlusion
   - Therapeutic embolization of
      - Intracranial or intraspinal tumors
      - Arterio-venous malformations
      - Traumatic fistulas (sinus cavernosus fistula)

B) Recanalizing techniques
   - Local intraarterial fibrinolysis of brain-supplying vessels (LIF)
      - Vertebro-basilar system
      - Internal carotid artery
      - Middle cerebral artery
   - Percutaneous transluminal angioplasty of brain-supplying vessels
      - Stenoses at the orifice of the vertebral artery
      - Subclavian artery stenosis with manifest steal
      - Postoperative strictures of the internal carotid artery

C) Intraarterial drug application
   - Short-acting barbiturate
   - Cytostatic drugs

ments CA 1000 with double floppy device or the Disa 1500. A Schwarzer 2-channel frequency analysis system and a Delalande Doppler-sonographic system were used for EEG and ultrasound monitoring.

History

In 1974 Serbinenko described different catheter methods for the occlusion of intracranial arteries and discussed the possibility of using those techniques for experimental studies. In 1981 we described the monitoring of continuous online EEG recordings during reversible balloon occlusions of the internal carotid artery in neurosurgical patients, in order to determine whether the internal carotid artery could be sacrified during the forthcoming operation. This was a variation of intraoperative reversible clamping methods used in anesthetized patients monitored by EEG, and of preoperative methods in which the internal carotid artery was compressed manually. Since the patients were awake, it was possible to assess additional factors such as motor function and language. The atraumatic full occlusion was demonstrated under X-ray control, and last but not least the occlusion could be resolved immediately (Hacke et al. 1981). Later we added SEP techniques in which loss of the cortical P25-N36 complex was demonstrated in patients who would not tolerate a permanent occlusion of the ICA (Hacke et al. 1983). While the C7-potential remained stable, the N20 cortical component showed in those cases a decrease of amplitude. On the other hand, in some patients it was possible to demonstrate preoperatively that occlusion of the internal carotid artery did not cause hemodynamic deficits.

In 1982 our group introduced the local intraarterial fibrinolytic therapy of brain supplying arteries, initially in basilar occlusion (Zeumer et al. 1982) which was monitored by both serial SEP and BAEP recordings (Hacke et al. 1982). Berenstein and his colleagues used intraoperative monitoring especially during embolization procedures of spinal cord tumors or arteriovenous malformations (AVM). During the percutaneous transluminal angioplasty of subclavian artery stenoses with manifest subclavian steal syndrome and vertebral artery stenoses the CW-ultrasound Dopplersonography was introduced as a monitoring method (Ringelstein and Zeumer 1984).

Results

Usually intraoperative monitoring is performed in order to obtain information whether the function of certain structures at risk remains stable or is alterated, which may lead to changes in the therapeutic or diagnostic procedure (Brown and Nash 1979; Grundy 1983; Hacke 1984a, b; McCallum and Bennett 1975; Nash et al. 1977; Raudzens 1982; Spielholz et al. 1979; Worth et al. 1982). Sometimes one can demonstrate normalization of initially impaired functions (Berenstein et al. 1984; Hacke et al. 1982), which may correlate with postoperative outcome. On the other hand, postoperative neurological deficits have been described in patients in whom the SEP monitored was stable throughout the operation (Berenstein et al. 1984; Spielholz et al. 1979). All those studies describe a more passive, observational approach to monitoring. In this context we want to point out that in intraoperative monitoring during interventional neuroradiology it is also difficult to characterize the changes of potentials that indicate truly persistent disturbances. Sometimes it takes more than 15 min for changes of potential during interventional neuroradiology to be fully reversed (Hacke et al. 1983). Therefore, for the sake of safety we accept each change of a potential as a warning to interrupt a procedure if possible. We monitored two patients during embolizing procedures in whom occlusion resulted in loss of the SEP which was followed by a gross postoperative deficit. This was because embolization is irreversible and the warning from the SEP came too late.

During interventional neuroradiology one is enabled to take additionally a more active and experimental approach to monitoring, by using different or multiple catheter techniques such as reversible balloon occlusion and intraarterial drug application. We feel that intraoperative monitoring has provided interventional neuroradiology new information of two types:

a) Identification of the functional territories of single or multiple feeding vessels
b) New insights into hemodynamics and the establishing of new sufficient collaterals

a) Identification of the Functional Territories of Single or Multiple Feeding Vessels

The basic idea of identification of functional territories is very simple: the reversible occlusion of a vessel that may be at risk during an intervention can be monitored with the same method as the operation itself. If the function remains stable for a certain time, permanent occlusion will presumably be tolerated.

There are modifications of this simple approach that may lead to decisions concerning the procedure of the planned final maneuver. For example, Berenstein and his colleagues (1984) performed "provocative" occlusion tests to identify the effects of a planned permanent embolization. In one of their cases a high flow AVM was fed from both anterior and posterior spinal arteries. The amplitude of the SEP diminished abruptly during balloon occlusion of the anterior spinal artery, while occlusion of the feeder from the posterior spinal artery resulted in increased amplitude of the potential. Final embolization, therefore, was performed through the posterior spinal artery, the anterior spinal artery was preserved and the patient showed clinical improvement postoperatively. In this case intraoperative monitoring helped to identify which of the feeding arteries could be used for embolization without clinical risk. Sometimes the diminution of SEP amplitude during provocative occlusions may indicate that occlusion enhances intraspinal steal phenomena, indicating that permanent embolization could not be helpful to the patient. Balloon occlusions are also helpful in identifying collateral blood flow, or conversely may indicate whether collaterals at risk are essential to the patient. This will be illustrated by one of our monitored cases (Fig. 1). In this sixty-year-old female patient the internal carotid artery was injured during a middle ear operation. The defect could not be fixed intraoperatively. The question was raised, whether the patient would tolerate permanent ligature of the internal carotid artery both at the neck and intracranially at the base of the skull, or if an extra-intracranial bypass was necessary. Therefore the first reversible balloon occlusion of the internal carotid artery was performed with the balloon placed precisely over the hole in the artery. There were no changes in the SEP monitoring. Therefore we decided to use a detachable balloon to perform the permanent occlusion of the artery right over the hole. Later a catheter was moved into the vertebral artery

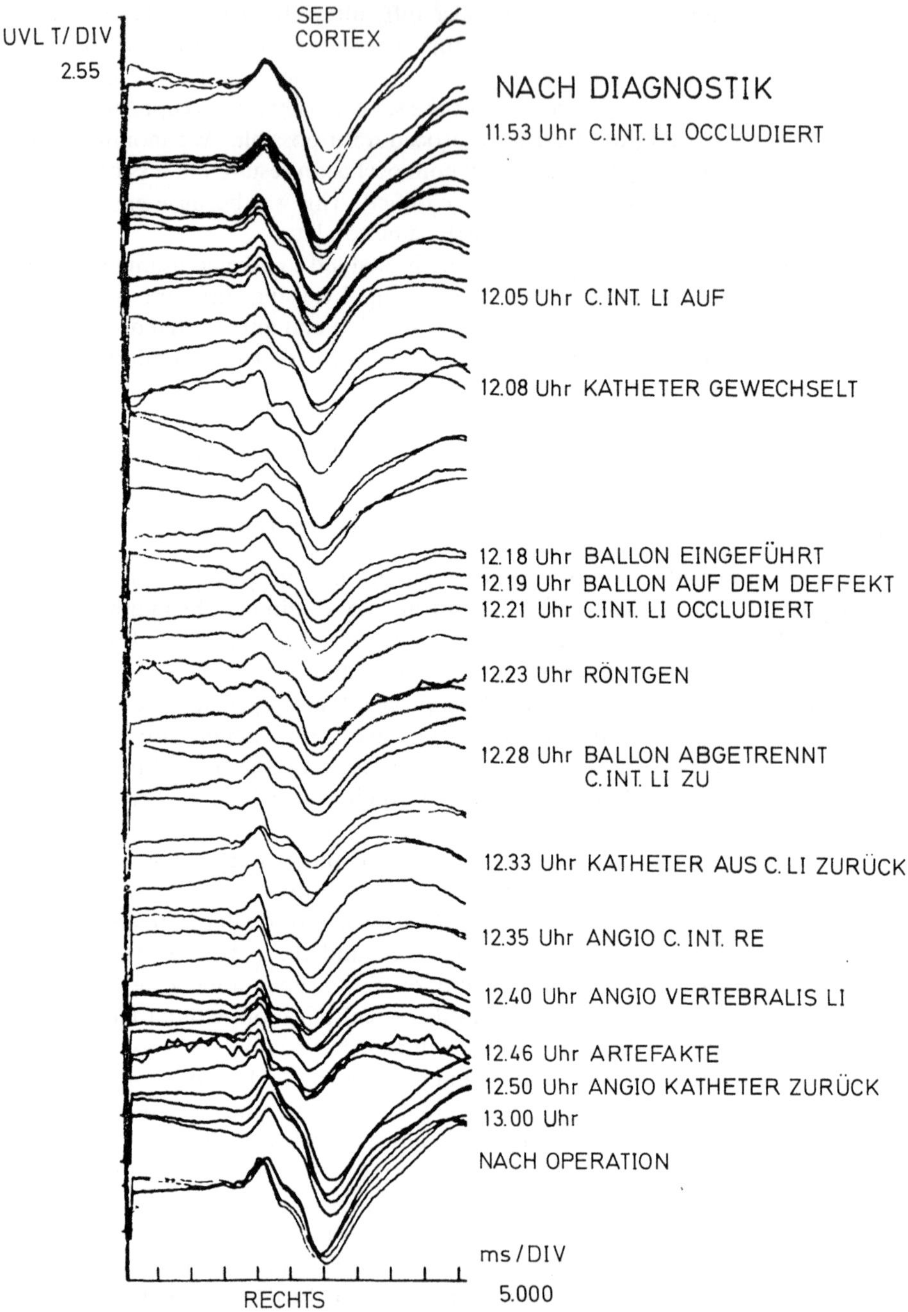

Fig. 1. SEP monitoring during reversible and permanent balloon occlusion of the left intracranial carotid artery. See text

and the amplitude of the median nerve SEP decreased significantly, indicating that the posterior communicating artery was essential for collateralization of the hemispheric circulation. This indicated that an extra-intracranial bypass would have been necessary if the balloon occlusion had not been performed. Additionally, reversible occlusion of the contralateral internal carotid artery and the external carotid artery showed the significance of the anterior communicating artery as well as the ophthalmic collateral.

Selective drug injection can be used to identify functional territories of an artery. The Wada-test (Wada and Rasmussen 1960) can be performed to identify whether the left hemisphere is speech-dominant in a patient. When the median nerve SEP was monitored, the following characteristic changes were observed: one minute after injection of sodium amytal (a short acting barbiturate) the SEP was significantly attenuated and subsequently vanished. Reappearance and normalization of the potential occurred eight minutes after injection. We think that this method could also be used intraoperatively in order to identify whether a vessel at risk leads to changes in the monitored potential after a provocative injection of sodium amytal.

In a separate trial we administered BCNU intraarterially for treatment of brain tumors. In order to try to avoid retinal damage the tip of the catheter was placed as close as possible to the tumor-supplying intracerebral vessel, at least above the origin of the ophthalmic artery. The calibrated-leak balloon catheters might have caused a partial occlusion of smaller brain vessels. Figure 2 shows the SEPs monitored during intraarterial application of BCNU with a catheter, which was placed initially at the trifurcation of the middle cerebral artery (b). Severe attenuation of the SEP occured before BCNU was administered (c). After retracting the catheter into the carotid siphon the monitored SEP normalized and there were no changes during injection of the drug (d). Monitoring is therefore useful in distinguishing between toxic and hemodynamic effects of the treatment.

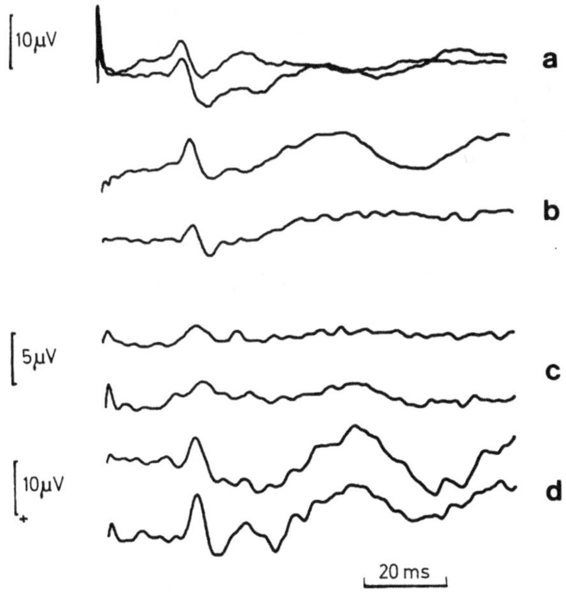

Fig. 2a–d. SEP monitoring during intraarterial application of BCNU. **a** Baseline potentials, **b–d** see text

b) New Insights into Hemodynamics and the Establishing of New Sufficient Collaterals

During monitoring of percutaneous transluminal angioplasty involving the subclavian artery or the vertebral artery, new insights into the hemodynamics and the establishment of new sufficient collaterals have been obtained. Balloon dilatation of the subclavian artery in the subclavian steal syndrome is very well monitored by CW-Doppler-sonography. Doppler-sonography reveals the fact that the reestablishing of an orthograde flow in the dependent vertebral artery after dilatation is delayed. This circumstance serves as a protection for possible embolizations from the stenosis into the brain (Ringelstein and Zeumer 1984).

Doppler-sonography also offers useful monitoring help during the dilatation of vertebral artery stenosis. In addition, the BAEP can be monitored. Figure 3 shows

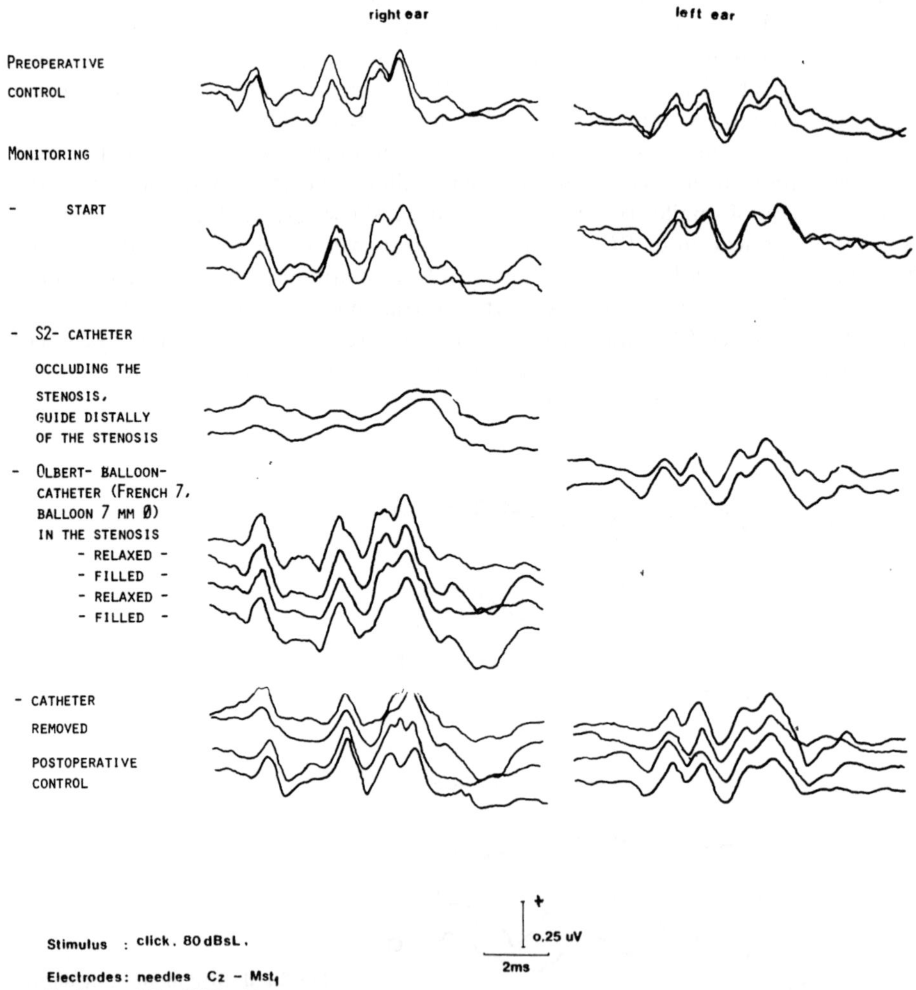

Fig. 3. BAEP monitoring during PTA of severe vertebral artery stenosis. See text

an example of BAEP monitoring during percutaneous transluminal angioplasty of vertebral artery stenosis (the contralateral vertebral artery was occluded!). Before the actual dilatation can be performed, one has to place a guide catheter through the highly stenotic part of the vertebral artery. This usually results in a complete cessation of blood flow, which can be demonstrated by Doppler-sonography. In these cases, as a rule, we find a dramatic alteration of the BAEP which may include the first and second peak or the inter-peak-latency. Such alterations are usually reversible within two minutes, but may persist for as long as 15 min before normalisation occurs. In the example illustrated in Fig. 3, after the first dilatation the BAEP was reestablished. During subsequent dilatations no new alterations of the BAEP occurred, although complete cessation of blood flow resulted from each new dilatation procedure. This circumstance implies that after the first complete arrest of the blood flow sufficient retrograde flow into the basilar artery must have been established, resulting in immediate blood supply during further occlusion of the remaining vertebral artery.

Conclusions

In our opinion, intraoperative monitoring during interventional neuroradiology can give information concerning functions at risk during interventional procedures, as well as providing new insights into hemodynamic mechanisms and the identification of functional territories of vessels at risk. The use of different balloon catheter techniques and administration of intraarterial drugs may also provide new information of relevance to other intraoperative monitoring methods.

References

Berenstein A, Young W, Ransohoff J, Benjamin V, Merkin H (1984) Somatosensory evoked potentials during spinal angiography and therapeutic transvascular embolization. J Neurosurg 60: 777–785

Brown RH, Nash CL (1979) Current status of spinal cord monitoring. Spine 4:466–470

Grundy BL (1983) Intraoperative monitoring of sensory evoked potentials. Anesthesiol 58:72–87

Hacke W, Zeumer H, Ringelstein EB (1981) EEG controlled occlusion of the internal carotid artery. Neuroradiology 22:19–22

Hacke W, Berg-Dammer E, Zeumer H (1982) Evoked potential monitoring during acute occlusion of the basilar artery and selective local thrombolytic therapy. Arch Psychiatr Nervenkr 232: 541–548

Hacke W, Zeumer H, Berg-Dammer E (1983) Monitoring of hemispheric and brainstem functions with neurophysiological methods during interventional neuroradiology. Am J Neurorad 4: 382–384

Hacke W (1984a) Methodische Aspekte intraoperativer Überwachung evozierter Potentiale. Z EEG EMG 15:127–131

Hacke W (1984b) Neuromonitoring. J Neurol 232:125–133

Lueders H, Gurd A, Hahn J, Andrish J, Weiker G, Klemm G (1982) A new technique for intraoperative monitoring of spinal cord function. Spine 7:110–115

McCallum JE, Bennett MH (1975) Electrophysiological monitoring of spinal cord function during intraspinal surgery. Surg Forum 26:469–471

Nash CL, Lorig RA, Schatzinger LA, Brown RH (1977) Spinal cord monitoring during operative treatment of the spine. Clin Orthop Rel Res 126:100–105

Raudzens PA (1982) Intraoperative monitoring of evoked potentials. Ann NY Acad Sci 388: 308–325

Ringelstein EB, Zeumer H (1984) Delayed reversal of vertebral artery blood flow following percutaneous transluminal angioplasty for subclavian steal syndromes. Neuroradiology 26:189–198

Serbinenko FA (1974) Balloon catheterization and occlusion of major cerebral vessels. J Neurosurg 41:125–145

Spielholz N, Benjamin MV, Engler GL, Ransohoff J (1979) Somatosensory evoked potentials during decompression and stabilisation of the spine. Methods and findings. Spine 4:500–505

Wada J, Rasmussen T (1960) Intracarotid injection of sodium amytal for the lateralization of cerebral speech dominance. J Neurosurg 17:266–282

Worth RM, Markard ON, DeRosa GP, Warren CM (1982) Intraoperative somatosensory evoked response monitoring during spinal cord surgery. In: Courjon J, Mauguière F, Revol M (eds) Advances in neurology, vol 32: Clinical applications of evoked potentials in neurology. Raven Press, New York, pp 367–373

Zeumer H, Hacke W, Kolmann HL, Poeck K (1982) Lokale Fibrinolysetherapie bei Basilaristhrombose. Dtsch Med Wochenschr 197:728–731

Zeumer H, Hacke W, Ringelstein EB (1983a) Local intraarterial thrombolysis in vertebrobasilar thrombo-embolic disease. Am J Neurorad 4:401–404

Zeumer H, Ringelstein EB, Hacke W (1983b) Gefäßrekanalisierende Verfahren der interventionellen Neuroradiologie. Fortschr Röntgenstr 139:467–475

Monitoring — Spinal Evoked Potentials

Experience of Epidural Spinal Cord Monitoring in 410 Cases

S. J. Jones[1], L. Carter, M. A. Edgar, T. Morley, A. O. Ransford, and P. J. Webb

Summary

This account describes a retrospective analysis of epidural spinal cord monitoring results in 410 patients undergoing posterior fusion and instrumentation for scoliosis. None developed serious neurological complications postoperatively but 9 (2.2%) showed mild defects. Out of 24 in whom the monitor indicated a transient or sustained conduction impairment (>60% amplitude reduction) 6 had problems postoperatively, representing a minimal correct detection rate of 1.5% and a maximal false positive rate of 4.4%. Of the 3 remaining cases, no potentials could be recorded from the appropriate limb(s) at any time in 2 who experienced mild neurological exacerbation due to root or cord lesions, and one patient in whom monitoring was uneventful developed signs of a lumbo-sacral root lesion.

Introduction

The preliminary results have been published (Jones et al. 1983) of an exploration of sensory pathway monitoring techniques in 138 patients undergoing correction of spinal deformities, mainly by posterior fusion and Harrington distraction instrumentation. The preferred method was that of recording conducted sensory potentials from the epidural space in the upper thoracic or lower cervical region, following stimulation of the posterior tibial nerve in the popliteal fossa. No serious neurological complications were encountered in these cases, but 3 had mild postoperative deficits after the monitor had indicated complete or partial conduction failure, with some recovery after release of the distraction force. Epidurally-recorded responses were more stable than similar potentials recorded from the vertebral bone, and were preferred to cortical potentials recorded from the scalp on the grounds of lesser susceptibility to anaesthetic factors, faster rate of feedback and the capability of assessing conduction in tracts other than and in addition to the posterior columns. A further advantage was the relatively high and restricted frequency recording bandwidth required.

The identical method has now been applied to 410 patients undergoing orthopaedic spinal surgery performed by 2 independent teams of surgeons at the Royal National Orthopaedic Hospital, London. This report summarizes the findings with regard to the incidence of transient and sustained conduction defects detected, and how these were related to neurological symptoms manifested postoperatively.

[1]Medical Research Council, The National Hospital for Nervous Diseases, London WC1N 3BG, Great Britain

Spinal Cord Monitoring
Ed. by J. Schramm and S. J. Jones

Patients and Methods

The patients were unselected cases of scoliosis or kyphoscoliosis undergoing posterior fusion with Harrington or Luque instrumentation. A wide variety of osteogenic and neurogenic conditions were included, with or without pre-existing neurological impairment, but most were classified as "idiopathic" and were neurologically normal.

The stimulus was a square-wave constant voltage impulse of duration 0.2 ms, delivered transcutaneously at 20/s to the posterior tibial nerve in the popliteal fossa on either side. The intensity (usually 100 V) was sufficient to induce contraction of the calf muscles and plantar flexion of the foot. A monopolar recording electrode (3 mm of stainless steel tubing, diameter 0.9 mm, at the tip of an autoclavable lead of similar diameter) was introduced to the dorsal epidural space through the midline interspinous ligament at the level of the rostral Harrington hook, and advanced cephalad by 1 or more vertebrae. The reference electrode was a stainless steel needle inserted into adjacent muscle. The amplifier bandpass was from 200 Hz to 2 kHz and a variable number of responses were averaged with 500 sampling points over an epoch of 30 ms. Recordings were obtained continuously throughout the operation, from exposure of the vertebrae until final closure (at least 20 min after the last application of distraction force). Measurements were taken of the latency of the initial positive deflection and the amplitude of succeeding negative peaks. Usually it was only necessary to record a measure of overall amplitude, but amplitude changes involving individual components were sometimes observed.

Results

The waveform and properties of conducted spinal cord potentials following tibial nerve stimulation have been described previously (Jones et al. 1982). The characteristic waveform recorded at low cervical level in neurologically normal patients consisted of 3 negative peaks at latencies of 14–22 ms, 0.5–6.0 μv in amplitude (Fig. 1). The incidence of failure due to technical factors was initially high, but eventually fell to a stable level of approximately 3%. No neurological deficits were observed postoperatively in any patient for whom monitoring was aborted for technical reasons. Effective monitoring was also impossible on account of gross degradation or absence of spinal potentials in 6 patients (1.5%) with marked pre-existing neurological deficits due to Friedreich's ataxia, hereditary motorsensory neuropathy or myelomeningocele. One of the latter developed a slight exacerbation of symptoms postoperatively.

In the majority of the remaining cases the response was easily recorded and remained stable throughout the procedure, with no more than 30% fluctuation in amplitude before distraction. Small changes were often attributable to slight movement in the position of stimulating and/or recording electrodes. Periods during which monitoring was impossible due to electro-cautery or excessive pick-up of "noise" were acceptably short in all but a very few cases.

The most frequently observed alteration in the waveform between pre- and post-distraction records was a latency increase of 0.2–1.0 ms (Fig. 1). This, however, was usually a gradual change occurring throughout the operation and sometimes reversed

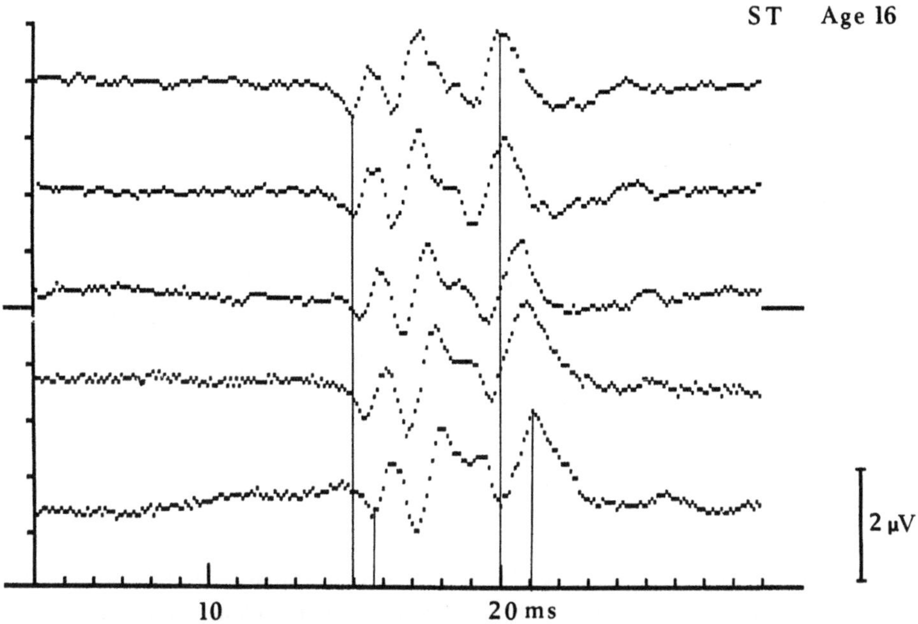

Fig. 1. Characteristic 3-component waveform (positive down) recorded epidurally at low cervical level following stimulation of the posterior tibial nerve at the popliteal fossa in a neurologically normal adolescent idiopathic scoliosis patient. A gradual increase in latency took place throughout the operation, unrelated to the magnitude of applied forces and without postoperative sequelae

shortly before closure. There were no associated neurological defects postoperatively, and the phenomenon was considered to be a benign slowing of conduction possibly due to cooling of the exposed vertebral column.

No more than ±30% change in amplitude was observed between pre- and post-distraction records in the great majority of cases. Figure 2 illustrates the distribution of overall amplitude changes between first and last (closure) recordings in 767 monitored limbs. The amplitude of the response fell by more than 10% in a slightly greater number than those in which there was a >10% amplitude increase (239 compared with 178 limbs) and the median change was −10%. Amplitude decrements of more than 10% were proportionately more frequent in the initial series of cases for whom a simple 2-point Harrington rod fixation technique was used (Jones et al. 1983), than in later cases for whom the correction was applied more gradually using distraction and compression rods with multiple fixation points. This suggests that a small degree of amplitude reduction may sometimes signify minor conduction impairment, insufficient to cause overt neurological defects but requiring close and continuous monitoring. Waveform changes without overall loss of amplitude were occasionally seen. Where these could not be definitely ascribed to a shift in the position of the recording electrode, particular attention was paid to the monitor and more significance attached to any subsequent amplitude loss.

A criterion, arrived at empirically, for significant deterioration of the response calling for interruption of surgery was an amplitude reduction of more than 60%,

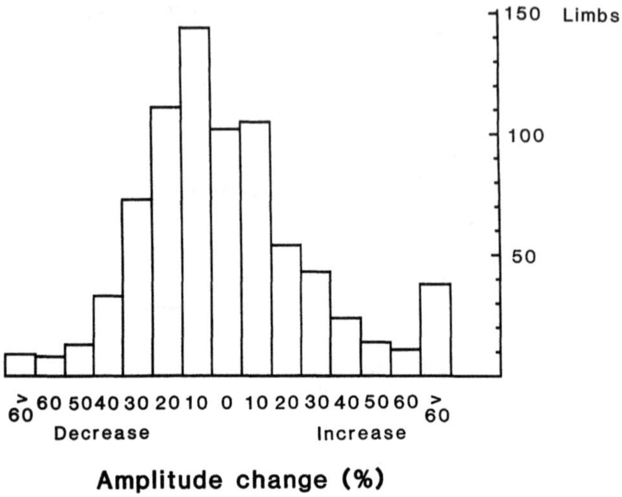

Fig. 2. Distribution of amplitude differences between initial and final (closure) recordings in 767 monitored limbs (410 cases)

affecting one or more components. This was seen in 24 cases (5.9%), with subsequent return to normal (either spontaneously or following release of distraction) in 16 and partial or no recovery in 8. Three of the former and three of the latter had mild postoperative signs (2) or symptoms (4) of spinal cord impairment. There was therefore a higher incidence of confirmed lesions among cases in whom spinal potentials failed to recover fully (38%), than among those in whom the electrophysiological impairment was transient (19%). The overall incidence of confirmed lesions in patients with electrophysiological deterioration was 25%. One patient in whom monitoring had been uneventful developed a left hemiparesis due to cerebro-vascular insufficiency 24 h later, and subsequently died. Two patients developed mild postoperative symptoms and/or signs attributable to lumbo-sacral root involvement. In one of the latter the impairment was unilateral, and no conducted potentials could be recorded at any time with stimulation on the affected side. In the other, waveforms which were initially poorly defined showed no significant change. Including these 2 cases and one of myelomeningocele mentioned above, a total of 9 patients (2.2%) experienced neurological exacerbation, detected electrophysiologically in 6. Perhaps more importantly, there were 18 patients who remained neurologically intact in spite of indications that some problem may have occurred. Some of the latter could probably be regarded as "false positives", but the maximal incidence of such was only 4.4% and not a great deal higher than the minimal "correct detection" rate of 1.5%.

Discussion

If statistical proof were needed of the effectiveness of spinal cord monitoring in preventing iatrogenic conduction impairment during spinal surgery it would be necessary to study more than 1000 patients in 2 parallel groups, each receiving identical treatment from the same surgical team in the same environment, with or without the monitoring facility. Whether or not such a study is justifiable on ethical grounds, it is clearly impractical since surgical techniques are subject to a continuous process of elaboration and refinement. For the same reason it is unrealistic to attempt to compare the 410 cases of the present study with, for example, the preceding 410 operated on by the same surgeons in the same hospital. It is only possible to observe that, in comparison with data collected several years ago from a large number of centres (MacEwen et al. 1975), the zero incidence of postoperative paraplegia observed in the present study is encouraging and suggests that spinal cord monitoring may be ethically and financially worthwhile. It is arguable that this outcome may have been largely due to improvement of surgical techniques, but in certain cases it does seem likely that severe complications would have arisen were it not for the fact that cord impairment was detected at an early stage.

It is clearly unsatisfactory that the criterion of 60% amplitude reduction, indicating that the risk of postoperative neurological impairment had risen from 1.5% to 25%, is a purely empirical figure encompassing trivial changes due, for example, to movement of the recording electrode relative to the spinal cord, in addition to non-trivial changes due to failure of conduction in a certain proportion of afferent fibres. Conduction block in as few as 10% of fibres is still a circumstance to be avoided if at all possible, although there may be only a small risk of overt clinical impairment. With this limitation in mind it would be useful to develop an electrode which could be held in closer contact with the dura throughout the operation.

In the patient group as a whole the maximal incidence of "false positives" was 4.4%, compared with a minimal "correct detection" rate of 1.5%. The former may well have included cases in whom incipient clinical defects were completely averted by moderation of the distraction force. Also not truly to be counted as "false positives" are those in whom the amplitude reduction was due to a genuine but temporary conduction block, possibly insufficient to manifest itself in symptoms and therefore requiring no change in surgical procedure. It is important from the point of view of the acceptability of a monitoring technique that the incidence of "false positives" be not too high, and the authors are satisfied that no such limitation is inherent in the present method.

Of even greater importance, however, is that the incidence of "false negatives" be kept to a minimum. Minor neurological exacerbation occurred in one case in spite of apparently satisfactory monitoring. This was attributable to a root lesion at L5 or S1, which might have been missed through the sensory volley entering the spinal cord at more rostral level. Monitoring of sacral segments and roots can be achieved by epidural stimulation of the cauda equina, and this technique may be particularly valuable in cases, for example, of severe spondylolisthesis.

Although no late-onset spinal cord complications were encountered in the present series, it has been the experience of many surgeons that defects may occasionally be manifest some hours or days after surgery. It is unlikely that such problems would ever be reliably predicted in per-operative recordings, although the danger might be signalled by labile responses, highly sensitive to surgical manipulation. One might advocate continuation of monitoring during the recovery period, but on the one hand the constant presence of a fairly intense electrical stimulus is likely to be poorly tolerated by an unanaesthetized patient, and on the other it is unlikely that epidural recordings would signal the onset of impairment very much in advance of the report of overt symptoms. Ensuring adequate oxygenation in the immediate postoperative period might go some way towards alleviating the risk of late-onset complications.

Two further patients suffered slight exacerbation of neurological symptoms postoperatively, one due to a lumbar root lesion which may have been sustained early-on in the procedure since no potentials could be recorded at any time with stimulation on the involved side. The other was a case of myelomeningocele with partially preserved lower limb function although, once again, no conducted potentials could be recorded even at the start of the operation. The present technique seems to be of limited value where there is significant impairment of sensory conduction preoperatively, but it may be that usable recordings could be obtained more often with the stimulus delivered directly to the spinal cord or cauda equina.

The precise mechanism whereby cord conduction is impaired in some patients undergoing correction of scoliosis is still unknown but is likely, in the majority, to involve local ischaemia of a few segments. The time course of changes observed in the present study, with amplitude loss following application of distraction force usually by more than 10 min, is compatible with an ischaemic mechanism. Cord anoxia is more likely to occur during hypotensive surgery, and it is here that monitoring becomes most imperative. In at least one case described above, however, the amplitude loss was immediate and followed percussive trauma rather than distraction. In such circumstances a different mechanism might be responsible, possibly analogous to that of spinal shock.

As a final point of discussion it is pertinent to reassess the position of the "wake-up" test. This provides a once-off, qualitative evaluation of motor pathway function and is unlikely to have detected some of the mild, mainly sensory deficits demonstrated in the present series. Furthermore, an arterio-vascular occlusion causing partial conduction failure during hypotensive surgery might be temporarily relieved when blood pressure is raised to perform a "wake-up" test, and might then recur after the patient is re-anaesthetized. The possibility that occlusion of the anterior spinal artery may cause damage confined to antero-lateral structures of the cord has frequently been raised, but has not been observed by the present authors. Were it to be found that an anterior cord syndrome is not reflected by any significant change of sensory potentials, a technique such as that involving recording of the efferent spinal volley from the epidural space following stimulation of the motor cortex through scalp electrodes (Boyd et al. 1985) may prove more quantifiable and easy to administer than the "wake-up" test.

Acknowledgement. We are grateful to Action Research for the Crippled Child for the provision of an equipment grant.

References

Boyd SG, Cowan JMA, Rothwell JC, Webb PJ, Marsden CD (1985) Monitoring spinal motor tract function using cortical stimulation: A preliminary report. In: Schramm J, Jones SJ (eds) Spinal cord monitoring. Springer, Berlin Heidelberg New York Tokyo, pp 227–230

Jones SJ, Edgar MA, Ransford AO (1982) Sensory nerve conduction in the human spinal cord: epidural recordings made during scoliosis surgery. J Neurol Neurosurg Psychiat 45: 446–451

Jones SJ, Edgar MA, Ransford AO, Thomas NP (1983) A system for the electrophysiological monitoring of the spinal cord during operations for scoliosis. J Bone Joint Surg 65-B: 134–139

MacEwen GD, Bunnell WP, Sriram K (1975) Acute neurological complications in the treatment of scoliosis: a report of the scoliosis research society. J Bone Joint Surg 57-A: 404–408

An Assessment of the Use of Spinal Cord Evoked Potentials in Prognosis Estimation of Injured Spinal Cord

T. Tamaki[1], H. Takano, K. Takakuwa, H. Tsuji, T. Nakagawa, K. Imai, and S. Inoue

Summary

The value of spinal cord evoked potentials for intraoperative prediction of the prognosis following acute or chronic spinal cord impairment was investigated in clinical cases. Simply by observing the pattern and amplitude of the potentials, we could not predict the prognosis significantly. Comparison of amplitudes in consecutive recordings, however, provided clinically useful information.

Conclusions derived from the present study are as follows:

1. The spinal evoked potential was not recorded in cases of neurologically complete spinal cord lesion.
2. Neurological improvement could be anticipated in cases who showed an increment in amplitude.
3. Patients showing no change in amplitude still had a certain possibility of improved neurological symptoms.
4. Decreased amplitude or complete disappearance of the potential were found to indicate postoperative neurological deterioration.
5. Complete disappearance of the potential was indicative of severe spinal cord dysfunction.

Introduction

The purpose of this paper is to evaluate the feasibility of the evoked spinal cord potential as an indicator to predict the neurological prognosis intraoperatively in patients with spinal cord lesions. Ideally, one might be able to determine the prognosis simply by analyzing the wave pattern and amplitude of the evoked potential, but, as is discussed subsequently, simple analysis of the evoked potential could offer no appropriate information to indicate the prognosis of the jeopardized spinal cord. However, periodical observation of the evoked potential during the course of the surgical intervention to the spinal cord or spine offered fairly reliable information. In another words, the alteration in amplitude was considered to be a sign which had prognostic value.

Material and Methods

The evoked spinal cord potentials following direct stimulation of the spinal cord were recorded from either epidural or subarachnoid space. Basically, the stimulating

[1]Department of Orthopaedic Surgery, Faculty of Medicine, Toyama Medical and Pharmaceutical University, 2630 Sugitani, Toyama 930-01, Japan

Spinal Cord Monitoring
Ed. by J. Schramm and S. J. Jones

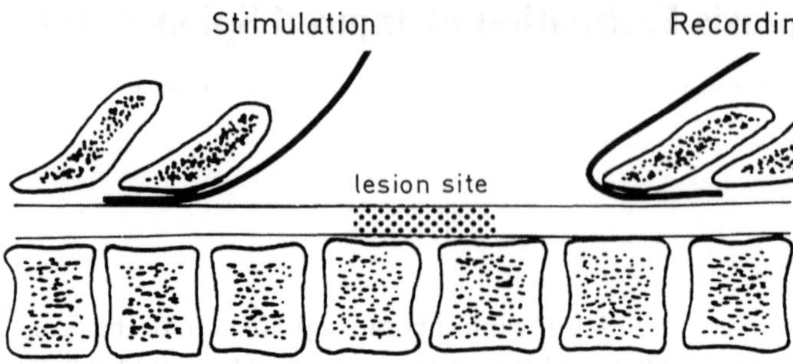

Fig. 1. If the recording of potentials or the introduction of electrodes was difficult, electrodes were inserted into the epidural or subarachnoid space from the site of laminectomy or opened dura mater

electrode was located cephalad to the site of spinal cord lesion in the epidural space, and the recording electrode was introduced into the subarachnoid space at the caudal site following our standard technique described elsewhere (Tamaki et al. 1977, 1981, 1984). For cases in whom there was difficulty in recording the evoked potential, the electrodes were inserted into epidural or subarachnoid space utilizing the site of laminectomy and opened dura mater (Fig. 1).

Potentials recorded from the incompletely impaired spinal cord were generally low in amplitude and atypical in wave pattern (Fig. 2). Therefore findings with regard to the neural structures responsible for each component of the potential, which have been obtained by many researchers (Harada 1983; Imai 1976; Tamaki 1979; Toyoda and Kanda 1984; Tsuyama et al. 1978) using neurologically normal animals, were not applicable to the pathological potentials of the present study.

In neurologically normal subjects, approximately 50% degradation of the initial spike wave was considered to be the alarm signal indicating hazardous effect on the spinal cord (Imai et al. 1984; Tamaki et al. 1984). However, in the circumstances of previously impaired spinal cord conduction, a different criterion should be established. In this study, 10% enhancement or degradation was arbitrarily defined as the minimal standard to indicate a significant change in amplitude.

Since we started to employ advanced electronic hardware such as the DISA 1500 system or the Medelec MS6 and MS92 in the past 5 years, 79 neurologically impaired spinal cords have been examined intraoperatively. They comprised 39 cases of spinal cord tumor, 9 of spinal cord trauma, 5 of ossification of the posterior longitudinal ligament, 5 of syringomyelia, 4 of congenital scoliosis, 3 of cancerous metastasis to the vertebra, 2 of ossification of the yellow ligament, 2 of achondroplasia and 10 of miscellaneous pathological conditions of the spinal cord.

Results

No spinal evoked potential was observed in 9 cases of complete spinal cord injury. Among 70 cases with incompletely impaired spinal cord conduction, no evoked spinal potential was recorded in three cases. Technical failure was encountered in 2 cases but no cause was found in 1. For 59 cases out of 67 in whom the evoked spinal

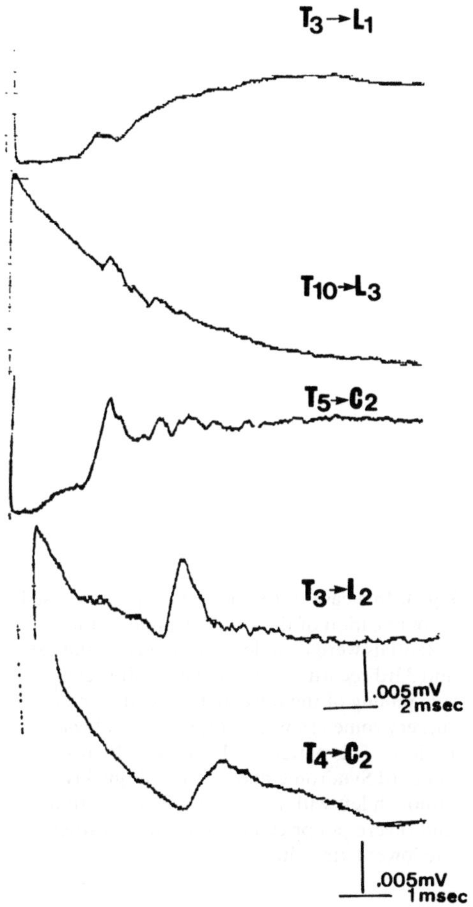

Fig. 2. Spinal cord evoked potentials in patients with incomplete spinal cord lesions. Potentials were low in amplitude and atypical in wave pattern. No close correlation between wave configuration and prognosis was revealed

cord potential was successfully recorded, the potentials were used as a tool to monitor spinal cord function intraoperatively. Among these cases, no change in amplitude was observed in 16. Neurological examination of these patients 4 weeks after the operation revealed unchanged neurological symptoms in 10, improved neurological status in 5 and neurological deterioration in 1. In the last case, the additional symptoms were caused by a lesion to the cauda equina which was not monitored. A conclusion derived from this analysis was that a patient with no change in amplitude during surgery still had a certain probability (33%) of improved neurological status.

Increments in amplitude ranging from 16% to 280% were observed in 33 cases; postoperatively, 31 of these improved in neurological status. In a case of intramedullary spinal cord tumor which showed a temporary degradation in amplitude of 19% coincident with the maneuver of tumor exstirpation, about 40% enhancement was observed at the end of surgery. The neurological status of this case was similar to the preoperative status; namely slight hypalgesia on the left side below C_4, impairment of vibration sense in the right lower extremity and motor weakness in the right extremities. In another case of metastatic cancer of the vertebra, the evoked poten-

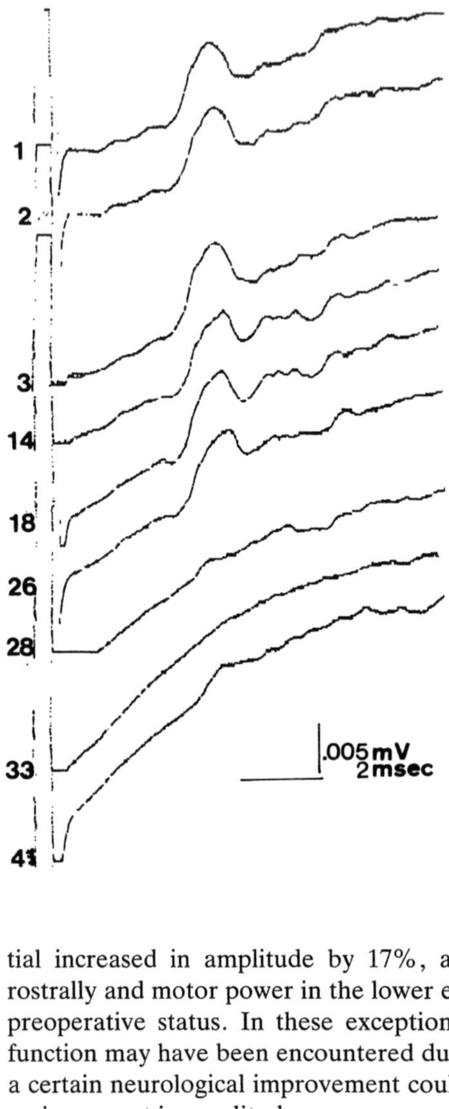

Fig. 3. In a case of thoracic myelopathy caused by ossification of the yellow ligament, the potentials were completely flattened in the 28th and 33rd recording, immediately after excessive pulling of the dura mater. At the end of surgery some recovery of the potential was noticed. This patient woke up with Brown-Séquard Syndrome and recovered quickly, although left with hyperesthesia in his trunk and severe pin prick and touch hyperesthesia in the lower extremities

tial increased in amplitude by 17%, although the zone of hypalgesia expanded rostrally and motor power in the lower extremities was degraded compared with the preoperative status. In these exceptional two cases, factors affecting spinal cord function may have been encountered during the postoperative course. In any event, a certain neurological improvement could be anticipated in those cases who showed an increment in amplitude.

Reduction of amplitude or complete flattening of the waveform was observed in 10 cases. Seven cases in this group demonstrated certain neurological sequelae in addition to the symptoms exhibited preoperatively. Conversely, in two cases with 10% degradation of amplitude, neurological improvement was observed. In the last case of this group, neurological symptoms were unchanged postoperatively. In four cases of the above mentioned group there was complete flattening of the wave form during the course of surgery, whereas three cases showed a recovery of small amplitude at the end of the operation. One case, who did not exhibit any recovery of the evoked potential, woke up with complete paraplegia and no recovery has been observed since. Three cases with small potentials at the end of surgery have shown improvement in neurological status, but 2 cases remain at the level of incomplete paraplegia, needing sticks to walk. The last case showed a remarkable recovery from a Brown-Séquard type of spinal cord injury postoperatively, although a zone of hyperesthesia remained in his trunk and high grade touch and pin prick hyperesthesia in the lower extremities (Fig. 3).

Carefully reviewing the clinical cases studied here from the point of view of neurological prognosis, no definite correlation was revealed between the static wave configurations and the postoperative clinical course. However, even small potentials, if they increased in amplitude during the surgical intervention, were a good sign that the patient would show a neurological recovery postoperatively.

Discussion

In the absence of recent relevant experimental work, we did not have any reliable neurophysiological means of predicting the possibility and extent of the functional recovery in patients with incompletely impaired spinal cord function. Analysis of the evoked spinal cord potential is one method which has the potentiality of answering the above mentioned demand of clinicians, but this was not always possible in patients with spinal cord dysfunction during the early stage of our clinical application of this method. This may be one of the reasons why conclusive studies have not been published on this matter, excepting the reports of Tsubokawa et al. (Tsubokawa and Moriyasu 1974; Tsubokawa 1978). He and his colleague observed that the prognosis in patients with incomplete spinal cord injury was predicted by studying not only the spinal SEP following peripheral nerve stimulation but also the potentials evoked by spinal cord stimulation. They have also reported that patients with enhanced evoked potentials after spinal cord cooling had a good prognosis, in contrast to patients with no evoked potentials in whom the prognosis was poor. In the present study, the prognosis of the cases who showed amplitude increment was good and similar to the findings of Tsubokawa et al., although a certain degree of recovery was observed in 33% of the patients who did not exhibit any change in amplitude until the end of surgery. These cases illustrate the difficulty in estimating the postoperative recovery course of the neural tissue.

As the present study revealed, modern advanced electronic hardware could increase the likelihood of recording evoked potentials from cases of incomplete spinal cord lesion. Therefore, the possibility of determining the prognosis of the patient intraoperatively has been increased. This also implies that each step of the operative procedure can be easily approved by the recording of such potentials, and this will contribute greatly to the development of surgical techniques.

Even with advanced hardware and techniques, it was not possible to obtain sufficient information from the static analysis of wave configurations, as performed in ECG analysis. In other words, only by comparison of the wave configurations in successive recordings could information relevant to the neurological prognosis be obtained. It remains controversial at present why it should be possible to estimate the plasticity or functional recovery of the spinal cord by observing evoked potentials which are conducted through barely surviving neural tissue. An increment of the potential might be interpreted as an early sign of favorable effect of the surgical maneuver on the spinal cord, such that, if the effect is maintained postoperatively, the recovery of function may continue to a certain level. The estimation of this terminal level of recovery has been an insistent demand of clinicians, but as yet we have not yet obtained any reliable information on this question.

References

Harada Y (1983) Study on experimental spinal cord injuries by distraction using spinal cord evoked potentials (in Japanese). J Jpn Orthop Ass 57:685–701

Imai K, Kobayashi H, Nakagawa T, Inoue S, Tamaki T (1984) The experiences and analysis of the spinal cord monitoring during surgery. In: Homma S, Tamaki T (eds) Fundamentals and clinical application of spinal cord monitoring. Saikon Press, Tokyo, p 211

Imai T (1976) Human electrospinogram evoked by direct stimulation on the spinal cord through epidural space (in Japanese). J Jpn Orthop Ass 50:1037–1056

Tamaki T, Kobayashi H, Yamane T, Egashira T, Tsuji H, Inoue S (1977) Clinical application of evoked spinal cord action potential (in Japanese). Seikeigeka 28:681–689

Tamaki T (1979) Basic analysis of evoked spinal cord potential elicited by direct spinal cord stimulation. In: Nash CL (ed) Proceedings of the Spinal Cord Monitoring Workshop, Data Acquisition and Analysis, Cleveland, Case Western Reserve University, p 17

Tamaki T, Tsuji H, Inoue S, Kobayashi H (1981) The prevention of iatrogenic spinal cord injury utilizing the evoked spinal cord potential. Int Orthop 4:313–317

Tamaki T, Noguchi T, Takano H, Tsuji H, Nakagawa T, Imai K, Inoue S (1984) Spinal cord monitoring as a clinical utilization of the spinal evoked potential. Clin Orthop 184:58–64

Toyoda A, Kanda K (1984) Origins of spinal cord potentials evoked by stimulation of the cat spinal cord. In: Homma S, Tamaki T (eds) Fundamentals and clinical application of spinal cord monitoring. Saikon Press, Tokyo, p 99

Tsubokawa T, Moriyasu N (1974) The clinical evaluation of spinal electrogram in spinal cord injury (in Japanese). Rinshonoha 16:398–409

Tsubokawa T (1978) The spinal cord evoked potentials as an indicator for diagnosis of spinal cord injury (in Japanese). Rinshonoha 20:236–243

Tsuyama N, Tsuzuki N, Kurokawa T, Imai T (1978) Clinical application of spinal cord action potential measurement. Int Orthop 2:39–46

Monitoring Spinal Motor Tract Function Using Cortical Stimulation: A Preliminary Report

S. G. Boyd[1], J. M. A. Cowan, J. C. Rothwell, P. J. Webb, and C. D. Marsden

Summary

We describe the preliminary findings of a study in which descending spinal cord potentials were recorded intraoperatively following single stimuli given to the scalp over the motor cortex. This might be a useful adjunct to sensory spinal cord monitoring, providing a rapid assessment of motor pathway function at times when the spinal cord is most at risk.

Introduction

A new technique for studying motor pathway function by stimulation of the cortex through the intact scalp has recently been assessed in several centres in London (Merton et al. 1982). The cortex is stimulated by applying electrodes to the skin, with the cathode at the vertex and the anode 7 cm laterally, approximately over the hand area of the motor cortex. The stimulator has a low output impedance and is capable of delivering a single pulse with a maximum output of 750 V over a period of 100 μs. In the conscious subject the latency to onset of the evoked EMG activity in a peripheral muscle can be measured. Percutaneous stimulation of the spinal cord is also possible with appropriate electrode placement (Marsden et al. 1982). Applications of the technique have included demonstration of abnormalities of central motor pathway conduction in Multiple Sclerosis (Gowan et al. 1984) and the investigation of spinal cord and cauda equina lesions in patients with incontinence (Snooks and Swash, to be published).

There is some clinical and experimental evidence that sensory and motor spinal tracts may be affected differentially during surgery to the spinal column (Bennett 1983; Levy 1983). For this reason, in addition to monitoring sensory pathway function, we have adapted this technique of motor cortex stimulation to monitor function in cortico-spinal pathways during spinal surgery. Eight patients aged 12–16 years who were undergoing surgical correction of scoliosis have been studied. None had neurological signs secondary to the scoliosis or any pyramidal tract lesion. Patients with epilepsy were excluded. Bipolar recording electrodes were inserted into the epidural space during the operation, when the spinal processes had been exposed, and recording from oesophageal electrodes was also attempted in some patients.

[1] Department of Clinical Neurophysiology, The Hospital for Sick Children, Great Ormond Street, London WC1N 3JH, Great Britain

Spinal Cord Monitoring
Ed. by J. Schramm and S. J. Jones

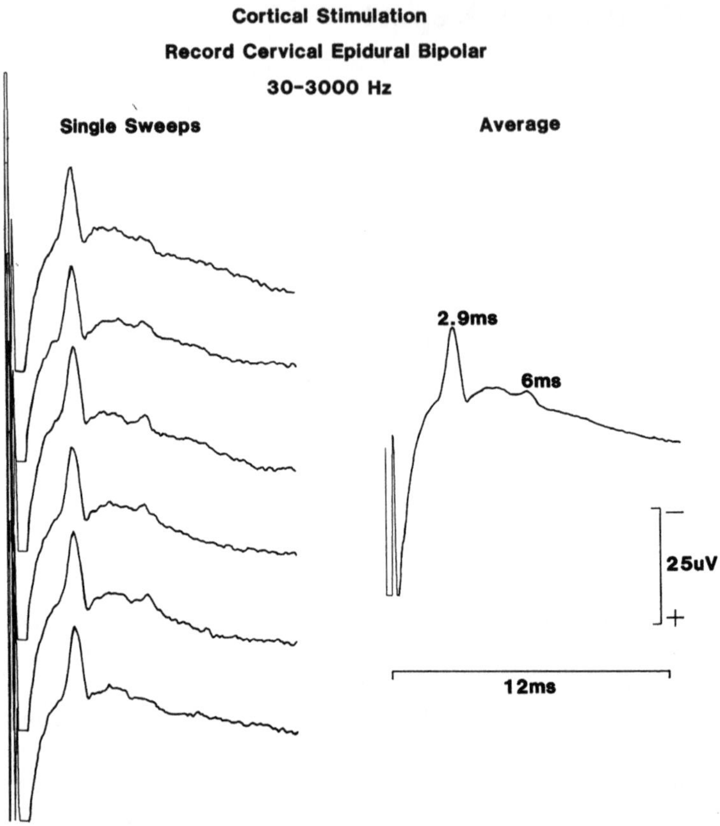

Fig. 1. Consecutive motor responses to single scalp stimuli demonstrate the stability and consistency of the main negative component. The small later components are detectable with single sweeps but are more clearly seen when the six responses have been averaged

Methods

The spinal sensory evoked potentials were recorded according to techniques described by Jones et al. (1982). Motor cortex stimulation was performed with the technique of Merton et al. (1982) and responses to single stimuli were amplified and displayed using either a Nicolet CA 1000 system (bandpass 30–300 Hz) or a Digitimer D 200 (bandpass 50–3000 Hz). Signals were also recorded on magnetic tape for further analysis.

Results

Responses to single stimuli could be recorded from epidural electrodes at all levels of the spinal cord and from the oesophagus in some patients. Responses to consecutive stimuli at rates of around 2 every 5 s were stable and consistent (Fig. 1). With increasing intensity of stimulation there was an increase in peak amplitude and a

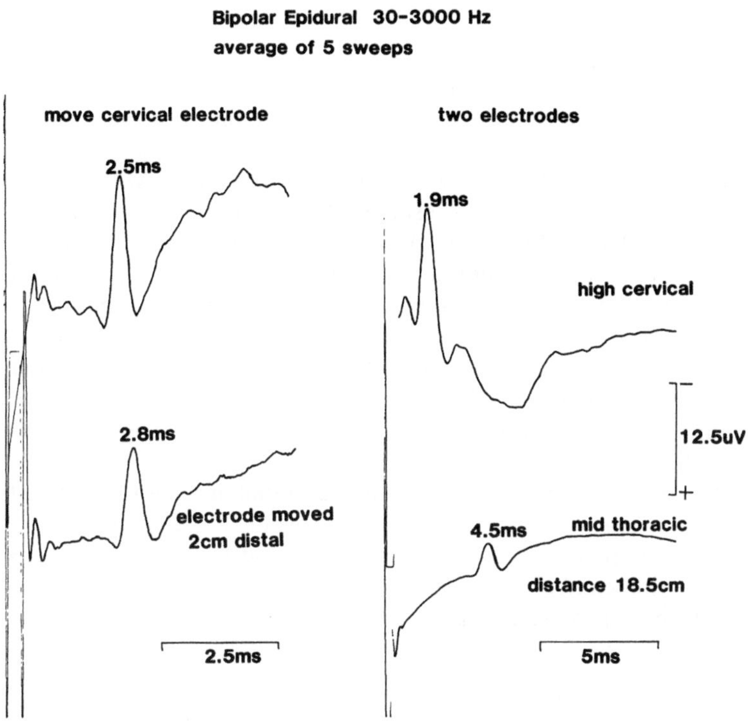

Fig. 2. Responses to motor cortex stimulation: on the *left,* note the difference in amplitude and latency of the motor response when the recording electrode is moved distally. On the *right,* the response to the same stimulus has been recorded from 2 sites. The amplitude of the response in thoracic segments of the cord is always smaller than in cervical segments

decrease in peak latency of the main negative response. At higher intensities of stimulation some later components were sometimes detected (Fig. 1) though these were of variable morphology in different patients.

The amplitude of the response in thoracic segments of the cord was always smaller than that seen in cervical segments (Fig. 2). The amplitude of the response in the cervical cord could vary, even over short distances (Fig. 2), as would be expected from the anatomical distribution of the cortico-spinal tracts. It was not possible to determine the exact level at which the recording electrodes lay and this may account for the range of response amplitudes from cervical recordings in different patients.

Where a single recording electrode was moved by a short, measured distance or where two recording electrodes were used (Fig. 2), it was possible to estimate motor conduction velocities in the cortico-spinal tracts which were found to range from 50–75 m/s (Boyd et al., to be published). Values for sensory conduction velocities were also measured in some of these patients and varied from 70 to nearly 90 m/s.

Discussion

The technique of motor cortex stimulation has demonstrated changes in function in spinal motor pathways which correlated with neurological findings in one further case studied. This was an 8-year-old child who had developed paraplegia following previous spinal surgery. Epidural electrodes were placed above and below the site of the mid-thoracic lesion and although a motor response could be recorded from above the lesion, no response was detected from the electrode placed below the lesion.

In cat experiments, Bennett (1983) noted that a modest degree of spinal cord ischaemia could produce profound changes in function of descending pathways which was not accompanied by changes in sensory evoked potentials. Such findings as well as some anecdotal clinical reports (Levy 1983) encourage us to believe that the monitoring of motor pathways may be an important adjunct to peroperative sensory pathway monitoring, and may detect impairment at an earlier stage than has been possible previously.

Although only a few patients have so far been studied, initial findings suggest that the monitoring of spinal motor pathway function using this technique may be clinically useful. The fact that responses to single stimuli can be easily recognised means that motor pathway function can be rapidly and repeatedly assessed at times during the operations when the cord is most at risk.

References

Bennett MH (1983) Effects of compression and ischaemia on spinal cord evoked potentials. Exp Neurol 80:508–519

Boyd SG, Cowan JMA, Marsden CD, Rothwell JC, Webb PJ (to be published) Direct estimation of cortico-spinal tract conduction velocity in man. J Physiol

Cowan JMA, Rothwell JC, Dick JPR, Thompson PD, Day BL, Marsden CD (1984) Abnormalities in central motor pathway conduction in multiple sclerosis. Lancet 2:304–307

Jones SJ, Edgar MA, Ransford AO (1982) Sensory nerve conduction in the human spinal cord: epidural recordings made during scoliosis surgery. J Neurol Neurosurg Psychiat 45:446–451

Levy WJ (1983) Spinal evoked motor potentials. J Neurosurg 59:912–913

Marsden CD, Merton PA, Morton HB (1982) Percutaneous stimulation of spinal cord and brain: pyramidal tract conduction velocities in man. J Physiol 328:61

Merton PA, Hill DK, Morton HB, Marsden CD (1982) Scope of a technique for electrical stimulation of human brain, spinal cord and muscle. Lancet 2:597–600

Snooks SJ, Swash M (to be published) Electrophysiological methods for investigating incontinence. Electroencephalogr Clin Neurophysiol

Spinal Evoked Potentials in Infants with Myelomeningocele

T. NAKAGAWA[1], K. IMAI, M. MURAKAMI, S.-I. INOUE, M. MAIE, T. YAMANE, and T. YAMASHITA

Summary

Spinal SEPs were recorded in 8 infants with myelomeningocele, in response to posterior tibial, peroneal and femoral nerve stimulation. Small, polyphasic potentials could be recorded from the neural plaque in all cases studied. Spinal SEPs were frequently absent, except with stimulation of the femoral nerve. Spinal SEPs generally correlated well with clinical sensory levels, but were obtained from levels lower than the sensory level in 2 cases.

Introduction

In the management of a newborn infant with myelomeningocele, an early and accurate assessment of the neurological condition is necessary. Hunt et al. (1973) emphasized the importance of the sensory level in assessing the neurological deficits. However, neurological examination is difficult and inaccurate in the neonatal period. Therefore, the value and importance of electrophysiological investigations in myelomeningocele neonates have been emphasized. Duckworth et al. (1976) used somatosensory evoked potentials (cortical SEPs) as a method for sensory examination of 18 infants and children with myelomeningocele. In the majority of patients, cortical SEPs correlated with clinical sensory levels. Reigel et al. (1976) studied 7 infants during surgery for thoraco-lumbar myelomeningocele using cortical SEPs, and evoked potentials from the neural plaque to demonstrate the residual function of the plaque and distal nerve.

In this study, we have attempted to use spinal SEPs to determine if they could be of any clinical use in the neurological assessment of infants with myelomeningocele.

Material and Methods

We studied 8 infants with myelomeningocele, 3 males and 5 females, ranging in age from 1 day to 5 weeks. One patient (case No. 2) was treated with nursing and symptomatic treatment and investigated within 12 h of birth. Seven patients were operated on for repair of myelomeningocele. All 8 patients were given a careful neurological examination at admission (Table 1).

[1] Department of Orthopaedic Surgery, School of Medicine, Chiba University, 1-8-1 Inohana, Chiba 280, Japan

Spinal Cord Monitoring
Ed. by J. Schramm and S. J. Jones

Table 1. Clinical material (1981. 3 ~ 1984. 6)

Case	Sex	Age (yrs. + mos.)	Time before closure operation	Neurological levels	
				Motor[a] (down to)	Sensory (down to)
1. K.G.	M	3 + 3	8 hours	S_{1-2}	L_5
2. F.I.	M	Died	*	L_{3-4}	L_3
3. C.I.	F	2 + 3	2 weeks	L_{1-2}	L_3
4. M.S.	F	1 + 9	8 hours	S_{1-2}	L_5
5. Y.M.	F	1 + 8	5 weeks	T_{12} (R) L_5 (L)	T_{11} (R) S_1 (L)
6. Y.M.	M	0 + 9	6 hours	L_5	L_5
7. C.I.	F	0 + 3	6 hours	L_{1-2}	L_3
8. F.S.	F	0 + 1	30 hours	L_{3-4}	L_1

[a] Classification by Sharrard (1964)
* Evoked potentials were recorded within 12 h of birth

After induction of anesthesia, disc surface electrodes were placed over the tibial nerve at the ankle, the peroneal nerve in the popliteal fossa and the femoral nerve below the inguinal ligament for stimulation.

The stimulus was a rectangular pulse of 0.2 ms duration, delivered at a rate of 3 or 10 Hz. The intensity of stimulation was adjusted to produce a weak muscle twitch. Evoked potentials following stimulation of each nerve were amplified with band-passes 200 Hz–2 kHz, 20 Hz–10 kHz (Medelec MS92) and 32 Hz–3.2 kHz (Medelec MS6). 100–1000 responses were summated with 20 or 50 ms analysis time. To record evoked potentials from the neural plaque, the bipolar catheter electrode designed by Tamaki et al. (1984) (Unique Medical Co., Ltd., Tokyo) was first placed over the neural plaque, and then introduced into the subarachnoid space rostral to the plaque during surgery.

Results

Results following stimulation of the peroneal nerve and the posterior tibial nerve are shown in Table 2.

Evoked Potentials from the Neural Plaque

In 7 patients, evoked potentials from the surface of the neural plaque were recorded following stimulation of the posterior tibial nerve and/or the peroneal nerve. Evoked potentials following posterior tibial nerve stimulation were present in 6 out of 7 patients. In all 7 patients, evoked potentials to peroneal nerve stimulation were present. In 3 out of 7 patients, evoked potentials in response to femoral nerve stimulation were studied but no responses could be recorded. The wave form of evoked potentials from the neural plaque was small in amplitude and polyphasic in pattern (Fig. 1).

Table 2. Results

Case	Neurological levels		Evoked potentials from neural plaque		Spinal evoked potentials	
	Motor (down to)	Sensory (down to)	Posterior tibial n.	Peroneal n.	Posterior tibial n.	Peroneal n.
1	S_{1-2}	L_5	+	+	+	+
2	L_{3-4}	L_3	−	+	/	/
3	L_{1-2}	L_3	+	+	−	−
4	S_{1-2}	L_5	+	+	+	+
5	T_{12} (R) L_5 (L)	T_{11} (R) S_1 (L)	/	/	− (R) + (L)	+ (R) + (L)
6	L_5	L_5	+	+	+	+
7	L_{1-2}	L_3	+	+	−	−
8	L_{3-4}	L_1	+	+	+ (R) − (L)	− (R) − (L)

+: Evoked potentials were recorded
−: No response

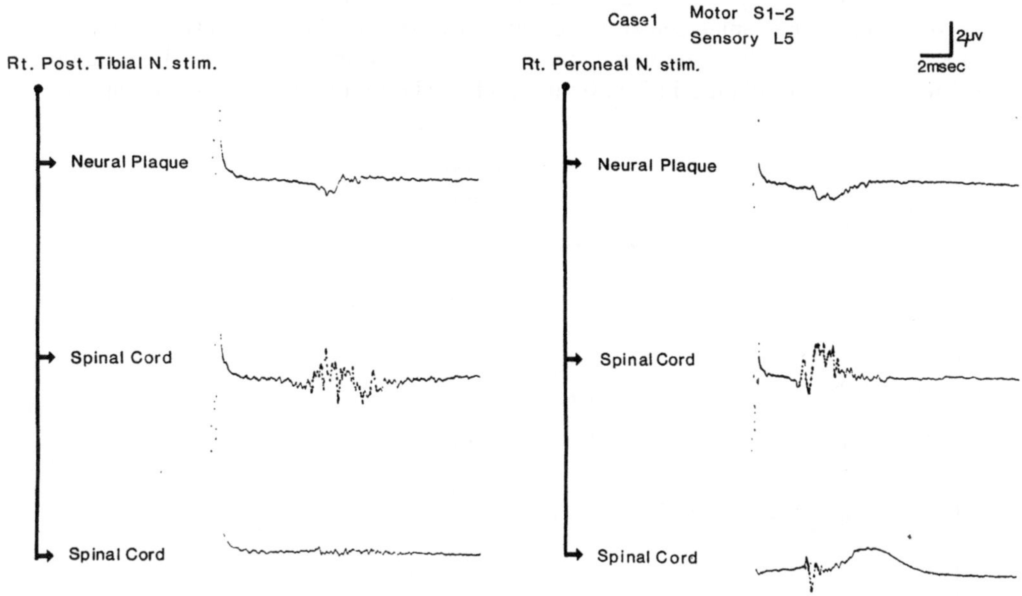

Fig. 1. Myelomeningocele. Case 1: Spinal SEPs recorded from the neural plaque and the spinal cord following posterior tibial and peroneal nerve stimulation in a case of myelomeningocele, showing small amplitude and polyphasic pattern

Spinal Evoked Potentials

Spinal SEPs were studied in 7 patients. Evoked potentials following stimulation of the posterior tibial nerve were present in 5 patients out of 7. In 2 out of 5 patients, spinal SEPs could not be observed unilaterally. Evoked potentials following peroneal nerve stimulation were present in 4 out of 7 patients. Recorded spinal SEPs were small in amplitude and tended to be polyphasic (Fig. 1). Spinal SEPs after stimulation of the femoral nerve were studied in 3 out of 8 patients, and present in all 3. The wave form of spinal SEPs evoked by femoral nerve stimulation consisted of an initial spike wave followed by slow waves.

Correlation Between Spinal SEPs and Clinical Sensory Levels

There was some disagreement between the motor and sensory levels observed. However, in 5 patients (case Nos. 1, 3, 4, 6 and 7), spinal SEPs fully correlated with clinical sensory levels. In patient No. 5 with asymmetrical neurological deficits, spinal SEPs following stimulation of the left lower limb were consistent with the clinical sensory level, but spinal SEPs following stimulation of the other side were inconsistent with clinical evaluation. In patient No. 8 spinal SEPs following stimulation of the right posterior tibial nerve were inconsistent with clinical sensory levels.

Discussion

With electrophysiological testing, Reigel et al. (1976) demonstrated that peripheral nerve afferents and somatosensory afferent pathways rostral to the neural plaque are intact in patients with myelomeningocele. In our study, evoked potentials could be recorded from the plaque in 7 patients, and marked temporal dispersion was ob-

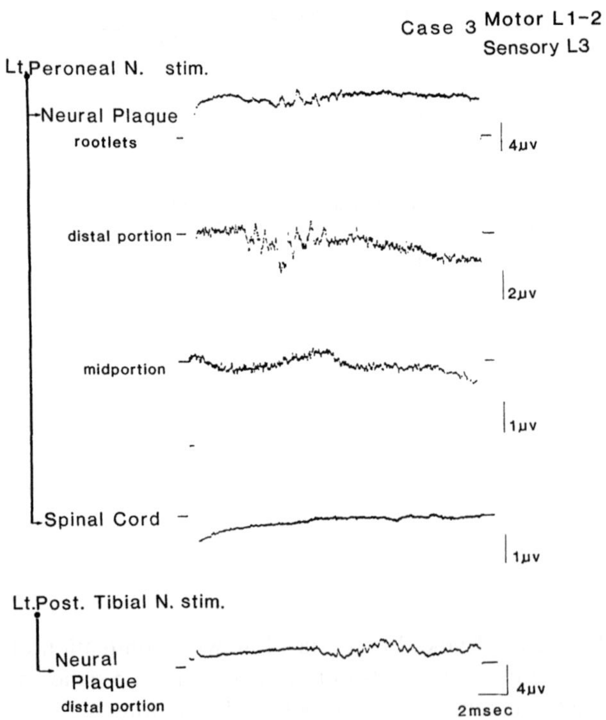

Fig. 2. Myelomeningocele. Case 3: Evoked potentials recorded from the neural plaque, showing marked temporal dispersion. No spinal cord potentials could be recorded at more rostral level

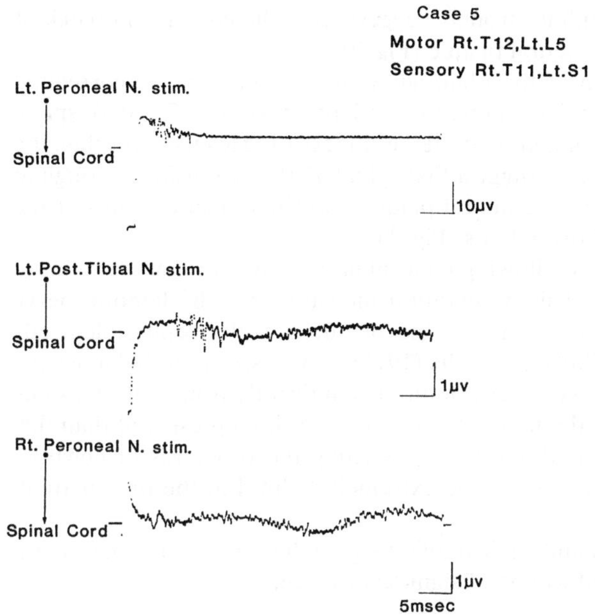

Fig. 3. Myelomeningocele. Case 5: Present peroneal and posterior tibial nerve spinal SEPs on the left, but absent peroneal nerve responses on the right, indicating residual dorsal column function on the left

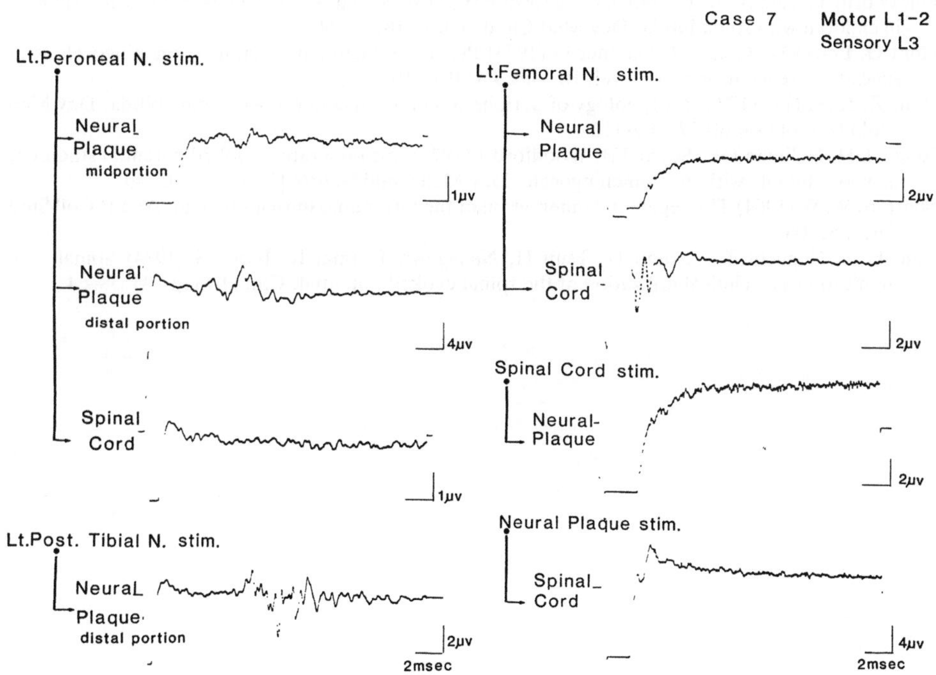

Fig. 4. Myelomeningocele. Case 7: Normal wave form of femoral nerve spinal SEPs, indicating intact electrophysiological function rostral to the plaque

served in most patients. This finding strongly suggests that the conduction block of afferent volleys originates in the neural plaque (Fig. 2).

Spinal SEPs with stimulation of the tibial nerve and peroneal nerve correlated well with clinical sensory levels in 5 patients. In 2 patients (case Nos. 5 and 8), spinal SEPs were obtained following stimulation of nerves innervating levels lower than the clinical sensory level. These findings suggest that spinal SEPs are sensitive enough to evaluate the residual functions of the neural plaque and the dorsal columns of the spinal cord, particularly in newborn infants (Fig. 3).

The wave form of spinal SEPs following femoral nerve stimulation was similar to normal segmental spinal SEPs. This observation indicates that the femoral nerve afferents and the dorsal columns rostral to the plaque are electrophysiologically intact (Fig. 4). We agree with Rališ and Rališ (1972) that in spina bifida babies the mean diameter of the femoral nerve does not differ significantly from that of normal babies, and that the function of the femoral nerve is more often preserved than that of the sciatic nerve. If our method assists in predicting the prognosis of newborn infants with myelomeningocele, it would be extremely helpful in the proper treatment of patients.

Further electrophysiological and clinical follow-up studies are needed before it is determined whether this method will be of significant value.

References

Duckworth T, Yamashita T, Franks CI, Brown BH (1976) Somatosensory evoked cortical responses in children with spina bifida. Dev Med Child Neurol 18:19–24

Hunt G, Lewin W, Gleave J, Gaidner D (1973) Predictive factors in open myelomeningocele with special reference to sensory level. Br Med J 4:197–201

Rališ Z, Rališ HM (1972) Morphology of peripheral nerves in children with spina bifida. Dev Med Child Neurol [Suppl] 27:109–116

Reigel DH, Dallman DE, Scarff TB, Woodford J (1976) Intra-operative evoked potential studies of newborn infants with myelomeningocele. Dev Med Child Neurol [Suppl] 37:42–49

Sharrard WJW (1964) The segmental innervation of the lower limb muscles in man. Ann R Coll Surg Engl 35:106

Tamaki T, Noguchi T, Takano H, Tsuji H, Nakagawa T, Imai K, Inoue S (1984) Spinal cord monitoring as a clinical utilization of the spinal evoked potential. Clin Orthop 184:58–64

Intraoperative Spinal Cord Monitoring at Different Levels and with Varying Surgical Pathology

E. Valencak[1], A. Witzmann, and F. Reisecker

Summary

Cortical and cervical somatosensory evoked potentials were recorded in response to median nerve or direct spinal cord stimulation during surgery of the craniocervical junction, vertebral or posterior inferior cerebellar artery aneurysms, cervical cordotomy and traumatic and tumorous lesions of the cervical and thoracic cord. SSEP were sensitive to hemorrhage and manipulation of the brainstem, but were unaffected by tractonucleotomy or high cervical cordotomy for pain relief. In traumatic cord lesions, direct cord stimulation at various levels was used to determine the level of the sensory lesion, and to predict the likelihood of functional recovery.

Introduction

The introduction of microsurgical technique into the surgery of the spinal cord is only of minor importance from the point of view of assessment of spinal cord function. Intraoperative monitoring of spinal cord function sets the following demands: (1) No additional damage to the spinal cord should be caused by the monitoring procedure. (2) There should be no significant extension of the duration of the operation due to the method. (3) Monitoring of spinal cord function must lead to conclusions, which are important for the outcome of the operation and in this way important for the medical procedure chosen. (4) The method should allow exact assessment.

The use of the Somatosensory Evoked Potentials (SSEP) seems to satisfy these conditions now and in the future. The following range of applications has been chosen in the present study (Table 1): (1) Lesions in the craniocervical junction: aneurysms of the vertebral or posterior inferior cerebellar artery (PICA), high cervical cordotomy and tracto-nucleotomy according to Sjöqvist-Kunc in trigeminal neuralgia following multiple sclerosis, epipharyngeal carcinoma and other types of intractable pain due to malignancy. (2) Traumatic and tumorous lesions of the cervical medulla ranging from C2 to C7 level. (3) Traumatic lesions of the thoracic medulla (Table 2).

[1]Neurochirurgische Abteilung des Wagner-Jauregg-Krankenhauses, Wagner-Jauregg-Weg 15, A-4020 Linz

Spinal Cord Monitoring
Ed. by J. Schramm and S. J. Jones

Table 1. Application of intraoperative spinal evoked potentials in surgery of different anatomic regions ($n = 45$)

A) Cranio-cervical junction (pontine and high medullary regions, C2 included)	$n = 5$
B) Cervical spine (C7-C2)	$n = 4$
C) Thoracic spine	
1. Intraoperative spinal monitoring	$n = 26$
2. Postoperative "therapeutic" stimulation	$n = 10$

Table 2. Application of intraoperative spinal evoked potentials in various pathologies ($n = 45$)

Intractable pain in epipharynx-carcinoma	1
Trigeminal neuralgia (multiple sclerosis)	3
Cervical disc—C6/7	1
Fracture-dislocation C5/6	1
Intramedullary ependymoma (C7-C3)	1
Aneurysm of vertebral artery	1
Aneurysm of PICA	1
Traumatic lesion of thoracic spine with dislocated fragments and spinal compression	25
Cordotomy D2	1
Postoperative "therapeutic" stimulation (5 weeks)	10

Methods

1) Craniocervical Junction and Cervical Medulla

Stimulation was carried out in two ways:

a) Square waves of 0.2 ms duration were delivered at a rate of 3/s by means of bipolar stimulating electrodes (DISA 13 L 22) over the median nerve at the wrist, with a cathode 3 cm proximal to the anode. Stimulus intensity was just sufficient to produce a small twitch of the abductor pollicis brevis in the neuroleptanalgesic narcotised patient.

b) After the laminectomy and opening of the dura, the spinal cord was stimulated directly from the epidural space by means of a bipolar electrode (Type Cordis). The position of the stimulation was in most cases the C5 level. Stimulus intensity was 1.5 V in all cases. The other parameters were the same as in the median nerve stimulation procedure.

Recording was made at two positions: at the C7 level, using a pre-frontal reference (Fpz) and at the C3 or C4 position of the International 10/20 system, using a reference at the mastoid on either side. The recording electrodes were stainless steel needles (Type Siemens 2264554). 512 sweeps were averaged with a DISA-1500 EMG system.

2) Thoracic Spinal Cord

After posterolateral thoracotomy or a combined thoracoabdominal approach to D12 and L1 (Valencak 1979) an epidural bipolar electrode (Type Cordis) was placed on the ventral surface of the cord first above the lesion and the SSEP was recorded. Several supralesional levels were examined. After this procedure the infralesional area was evaluated in the same way. Infralesional evoked potentials are very important, especially in patients suffering from complete transverse spinal cord lesion, to prove stimulus conduction and furthermore "neuronal continuity". Stimulation of the spinal cord was performed with the dual object of recording the SSEP and observing motor activity in the lower extremities and the abdominal muscles.

In the immediate postoperative period, monitoring of spinal cord function as well as "therapeutic stimulation" was carried out. To attain this goal, at the end of the operation a bipolar electrode (Type Cordis) was positioned infralesionally in the ventral epidural space and led out of the wound. Stimulation started on the first day after the operation and was carried out for half-an-hour with one hour intervals. A Stimucord External Neural Stimulator Model 910 A was used. The intensity of the stimulation was chosen in such a way that the patient, suffering from incomplete transverse lesion of the spinal cord with paraparesis, showed a small motor response, which was tested by a physician. Patients suffering from a complete transverse motor lesion of the spinal cord with paraplegia felt a phantom-like sensation. Intensity of stimulation in these patients was chosen in such a way that the patient was just able to feel this sensation. We have, until now, stimulated the patients for up to five weeks.

Results

Lesions in the Craniocervical Junction

1) Aneurysms of the Vertebro-basilar System

Aneurysm of the Posterior Inferior Cerebellar Artery (PICA). During the dissection of the vertebral artery and especially of the PICA, significant latency delay of the N20 and P28 waves took place. At the end of the operation this latency delay was improved again.

Aneurysm of the Vertebral Artery. The aneurysm was located on the point of confluence with the basilar artery. Prior to operations the patient was classified as clinical grade IV according to Hunt and Hess. At the beginning of the operation the N13-wave was slightly delayed (15.2 ms). In our opinion this was on account of the presence of blood in the prepontine cistern, due to the subarachnoid hemorrhage. The Erb's point potential was normal, indicating unchanged peripheral impulse conduction. Five minutes after preparation and clipping of the aneurysm a further delay of N13 to 16.5 ms was observed, causing delayed conduction of the afferent impulse to the cortical generating structures so that the N20 wave showed latency delay also (Fig. 1). In contrast to that the central conduction time (N13–N20) was normal (6.0 ms), indicating that the lesion was to be found in the brainstem. At the end of

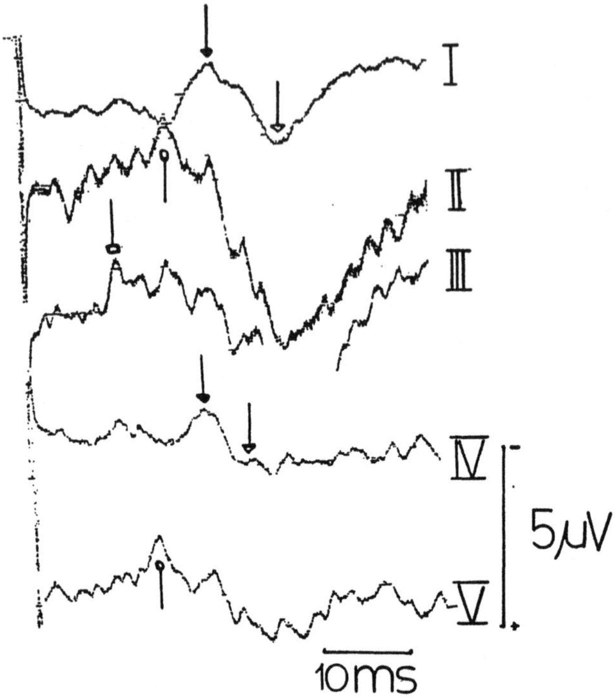

Fig. 1. Aneurysm of the right vertebral artery. Left median nerve stimulation. ↓ N20, ↓ P28, ↓ Erb's point potential, ♀ N13. *I, II, III,* 5 min after preparation and clipping of the aneurysm. All SSEP components except EP are pathologically delayed as a result of disturbance of the lower part of the brainstem, due to blood within the prepontine cistern as well as the operative manipulations on brainstem vessels. *IV, V,* 5 days after the operation. The patient is awake. The latencies of all components have diminished compared with the intraoperatively recorded responses

the operation direct stimulation of the spinal cord at the C5 level was carried out by means of a bipolar electrode. A large negative cortical evoked potential occurred after a latency of 9.6 ms, in addition to a spinal evoked potential after 6.8 ms. SSEP were repeated five days after the operation while the patient was awake (Fig. 1). At this time the N13 as well as the N20 potential latencies were normalized and the patient was in a good condition. This suggested that the brainstem irritation may have been a result of both blood within the prepontine cistern and the operative manipulation of brainstem vessels.

2) Tracto-nucleotomy and High Cervical Cordotomy

In three patients suffering from multiple sclerosis with consequent trigeminal neuralgia and in one patient suffering from epipharyngeal carcinoma with symptomatic trigeminal neuralgia, tracto-nucleotomy according to Sjöqvist-Kunc was carried out. In all these patients no change of the SSEP pattern occurred after the spinal cord incision (Fig. 2). In another patient suffering from malignant infiltration of the brachial and the cervical plexus (carcinoma of the thyroid gland), high cervical cordotomy was performed without change of the SSEP pattern after incision.

Lesions of the Cervical Medulla

In the patient with cervical disk herniation at C6/7 level the SSEP pattern was normal. The SSEP components of the patient with dislocation and fracture of the C5 body with a consequent quadriparetic syndrome showed a marked latency delay and

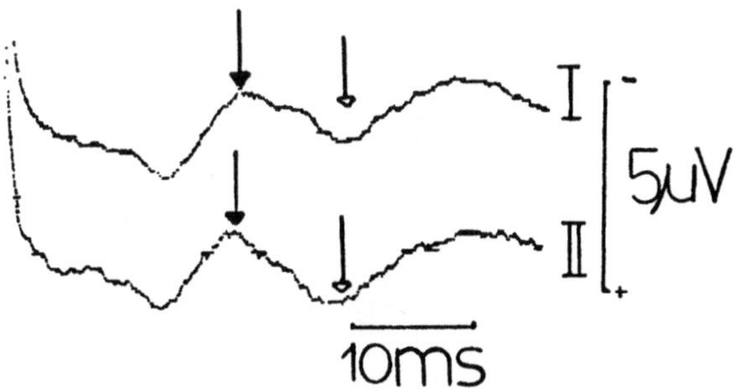

Fig. 2. Multiple sclerosis with consequent trigeminal neuralgia on the right side. Tracto-nucleotomy was carried out according to Sjöqvist on the right side. Right median nerve stimulation. ↓ N20, ⇣ P28. *I,* Scalp recorded potentials before the incision. *II,* Scalp recorded potentials after the incision. No change in latency and amplitude of the N20/P28 complex

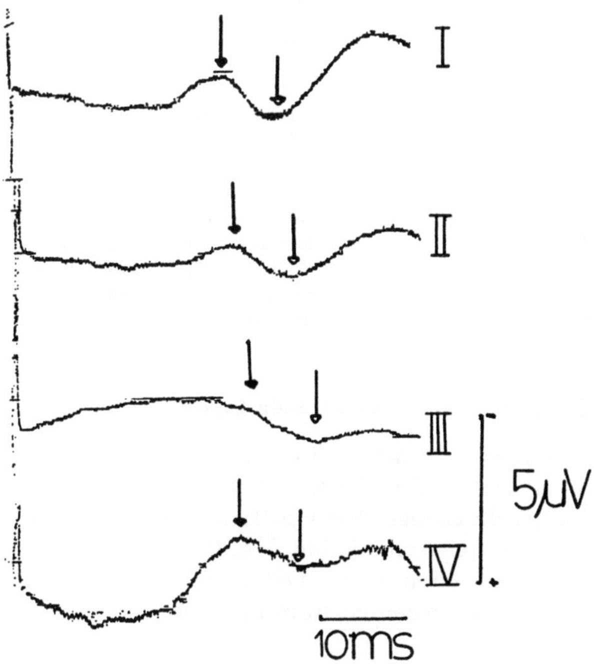

Fig. 3. Intramedullary ependymoma C4-C7. Intraoperative scalp recorded potentials at different stages of the surgical procedure. ↓ N20, ⇣ P28. *I,* At the beginning of the operation. The patient has been narcotised. *II,* After opening of the dura the SSEP pattern is almost unchanged, although latencies are slightly increased. *III,* During the final stages of the removal of the tumor some strong adhesions between the tumor and the spinal cord are to be resolved. The cortically recorded potentials become remarkably diminished and deformed. *IV,* 20 min after administration of 100 mg dexamethasone the latency-delay as well as the deformation and amplitude reduction show some tendency to recover

amplitude reduction, but no interruption of the neuronal continuity. In the postoperative period, improvement of the quadriparesis occurred to a very high degree. In one patient, operated on because of an intramedullary ependymoma (C4–C7), the intraoperative SSEP pattern showed a remarkable course: at the beginning of the operation prolonged latency of the N20 (23.4 ms) and the P28 (31.2 ms) components due to neoplasm was observed (Fig. 3). After opening of the dura a further increase of N20 and P28 latency took place. After a short period in which latencies were reduced, N20 and P28 became markedly deformed and reduced in amplitude during the final stages of the removal. By means of intravenous administration of cortisone the cortically generated potentials returned, but with considerable latency delay and amplitude reduction. Direct supra- and infralesional stimulation of the spinal cord confirmed the difference in neurophysiologic behaviour above and below the lesion.

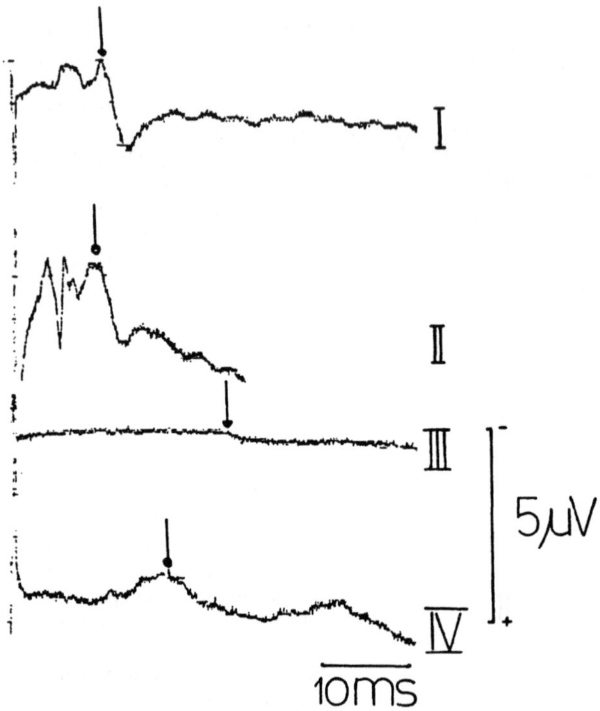

Fig. 4. Traumatic fracture of the vertebral body D12 with compression of the spinal cord. At the time of operation there was a complete motor transverse lesion of the cord with paraplegia. ↓ Scalp recorded potential, ↓ potential recorded at C7-level. *I, II, Supralesional* intraoperative direct stimulation of the spinal cord at D6 level. Cortical response with a negative peak at 11 ms followed by a positive peak at 14 ms (I). The spinogram reveals an initial spike wave followed by a biphasic potential with the negative peak at 9.5 ms and the positive peak at 13.8 ms (II). *III, IV, Infralesional* intraoperative direct stimulation of the spinal cord at L2-level. The scalp recorded potentials (III) as well as the potentials recorded at the C7-level (IV) show marked latency-delay, deformation and amplitude reduction

Lesions of the Thoracic Spinal Cord and of the Upper Cauda Equina

a) It was possible to define accurately the exact level of the lesion by recording supralesionally evoked potentials and moving the stimulating electrodes downwards, step by step. The pathologically changed SSEP pattern appeared suddenly and corresponded to the level of the spinal cord lesion (Fig. 4).
b) In order to assess the neuronal conduction of the medulla by means of infralesional stimulation, potentials may be recorded from the scalp and from the surface of C7 (Fig. 4).
c) In some cases with compression of the cauda equina due to fracture of parts of one or more vertebrae, no pathological sign could be found in the neurological examination. Yet the SSEP pattern may reveal latency delay, amplitude reduction and/or deformation in such cases.

Discussion

Lesions in the region of the craniocervical junction due to ruptured aneurysms of the vertebrobasilar system call for SSEP monitoring intraoperatively because of the dangerous disturbance of the microvascular supply of the brainstem due to blood in the prepontine cisterne. Studies of the venous part of vascularisation of the brainstem (Duvernoy 1975) as well as the arterial part (Duvernoy 1978) revealed a high degree of vascularity as well as close connection to the different brain structures. It was apparent to us that the disturbance of brainstem function appearing in severe subarachnoidal hemorrhage disappeared a few days postoperatively, because the blood was carefully washed out of the prepontine cisterne. This recovery of brainstem function was manifested, not only as an improved clinical condition but also as a return to normal latencies and amplitudes of the SSEP pattern. In our opinion, it is to be emphasized that evaluation of the CCT (Symon et al. 1984; Wang et al. 1984) is of importance, in addition to correlations between pathological changes of some SSEP components and the location of the brain lesion.

Intraoperative monitoring in the craniocervical region reveals a remarkable stability of the SSEP pattern; the SSEP components do not change after microsurgical incisions (for example tracto-nucleotomy according to Sjöqvist-Kunc), even when those are carried in the immediate neighbourhood of the gracile and cuneate nucleus.

This fact does not go without saying because, according to the studies of Albe-Fessard in 1966 and 1967, the so-called "spino-cervical" tract is part of the neospinothalamic tract. This tract is located in the immediate neighbourhood of the incision and belongs to the specific lemniscal somatosensory system. The axons of the laterally-positioned spino-cervical tract run to the lateral cervical nucleus at C1 level, as established by Rexed and Ström (1952). These findings were confirmed by Eccles et al. (1960). This suggests the conclusion that SSEP components are mediated through a very circumscribed pathway, which is not disturbed by microsurgical incisions even in the immediate vicinity. These neuroanatomical findings were first mentioned by Sjöqvist (1937) and later by Kunc (1962, 1965). The behaviour of SSEP pattern in the patient suffering from a large intramedullary ependymoma is astonishing. There is general agreement that SSEP are mediated by the dorsal columns. The marked disturbance of the spinal cord, and particularly of the dorsal columns, due to a large median microsurgical incision of the dorsal raphe with consequent removal of the tumor, caused latency delays, amplitude reductions and deformation of some components, but never caused total abolition. This corresponds to the good recovery in the postoperative course and emphasizes the importance of SSEP monitoring for postoperative prognosis. Such an assumption is confirmed by the studies of Ziganow and Rowed (1980) and Bricolo et al. (1976). These studies demonstrate good neurological recovery after spinal trauma, when SSEP components could be recorded.

In the thoracic and upper lumbar region only direct supra- and infralesional stimulation of the medulla was used. As mentioned above, the bipolar stimulating electrode was placed epidurally and ventro-laterally in the spinal canal. By this method we were able to stimulate not only the dorsal columns but also the motoneurons, producing muscle contractions in the lower extremities and in the abdominal muscle after the patient has been narcotised by means of neuroleptanalgesic drugs. Ventrolateral stimulation allows exact localisation of the level of disturbance also. Obviously, in some cases with a complete motor transverse lesion the SSEP pattern is very deformed and latencies are delayed, but consistent potentials can still be recorded. It is to be assumed that the dorsal columns and the pyramidal tract are electrophysiologically independent of each other. This assumption is confirmed by the findings of Levy (1983) and Owen et al. (1979). These findings prompted us to carry out "therapeutic stimulation" in 10 patients suffering from complete motor transverse lesions, but incomplete sensory transverse lesions, which had remained unchanged for a few months. Previously decompression of the medulla had been carried out via a transthoracic approach (Valencak 1979).

We could make the following observations: in many patients the sensory level of the lesion was lowered for up to six segments. Some patients revealed a small degree of motor activity and trophic ulcera disappeared. Most of the patients reported a feeling of warmness due to improved blood circulation in the paraplegic lower extremities. Sedgwick et al. (1983) made use of this phenomenon of "electrophysiologic sympathectomy" by means of epidural spinal stimulation in patients with vascular diseases of the lower extremities. Finally intraoperative examination of the neuronal conduction of SSEP components will be discussed briefly. In all patients for whom by means of the infralesional stimulation technique SSEP potentials could be recorded either over the scalp or over the surface of C7, recovery of neurological signs was later observed. Follow-up studies, however, revealed that there was a wide range of recovery from very good (able to walk without crutches) to poor, in which case motor function remained of no practical use to the patient.

References

Albe-Fessard D, Liebeskind J (1966) Origine des messages somatosensitifs activant les cellules du cortex moteur chez le singe. Exp Brain Res 1:127–146

Albe-Fessard D (1967) Organization of somatic central projection. In: Neff WD (ed) Contributions to sensory physiology, vol 2. Academic Press, New York

Bricolo A, Ore GD, da Pian R, Faccioli F (1976) Local cooling in spinal cord injury. Surg Neurol 6:101–106

Duvernoy H (1975) The superficial veins of the human brain. Springer, Berlin Heidelberg New York

Duvernoy H (1978) Human brainstem vessels. Springer, Berlin Heidelberg New York

Eccles JC (1960) Types of neurons in and around the intermediate nucleus of the lumbosacral cord. J Physiol (London) 154:89–114

Kunc Z (1965) Treatment of essential neuralgia of the 9th nerve by selective tractotomy. J Neurosurg 23:494–500

Kunc Z, Marsala J (1962) La localisation et la terminaison des voies afférentes des nerfs IX et X dans le bulbe. Acta Neurochir 10:512–522

Levy WJ (1983) Spinal evoked potentials from the motortracts. J Neurosurg 58:38–45

Owen MP, Brown HH, Spetzler RF (1979) Excision of intramedullary arteriovenous malformation using intraoperative spinal cord monitoring. Surg Neurol 12:271–276

Rexed B, Ström G (1952) Afferent nervous connection in the lateral cervical nucleus. Acta Physiol Scand 25:219–229

Sedgwick EM, Illis LS, Tallis RC (1983) Spinal cord stimulation in peripheral vascular disease. In: 1st Vienna International Workshop on Functional Electrostimulation. Vienna Bioengineering Laboratory

Sjöqvist O (1937) Eine neue Operationsmethode bei Trigeminusneuralgie: Durchschneidung des Tractus spinalis trigemini. Zbl Neurochir 2:274–281

Symon L, Wang AD, Costa e Silva IE, Gentili F (1984) Perioperative use of somatosensory evoked responses in aneurysm surgery. J Neurosurg 60:269–276

Valencak E (1979) Die vorderen operativen Zugänge zur Brust- und Lendenwirbelsäule. Acta Chir Austriaca 6:121–127

Wang AD, Cone J, Symon L, Costa e Silva IE (1984) Somatosensory evoked potential monitoring during the management of aneurysmal SAH. J Neurosurg 60:264–269

Ziganow S, Rowed DW (1980) The cortical somatosensory evoked potential in acute spinal cord injuries (Abstract). Symposium International. Applications Cliniques des Potentiels Evoqués en Neurologie, Lyon

Clinical Usefulness of Spinal Cord Evoked Potentials

H. Baba[1], I. Shima, K. Tomita, S.-I. Umeda, Y. Sawada, H. Yugami, U. Ugaji, K. Shinoda, N. Yoshimizu, S. Masuyama, K. Yonezawa, S. Tsuji, and S. Nomura

Summary

Spinal cord evoked potentials were recorded from the cervical or thoracic epidural space following spinal cord stimulation in 21 patients with cervical spondylotic myelopathy, 18 with ossification of the posterior longitudinal ligament and 7 with spinal cord injury due to fracture-dislocation. Thirteen cases were monitored during surgery. Amplitude or waveform abnormalities were detected in the majority of cases, and could often be used to determine the affected vertebral levels. Amplitude was sometimes greatly increased after decompressive surgery.

Introduction

Pathophysiological aspects of the research on spinal cord disease are increasing with modern recording equipment, but it is fair to say that much still remains to be done accurately to assess the damaged spinal cord. Since 1982, spinal cord evoked potentials have been studied in patients with various kinds of myelopathies.

One of the main purposes of the spinal cord evoked potential has been the correct determination of the vertebral level at which surgical intervention is indicated. Another has been the investigation of neurophysiological properties of the evoked potentials. There are several ways to monitor evoked responses, and the authors have employed the epidural recording method since the amplitude of the evoked response is usually large (up to 200 μV). The posterior tibial nerve is often used for stimulation, but the authors have directly stimulated the spinal cord epidurally in the thoraco-lumbar region in order to obtain a sufficiently large and stable evoked response. This report describes the electrophysiological characteristics of spinal cord evoked potentials in patients with cervical or thoracic myelopathy resulting from various spinal diseases, and the technique of electrodiagnosis of the affected levels in comparison with conventional neuroradiological techniques.

Material and Methods

Forty-six patients aged 30 to 74 years (mean 53 years) were investigated with respect to the difference between evoked reponse abnormalities and conventional neuroradiological findings. Twenty-one patients had cervical spondylotic myelopathy (CSM), 18 ossification of the posterior longitudinal ligament (OPLL), and 7 fracture-

[1] Department of Orthopaedic Surgery, School of Medicine, Kanazawa University, 13-1 Takaramachi, Kanazawa, 920, Japan

Spinal Cord Monitoring
Ed. by J. Schramm and S. J. Jones

dislocation of the spine with spinal cord injury. Thirty-seven cases were operated on and spinal cord evoked potentials were monitored during surgery in 13. In cases with OPLL, the neurological stage of myelopathy was more advanced than in cases with CSM, and 17 out of 18 required surgical intervention.

A bipolar epidural electrode with a 1 cm spacing placed in the thoraco-lumbar region, usually at T12–L1, was used for stimulation while the recording electrode was placed either in the thoracic or in the cervical region. The evoked response was mapped by stepwise placement of the recording electrode in a craniocaudal direction. The intensity of the stimulating current usually was set at just above the threshold to produce a minimal muscle twitch in the legs, but an intensity of 2 to 3 times threshold was often required to obtain a well defined evoked response. Rectangular impulses at 20 impulses per second and 0.2 ms duration were used for stimulation. Responses occurring within 50 ms following the triggered stimulus were amplified by a Medelec AA6 with a bandpass of 10 Hz to 10 kHz, and averaged 64 to 256 times with a resolution of 5 points per ms, by a Medelec DAV6 averaging computer.

Results

The conducted spinal cord evoked potentials (SEPs) recorded from the thoracic or cervical epidural space consisted of an initial spike wave (P1–N1) followed by slow negative-positive waves (P2–N2). Abnormal findings in spinal cord evoked potentials were defined as follows: latency increase at the injured vertebral segments; polyphasicity; amplitude reduction; positive-going waveform ("killed end" potentials). Seventy-two per cent of cases with CSM had various kinds of abnormal findings in spinal cord evoked potentials, whereas 89 per cent with OPLL and all cases with spinal cord injury had various abnormalities. The overall incidence of polyphasicity at involved vertebral segments was 43 per cent in CSM, 61 per cent in OPLL, and 14 per cent in spinal cord injury. Concerning the possibility of latency increase at the injured levels, this was difficult to identify since components with prolonged latencies might be concealed by normal components due to activity in slower fibers.

Twenty-nine cases out of 46 were appropriate for Hattori's classification of cervical myelopathy, and were divided into three groups: 2 with Type I, characterized as having only upper extremity involvement; 7 with Type II, with exaggerated deep tendon reflexes in the legs in addition to Type I symptoms; 20 with Type III, having sensory disturbance in the trunk and lower extremities besides Type II symptoms. In the Type III cases, electrophysiological diagnosis of the affected vertebral levels was possible in 70 per cent, whereas in the other types the diagnosis was relatively hard. The incidence of positive-going "killed end" potentials at the affected vertebral levels established by conventional neuroradiological pictures was 29 per cent in CSM, 78 per cent in OPLL and 43 per cent in spinal cord injury. In cases with OPLL, neurologically more advanced than those with CSM, the incidence of positive potentials was higher than in CSM. In the spinal cord injured patients, positive potentials were recorded from only 3 out of 7 patients, and no clear evoked potentials could be elicited in the other 4. Figure 1 shows typical positive potentials recorded from a case with an incomplete Brown-Séquard syndrome due to a herniated disc at C4-5.

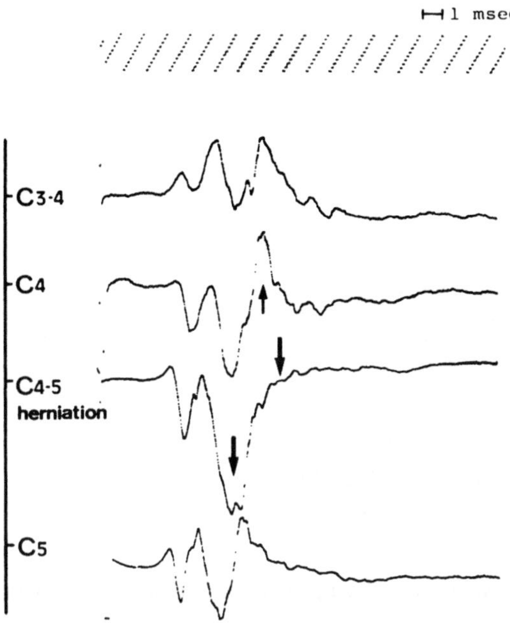

Fig. 1. Spinal cord evoked potentials recorded from a case with disc herniation at C4-5. Note typical positive deflection

Table 1. Peak-to-peak amplitude changes at the affected levels compared with neighboring levels (mapping of SEP amplitude). $n = 46$ responses

	CSM	OPLL	Spinal cord injury
Increased	14%	44%	14%
Decreased	43%	28%	57%
Unchanged	43%	28%	29%

Myelography demonstrated that the compressive filling defect was limited to C4-5, but a large negative potential following the second positive component was recorded from C4. It was assumed that the electrical conduction disturbance extended up to C4 in this case.

Table 1 shows a breakdown of 46 mapped responses, listing peak-to-peak amplitude changes at the affected levels compared with neighboring vertebral levels. In cases with CSM and spinal cord injury the amplitude at the involved levels was mostly decreased or unchanged. However, nearly half of the cases with OPLL (44 per cent) showed an increased amplitude at the involved levels.

Figure 2 shows the correlation between peak-to-peak amplitude changes and the blood pressure recorded during posterior decompression surgery for cervical OPLL from C2 to C7. SEP amplitude increased after the mean blood pressure was briefly reduced down to 50 mm Hg. In this case the amplitude increase (augmentation) was thought to be due to prolonged hypotension with a continuation of bleeding. The

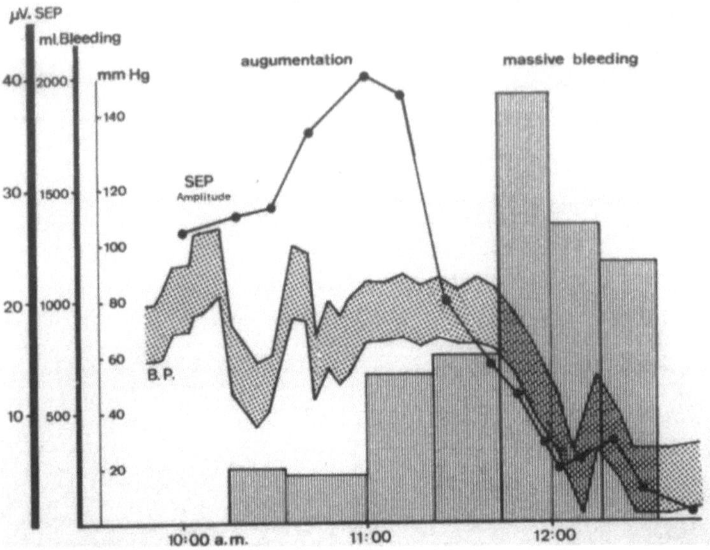

Fig. 2. Correlation between peak-to-peak SEP amplitude, blood pressure, and volume of bleeding during surgery

following rapid reduction of amplitude was thought to be due to massive bleeding which may have caused irreversible mal-circulation and pathological damage to the spinal cord. As mentioned before, compressive myelopathy may result in amplitude abnormality of the evoked response. However, one case with a disc herniation at C5-6 showed a large SEP amplitude at C5-6 and another with cervical OPLL showed a decreased amplitude at the level responsible for his myelopathy. Thus, it is difficult to relate amplitude changes to the determination of the vertebral levels responsible for myelopathy. Moreover, there was no relationship between amplitude increase or decrease, the type of the disease causing myelopathy or the type of myelopathy (Hattori) and the severity of the disease.

In one case with a burst fracture of the L2 vertebra associated with cauda equina injury, intra-operative monitoring of the evoked response was very successful. The preoperative peak-to-peak amplitude of the evoked response was less than 20 μV, but this increased to 120 μV after anterior decompressive surgery was completed.

Discussion

Spinal cord monitoring with epidural electrodes permits study and research on the spinal cord in healthy or diseased conditions. For cases in which the affected vertebral levels are hard to ascertain by conventional neuroradiological techniques, spinal cord evoked potentials provide a better opportunity for functional assessment.

In contrast to the posterior tibial nerve stimulation technique described by Jones et al. (1984), the authors have used the epidural stimulation and recording technique in order to obtain a larger and more easily recorded evoked response. Indications for the use of this technique, however, must be carefully considered an account of its invasiveness compared with other methods. McNeal et al. (1984) described a further method in which SEPs were recorded from a Kirschner wire firmly inserted into the spinous process, and demonstrated highly reproducible evoked responses in their series of scoliosis cases. In the present study, one case with multiple level CSM was monitored using this technique on account of failure of the epidural recording during surgery. The peak-to-peak amplitude of the evoked response before surgery was $10\,\mu V$, and this increased to $40\,\mu V$ after the laminoplasty (Kimura et al. 1984) was completed. The evoked response from bone was somewhat less well defined than that from the epidural space, but these records indicate an alternative possibility for intra-operative spinal cord monitoring.

Jones et al. (1984) emphasised the importance of amplitude analysis in their scoliosis correction series. In the present study, however, it was very hard to utilize mapping of SEP amplitude for determination of the affected vertebral levels in cervical or thoracic myelopathy. Identification of positive "killed end" potentials can be of importance in the neurophysiological assessment of electrical conduction in the injured spinal cord (Baba et al. 1985). Deecke and Tator (1973) found that positive potentials produced by the injury current were very sensitive indicators of spinal cord conduction disturbance. These positive potentials were a very useful and reliable parameter in the precise determination of the sites of functional conduction disturbance, and their incidence was 29 per cent in cases with CSM and 78 per cent with OPLL. Although these positivities were useful, further investigation is needed in order to explain the low incidence in CSM and to establish the correlation with morphological and/or physiological damage to the spinal cord.

In conclusion, spinal cord monitoring using epidural electrodes is one of the most useful diagnostic tools for the physiological determination of the vertebral levels responsible for myelopathy.

References

Baba H, Shima I, Nomura S (1985) Clinical application of cauda equina action potentials (in Japanese). Cent Jap J Orthop Traumat 27(4):21–23

Deecke L, Tator CH (1973) Neurophysiological assessment of afferent and efferent conduction in the injured spinal cord of monkeys. J Neurosurg 39:65–74

Jones SJ, Edgar MA, Ransford AO (1984) Epidural recording of spinal SEPs following posterior tibial nerve stimulation in man. Conduction properties of normal human spinal cord and changes occurring during scoliosis correction. In: Homma S, Tamaki T (eds) Fundamentals and clinical application of spinal cord monitoring. Saikon Publishing, Tokyo, pp 223–236

Kimura I, Oh-hama M, Shingu H (1984) Cervical myelopathy treated by canal-expansive laminaplasty: Computed tomographic and myelographic findings. J Bone Joint Surg 66-A:914–920

McNeal DR, Swank S, Satomi K, Passoff T (1984) Comparison of spinal evoked potentials recorded from bone and the epidural space. In: Homma S, Tamaki T (eds) Fundamentals and clinical application of spinal cord monitoring. Saikon Publishing, Tokyo, pp 77–86

Spinal Cord Monitoring in Neurosurgery in the United States

T. B. Ducker[1], R. W. Thatcher, D. S. Cantor, and R. McAlaster

Summary

The clinical use of evoked potentials to monitor spinal cord problems in the neurosurgery community of the United States is becoming more and more common. Unfortunately, it is currently utilized in less than 1% of our cases. Its utilization as a diagnostic tool, during vascular manipulation of the spinal cord, in spinal corrective operations and during intrinsic cord procedures have all been reported. Not only have the sensory functions been assessed, but more recently techniques to stimulate the brain without causing seizure activity and record motor responses down the spinal cord have been developed. Once all these techniques become more standardized, not only in nomenclature but also in wave form, then their utilization will be more widely accepted. In those patients who have minimal neurologic deficit with maximum complaints, the diagnostic utility of evoked potentials is respected. In those patients who have devastating neurologic problems, the extent of the pathology can be measured. The frailty of the vascularity to the spinal cord and its influence on spinal cord conduction is now more than ever appreciated. Soon, it will be easier to do certain operative manipulations with reassurance that the spinal cord will not be irreparably damaged.

Introduction

In the clinical practice of neurosurgery in the United States, electrophysiological recording from the spinal cord has been well tested but is not commonly used. For example, less than 1% of the cases treated by the neurosurgery community have had either pre-operative, intra-operative, or post-operative study of the conduction within the spinal cord. Reliance on the history, clinical examination, X-rays, computerized tomography, as well as the various myelographic techniques is the mainstay of clinical neurosurgery as it applies to treatment of disorders of the spinal cord and its nerve roots. Even though the actual testing is not common in the majority of cases, its clinical utilization has been well reviewed by many centers in the United States. Most of these centers are tied to Universities; however, there are select physicians in private hospitals that have also pursued the tests and use them in their practice. From all of this work there has been a considerable amount of literature recently, published not only in our local journals but in the three journals that concentrate on neurosurgery in the United States.

The use of spinal cord monitoring primarily by evoked potential testing can be subdivided into (a) diagnostic tests, (b) vascular manipulation of the spinal cord, (c)

[1]Neurosurgery, University of Maryland, 100 Cathedral Street, Annapolis, MD 21401, USA

Spinal Cord Monitoring
Ed. by J. Schramm and S. J. Jones

spinal corrective operations, and, (d) spinal cord decompressive and/or manipulative procedures. Each of these four separate categories can be discussed when one considers an overall view of spinal cord diagnostic testing by the current monitoring techniques.

Before proceeding with that, the importance of normative data cannot be overemphasized.

In each of the various laboratories across the United States, standards have been well worked out in healthy subjects. An overall view of normative data recently published by Lastimosa et al. (1982) is indicative of such a study. That particular study, carried out at the University of Kentucky, published representative wave forms of the various evoked responses, primarily concentrating on those seen after posterior tibial nerve stimulation at the ankle. The various wave forms that can be measured include: the popliteal fossa, negativity at 8 ms; at the L4 lumbar vertebra, negativity at 22 ms. These various latencies are dependent on the stature of the person. However, once the conducted volley enters the spinal cord the response amplitude and latency seem to be independent of the patient's height. In the wave form recorded from the scalp there seem to be rather consistent responses at 37 ms with a positive deflection and at 45 ms with a negative deflection. In many laboratories a negative response is represented by an upward deflection (the "traditional" method) and positive downward on the oscilloscope. On the other hand, much of the American literature is done with positive as an upward deflection (the "purist" method). Because of these variations, there is a considerable amount of confusion as to exactly what the wave form should or should not look like. This has led to such designations as P1, N1, P2, N2, wherein the P and N refer to positive and negative and the first numbers refer to the actual peak itself. On the other hand, such designations as P37 refer to a positive wave at 37 milliseconds. Unfortunately, much of the literature in this hemisphere is confused by the lack of consistency and standardization in not only the interpretation of the waves, but also the actual initials applied to the evoked potentials themselves. For example, while EP often means evoked potentials, the S in front of it can mean spinal, sensory, somatosensory, etc.

While there have been a considerable number of publications with various adaptations of normative data, the wide acceptance of these tests has been hindered because of the lack of uniformity in the presentations, nomenclature, wave forms, etc. Nevertheless, the clinical use of these tests has found its way into our pre-operative, intraoperative, and post-operative care of patients.

Evoked Potentials in Spinal Cord Disease

The application of evoked potentials as a diagnostic test in patients with spinal cord disease has been evaluated in a variety of disorders. The initial studies were commonly done in trauma in such centers as South Carolina (Perot 1973), and in Toronto (Rowed et al. 1978). Once these initial evaluations had been published, the utilization of evoked potentials in intrinsic cord disease, spinal cord compressive disorders, craniovertebral abnormalities, and evaluation of the damage to the cord post trauma became increasingly widespread.

At the same time that these diagnostic tests were being done in patients, basic laboratory work was being carried out to define exactly which components of the evoked response correspond to which tracts within the spinal cord. It was clear from all that the primary response of a cortical evoked potential after stimulation of a peripheral nerve was carried in the dorsal columns of the spinal cord. This applies to the major first positive and negative deflection, as well as various deflections after that. As time elapsed, it was clear that not only the dorsal columns participate in the overall wave form as seen at the scalp, but that certain aspects of the anterolateral columns, to which we ascribe pain, temperature, touch sensation etc., contribute to the longer latency responses. The third positive and third negative waves definitely seem to be influenced by the anterolateral columns as opposed to the dorsal columns. Long term experimental material reported by Simpson et al. (1983) confirms the same in animals.

In evaluating intrinsic cord disease, multiple sclerosis is the most common disorder for which a patient will be tested. Multiple sclerosis cases comprise a chronic population which is always available, and individual patients often exhibit change in their complaints without a drastic change in their neurologic findings. In this patient population, significant changes within the evoked potentials recorded from the scalp are often seen. In patients with nebulous complaints, a good healthy evoked potential is often reassuring. In our own clinical center patients with hysteria have been examined, primarily to reassure them that their complaint of numbness will clear because their cortex is indeed receiving the messages.

In spinal cord compressive disorders, the most common being cervical spondolytic myelopathy, the evoked potential is often altered. The primary positive and negative response may be greatly diminished. This first affects the amplitude of the wave form which is recorded following stimulation of peripheral nerves in the leg. Spondolytic myelopathy will alter the median nerve evoked potential much less. The alteration of wave form is directly in proportion to the patient's neurologic deficit. In patients where the leg amplitude is less than half that of the arm, this abnormality is a good indication of a posterior column disorder in a compressive myelopathy. Dorfman and coworkers (1980) were among the first to point this out and this comparison ratio has withstood the test of time in clinical studies. Often these spondolytic myelopathy patients will describe a diffuse numbness in the lower half of their body, extending up to the thoracic level, while clinical testing will show only minor diminution of the joint-position and vibratory senses. The evoked potentials can be quite abnormal, along with the finding of diffuse hyperreflexia and spastic movements in the lower extremities. Whether the myelopathy is spontaneous or partly induced by trauma, the abnormal recordings are basically the same. If there is a devastating neurologic deficit, the standard cortical evoked potential response is absent or severely attenuated.

Certain patients have abnormalities at the craniovertebral junction. Chronic malformation of the odontoid process with subluxation of C1 on C2 is the most common example of this kind of disorder. This can occur on a congenital, traumatic or rheumatological basis. Again this patient population will present with a diffuse hyper-reflexia and difficulties with rather vague sensory abnormalities. In these patients, the cortical evoked response is often abnormal.

In all patients, the overall conduction velocity should be in the range of 50–60 meters per second. If the conduction through the spinal cord to the cortex is delayed, and the first latencies are delayed to greater than 45 ms, then clinical pathology is present in the majority of cases. From all studies available, conduction time in the spinal cord is fairly constant in a normothermic healthy individual. Conduction time is only delayed for a specific wave form when there is hypothermia or certain anesthetics are employed.

In spinal cord trauma the evoked potential may be lost initially even for those animals or people in whom there is a good recovery (Ducker et al. 1978), but a rapid return with the appearance of a normal evoked response is indeed a good sign. This has been reported by Rowed et al. (1978) and has been confirmed by other centers such as New York (Young 1982). The presence of even a weak somatosensory evoked potential early in the recovery phase is definitely a favorable prognostic sign. It is a good predictor of return of function, and appears to be an even better predictor than a neurologic examination. However, after six to eight weeks, the neurologic examination itself is the best predictor of how well a patient is going to do. In some of those patients, there can be a favorable return of evoked potentials only for the neurologic deficit to remain rather severe.

Recently, not only sensory evoked potentials have been used, but motor responses as well. The recent studv by Levy et al. (1984) is a significant break-through in recording motor evoked responses. By placing cranial stimulating plates over the motor cortex, along with a plate in the mouth on the palate, and recording between the spinous processes over the spinal cord, evoked responses can be consistently produced. These can be readily recorded from the cervical, thoracic, and lumbar area. They correlate beautifully with a motor neurologic deficit, especially when the lesion is complete. Again, the conduction velocity is rapid, being greater than 60–70 meters per second. With the development of this newer test, published only late in 1984, it is now possible comfortably to record not only ascending impulses in the sensory system but also descending impulses which are most likely to be propagated in the motor system. By using both the motor and sensory stimuli, the exact extent of pathology within the spinal cord can now be defined. Figure 1 is an example of such a sensory examination wherein the significant spinal cord pathology begins at C6, and by study of the descending pathways we know the upper limit of the pathology to be clearly at C4.

Vascular Manipulation of the Cord

The vascularity to the spinal cord is important in the maintenance of conduction through the cord itself. The spinal cord seems to be unduly sensitive to any significant vascular manipulation. This seems to be over and above the usual ischemic effects we see from the cortical evoked potential while carrying out carotid surgery and stimulating the median nerve. The spinal cord responds adversely within seconds of any type of significant ischemic effect. As recently pointed out by Berenstein et al. (1984), angiography of the vessels feeding the spinal cord that only transiently robs the cord of its nourishment can drop the somatosensory evoked potential amplitude transiently, for 2–4 min, in over 85% of the cases when a major spinal medullo-radicular artery is injected. The classic example is the greater radicular artery, known as the Artery of Adamkiewicz. When this vessel is injected, there is a marked depression of evoked potentials for 2–5 min in the majority of cases. In studying

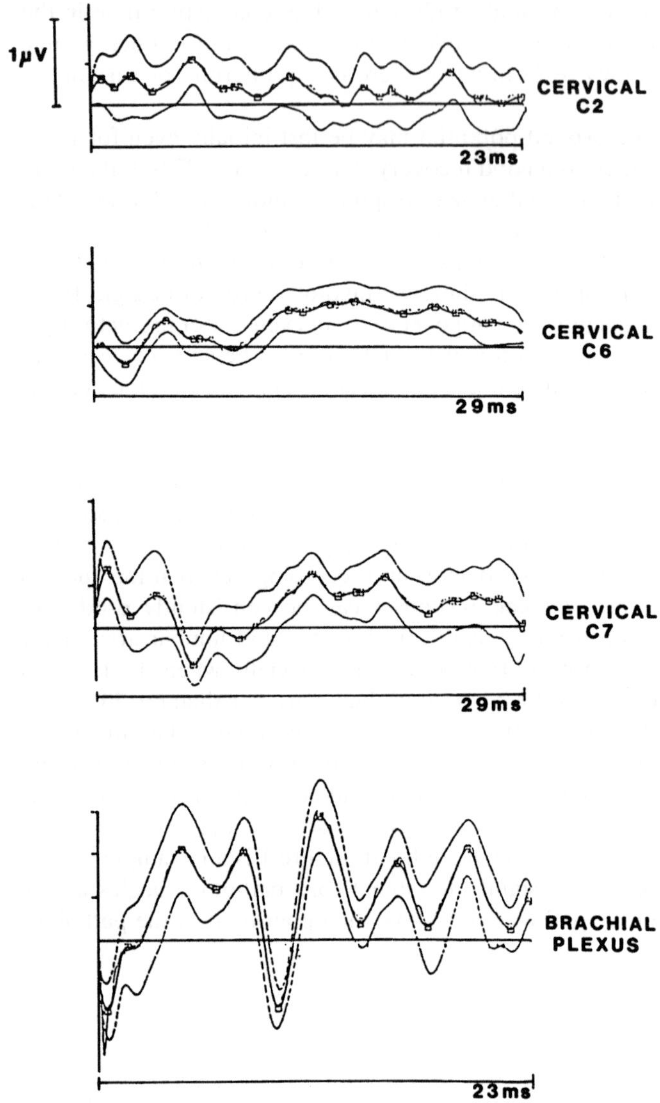

Fig. 1. Somatosensory recordings from leads placed over brachial plexus and spine. Patient R.L. is a 29-year-old white male who fell backwards off a boat dock, striking his back or neck on something in the water. Patient is complaining of severe neck pain and is unable to move arms or legs. Infarct at C4-C5

patients where there is vascular malformation of the cord or adjacent bones, injecting the vessel which feeds not only the vascular malformation and/or tumor but also the cord causes a depression of the evoked potential. This identifies the vessels at high risk in embolization. By knowing these vessels, the embolization procedures can be more carefully planned and periodically carried out to minimize the risk of a devastating neurologic deficit to the patient.

In patients who have huge aortic aneurysms, during the angiography to demonstrate the aneurysmal dilatation the evoked potentials may become markedly depressed. It is in these patients that the vascular supply to the spinal cord may be impaired. It is the same group of patients who run a high risk of being paraplegic following aortic replacement. In these patients with great vessel disease, it is often impossible to identify in the operating room the exact feeding vessel to the spinal cord. It may be that in time the important small feeding vessel will be identified and maintained so that neurologic function is sustained.

Vascular maintenance as a function of perfusion pressure, best measured by blood pressure, is rapidly becoming recognized as the single most important parameter that may influence spinal cord conduction. In patients with cervical spondylosis and impaired neurologic function, a drop of the blood pressure within the procedure without any other manipulation may erase what remaining wave forms of the evoked potentials that exist. This has been witnessed on several occasions in the operating room. As emphasized by Grundy (1982) cord compression and/or stretching with hypotension may erase a measurable evoked potential, which can readily be recovered by restoring a healthy perfusion pressure.

In certain patients the position of the neck may interfere with the vascularity of the cord and change the evoked potential. This has been seen in positioning the patients in the operating room, and does influence the choice of operative positions, such as a sitting position. In fact it has so influenced certain centers, such as our own, that patients with significant cervical myelopathy undergoing decompressive procedures are no longer placed in the sitting position. It is more important to maintain their blood pressure, maintain traction, and operate without loss of blood pressure in order to maintain perfusion in a compromised cord. Experimental studies in the past, on several occasions, have indeed confirmed this finding, for example Hukuda and Wilson (1972).

Spinal Corrective Procedures

In operations wherein there has been a correction of a spinal deformity, the measuring of sensory function as recorded by the cortical somatosensory evoked potentials has been found to be extremely useful in the evaluation of overall cord function. For some reason, even though the evoked potential is primarily mediated by the dorsal columns and this response is primarily a sensory response, maintenance of these evoked potentials is a good reflection of the overall health of the cord.

Vascular impairment should infarct primarily the anterior aspects of the cord, but if such occurs it seems that the evoked potential is also altered. In scoliosis surgery, where Harrington rods are used, it appears that one no longer needs to use the wake-up test in by far of the majority cases. By using balanced anesthesia of nitrous oxide, narcotics and Pancuronium and avoiding Halothane, Enflurane, etc., the evoked potentials are not suppressed, so that they are readily recorded throughout the procedure. In the mid and late 1970's, Engler et al. (1978) used the procedure in 55 patients without suffering a devastating neurologic deficit and avoided the problems associated with a wake-up test. Simultaneous with these studies Nash et al. (1984), who are reporting their recent data in this book, utilized the intra-operative monitoring technique not only in scoliosis but also in fractures, spondylosis, etc. They have pointed out the importance of maintaining blood pressure and perfusion pressure in order to maintain a satisfactory evoked potential. Maccabee et al. (1983) recorded evoked potentials not only from the scalp, but also from pins placed within the vertebrae. By studying evoked potentials both pre- and post-distraction, they were able accurately to assess cord conduction in all of their patients. In 2 of 43 patients, the evoked potentials were abolished during distraction. After easing of the distraction force, the evoked potential recovered. In each of these cases, the patient's clinical course was favorable. Maccabee et al. (1983), along with other authors, have pointed out not only the importance of hypotension, but also the possibility that hypothermia during a prolonged case may unfavorably alter the evoked potential.

Cord Decompressive Operations

In spinal cord decompressive procedures performed by the neurosurgical community in the United States, the use of evoked potentials is increasing but, as pointed out earlier, probably less than 1% of our patients are so monitored. In many of the patients the pre-operative status, including motor and sensory deficits, alters their evoked potentials so that they are abnormal when the patients are first seen. With this abnormal evoked potential, intra-operative recordings are more difficult. Reliable recordings can be obtained from scalp electrodes in only 75 to 85% of such patients. When these recordings are available, they do aid in the operative care. For example, cord arteriovenous malformations attacked through a posterior laminectomy can be treated by progressive occlusion of feeding vessels with assurance of good neurologic prognosis if a significant evoked potential is maintained. Because of the problems in obtaining consistent and reliable cortical evoked potentials in these neurologically handicapped patients, the neurosurgeons have turned to the techniques commonly used by our Japanese colleagues, which are reported in various chapters of this book.

Hahn et al. (1981) and others have relied on epidural recordings, obtained by placing a needle electrode between the spinous processes down under the lamina. With a reference electrode located in the skin over the recording electrode, the evoked potentials along the spinal column can be readily recorded. By placing the electrodes above and below the operative site, one can record the incoming volley and the volley propagated through the operative area. This technique has become more and more established and is described in our own clinical neurosurgery handbook.

At the same time in Boston, Macon and Poletti (1982) developed intra-operative epidural recording electrodes. Once the operation had been initiated and the laminectomy started, intra-operative electrodes were placed both rostral and caudal to the laminectomy site. These electrodes were separated by more than 30 mm. Their positions were confirmed by X-ray, so that the actual position and distance between them could be appropriately measured. Then lower extremity peripheral nerve stimulation was carried out so that there could be accurate recordings along the spinal cord. Using balanced anesthesia with Pavulon, the various evoked potentials could be consistently recorded. Macon and Poletti (1982), along with other authors, recorded the fact that on entering the lumbar spinal cord the evoked potential tended to be monophasic with a sharp negative wave. The more rostral (or the more cervical) the recording is, the more polyphasic is the wave form and the less the amplitude. The same authors felt there was a difference in the conduction velocity, being only approximately 40 m/s in the lumbar area but ascending to a speed of greater than 65–75 m/s in the cervical area. A change in conduction velocity has not been found to be always the case in our own laboratory, for we feel that a good healthy conduction velocity within the spinal cord is always greater than 60 m/s. However, the data from Boston still seems impressive.

The same authors, working with the clinical neurosurgeons, applied this intra-operative epidural evoked potential technique in cases of tumor, arteriovenous malformation and spondylosis, and also in pain procedures (Macon et al. 1982). While in 85% of these patients there was no significant change in the evoked potential and no significant change in the post-operative function, the fact remains that 15% of the patients had a marked alteration of the evoked potential along the spinal cord during operation. When there was an intra-operative change in these 15% of the patients, a post-operative neurologic deficit occurred in each case. Therefore, the spinal evoked potential accurately predicted the post-operative neurologic worsening. We have had the same experience, and the reliability of their technique has become more and more accepted.

Not only have there been epidural recordings, but more recently there have been intra-operative subdural recordings. As reported by Whittle et al. (1984), proper electrode development is possible such that they can be safely placed beneath the dura on the spinal cord itself and record highly reproducible responses to peripheral nerve stimulation. However, subdural recording does not seem to be necessary in the majority of cases.

References

Berenstein A, Young W, Ransohoff J, Benjamin V, Merkin H (1984) Somatosensory evoked potentials during spinal angiography and therapeutic transvascular embolization. J Neurosurg 60: 777–785

Brown RH, Nash CL (1984) Implementation and evaluation of intraoperative somatosensory cortical potential procedures and pitfalls. In: Homma S, Tamaki T (eds) Fundamentals and clinical application of spinal cord monitoring. Saikon Publishing, Tokyo, pp 373–384

Dorfman LJ, Perkash J, Bosley TM, Cummins KL (1980) Use of cerebral evoked potentials to evaluate spinal somatosensory function in patients with traumatic and surgical myelopathies. J Neurosurg 52: 654–660

Ducker TB, Salcman M, Lucas JT (1978) Experimental spinal cord trauma. II: Blood flow tissue oxygen, evoked potentials in both paretic and plegic monkeys. Surg Neurol 10: 64–70

Engler GL, Spielholz NI, Bernhard WN, Danziger F, Merkin H, Wolff T (1978) Somatosensory evoked potentials during Harrington instrumentation for scoliosis. J Bone Joint Surg 60-A: 528–532

Grundy BL (1982) Monitoring of sensory evoked potentials during neurosurgical operations: Methods and applications. Neurosurgery 11: 556–575

Hahn JF, Lesser R, Klein G, Lüders H (1981) Simple technique for monitoring intraoperative spinal cord function. Neurosurg 9: 692–695

Hukuda S, Wilson CB (1972) Experimental cervical myelopathy: effects of compression and ischemia on the canine cervical cord. J Neurosurg 37: 631–652

Lastimosa ACB et al. (1982) Lumbar spinal cord and early cortical evoked potentials after tibial nerve stimulation: Effects of stature on normative data. Electroencephalogr Clin Neurophysiol 54: 499–507

Levy WJ, York DH, McCaffrey M, Tanzer F (1984) Motor evoked potentials from transcranial stimulation of the motor cortex in humans. Neurosurgery 15:287–302

Macon JB, Poletti CE (1982) Conducted somatosensory evoked potentials during spinal surgery. Part I: Control conduction velocity measurements. J Neurosurg 57:349–353

Macon JB, Sweet WH, Ojemann RG, Zervas NT (1982) Conducted somatosensory evoked potentials during spinal surgery. Part II: Clinical applications. J Neurosurg 57:354–359

Maccabee PJ, Levine DB, Pinkhasov EJ, Cracco RQ, Tsairis P (1983) Evoked potentials recorded from scalp and spinous processes during spinal column surgery. Electroenceph Clin Neurophysiol 56:569–582

Perot PL (1973) The clinical use of somatosensory evoked potentials in spinal cord injury. Clin Neurosurg 20:367–381

Rowed DW, McLean JAG, Tator CH (1978) Somatosensory evoked potentials in acute spinal cord injury: Prognostic value. Surg Neurol 9:203–210

Simpson RK Jr, Blackburn JG, Martin FH, Katz S (1983) Long term effects of spinal cord lesions on the configuration of somatosensory evoked potentials. Part I: Alterations in waveform characteristics. Clin Electroenceph 14:142–151

Whittle IR, Johnston IH, Besser M (1984) Spinal cord monitoring during surgery by direct recording of somatosensory evoked potentials. J Neurosurg 60:440–443

Young W (1982) Correlation of somatosensory evoked potentials and neurologic findings in spinal cord injury. In: Tator C (ed) Early management of acute spinal cord injury. Raven Press, New York, pp 153–165

Atraumatic Cord Lesions

Conductivity in Elastically Stretched and Relaxed Spinal Cord Evaluated by Somatosensory Evoked Potentials

F. L. GLÖTZNER[1] and R. NETAL

Summary

The effect of neck flexion and extension on early components of the tibial nerve cortical evoked potential was studied in healthy subjects and in multiple sclerosis and scoliosis patients. No significant latency differences were demonstrated, and there was consequently no evidence of an acute decrease of cord conductivity produced by stretching of axons during flexion.

Introduction

Recently it has been postulated by Breig et al. (1982) that intramedullary tension influences axonal conductivity. According to this hypothesis tension will lead to a decrease and ultimately to a cessation of conductivity in the spinal cord. As the authors put it, "the soft medullary tissue near the lesion is exposed to adverse tension arising from the relative hardness and stiffness of the lesion itself". Breig (1984) therefore proposed a surgically maintained spinal cord relaxation, the so-called cervicolordodesis, whereby neurological deficits in multiple sclerosis and cervical myelopathy were seen to disappear. This biomechanical relaxation therapy was derived from detailed experimental studies on traumatized spinal cord (Breig et al. 1982). In addition the authors performed measurements on human cadaveric cord showing a difference of five to seven centimeters in length between maximum extension and full flexion of the vertebral column.

In order to test the above hypothesis, we investigated the conductivity of the spinal cord by means of somatosensory evoked potentials. The first part of the study deals with the question of conductivity changes in the intact cord during maximum extension and full flexion of the spine. The second part considers this question in patients with multiple sclerosis. Finally, in the third part the results of positional changes in scoliotic individuals are reported.

Methods

Somatosensory potentials were evoked in 32 healthy adults, comprising 15 men and 17 women. They were between 17 and 55 years of age, with 19 persons between 20 and 30 years and six older than 40. Five patients with multiple sclerosis—three men and two women— were between 24 and 60 years of age. Three of them were rated 6 on the scale of Kurtzke (1965), with assistance required for walking (inability to

[1]Neurologische Abteilung am Krankenhaus Rummelsberg, D-8501 Schwarzenbruck

Spinal Cord Monitoring
Ed. by J. Schramm and S. J. Jones

work without special provisions) one 5 and one young lady 2 (minimal disability). The progression rate varied between 0.22 in one and higher rates up to 0.5 in the others, i.e. the chronic progressive course of the disease predominated.

The electrophysiologic examination was performed in two different positions: maximum extension of the cervical spine in a comfortable supine position and maximum bending by bringing up the knees to the head in a kind of fetal posture. For some of the severely incapacitated patients maximum flexion or extension was impossible.

Somatosensory potentials were evoked by light electrical stimulation of the tibial nerve at the ankle, using a DISA Neuromatic 2000. Rectangular pulses of 200 microseconds duration at a frequency of 2 Hz were applied.

The stimulus intensity did not exceed 15 milliampères. Following the International 10-20 system, two small steel needle electrodes were inserted into the scalp. The active electrode was placed between C_z and P_z and the reference electrode at F_z. Amplification was performed with an upper frequency limit of 2 kHz and a lower limit of 2 Hz. 200 responses were processed in the averager. The amplitude of the responses rarely exceeded 5 microvolts.

Evoked responses were accepted if two successive trials showed only minor variations in waveform and variations of no more than 1–2 ms for the early components. Two trials were performed in either posture.

Results

Healthy Adults. 32 healthy adults were divided into two groups, one (16 subjects) with a body height of less than 175 cm and one (15 subjects) with a height of more than 175 cm. One person measured exactly 175 cm and was therefore excluded. The latency values in Table 1 are given as means and standard deviations. Standard deviations were larger in the taller group, and for SEP components with longer latencies. The results shown in Table 1 demonstrate the increased latencies for all early components in the taller individuals. There was no statistically significant difference between the latencies during maximum extension and maximum flexion. This was true for both groups—the smaller and the taller individuals.

Multiple Sclerosis. In five patients suffering from multiple sclerosis (Table 2) the early components of tibial nerve evoked potentials were recorded. Component N_1 could not be clearly discerned in all cases and was therefore omitted. Components P_1 and N_2 were analysed. With the exception of one value—N_2 in patient H.N.—all latencies increased slightly during bending of the spine. The maximum increase amounted to two milliseconds for component N_2 in patient H.F. These differences are not statistically significant (for P_1 $p<0.1$, 2-tailed) on account of the small size of the sample, but the tendency to increased latencies during flexion is suggestive. An example of the type of potentials recorded and the alteration induced by postural change is shown in Fig. 1, referring to patient B.K. in Table 2.

It may be noted in this context that no correlation was found between the degree of the disability and the latency of early components. However, absence of somatosensory potentials following tibial nerve stimulation was associated with high ratings

Table 1. Latency of tibialis evoked potentials in healthy adults during maximum extension and flexion of the spine

Stature	Component	Extension (ms)	Flexion (ms)
Under 175 cm ($n=16$)	N_1	31.8 ± 0.8	32.0 ± 0.6
	P_1	38.3 ± 0.9	38.3 ± 0.8
	N_2	47.3 ± 1.6	47.3 ± 1.6
Over 175 cm ($n=15$)	N_1	34.8 ± 1.4	35.2 ± 1.3
	P_1	40.9 ± 1.8	41.3 ± 1.7
	N_2	49.8 ± 2.0	50.0 ± 2.0

Table 2. Latency of tibialis evoked potentials in patients with multiple sclerosis during extension and flexion

Patient	Component	Extension (ms)	Flexion (ms)
H.K.	P_1	57.0	58.0
	N_2	67.8	68.0
H.N.	P_1	50.0	51.6
	N_2	58.6	58.4
B.K.	P_1	51.2	51.8
	N_2	59.2	60.4
H.F.	P_1	43.2	44.6
	N_2	54.0	56.0
I.K.	P_1	59.8	61.4

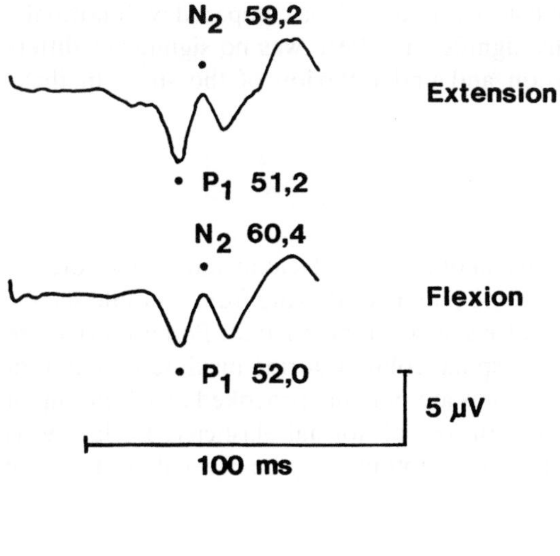

Fig. 1. Latency of tibialis evoked potentials in a patient (B.K.) with multiple sclerosis during extension and flexion

Table 3. Latency of tibialis evoked potentials in scoliotic and healthy individuals during maximum extension and flexion (see text)

Group	Com-ponent	Extension latency/height	Flexion (ms/m)
Healthy adults ($n = 32$)	P_1	22.9 ± 0.8	23.0 ± 0.8
	N_2	28.2 ± 1.0	28.3 ± 1.0
Scoliotic patients ($n = 7$)	P_1	24.2 ± 3.4	24.3 ± 3.2
	N_2	29.4 ± 1.7	29.8 ± 2.2

of 7 and 8 on the disability scale (wheelchair- or bed-ridden) in an additional group of 6 patients with multiple sclerosis not included in this study. The impossibility of evoking potentials in these patients might be primarily due to a particular susceptibility to flexor or extensor spasms.

Scoliosis. Seven patients with scoliosis underwent the same procedure as the two groups of persons already described (Table 3). The angle of dorsal scoliosis was found to range from 30 to 98 degrees. In five adolescents the values were between 30 and 40 degrees, in a young woman of 21 years it amounted to 62 degrees and in a 73-year-old woman to 98 degrees. All seven patients were female. The rigidity of the spine prevented maximum flexion in most cases.

In order to compare potentials of healthy adults with those of scoliotic patients, differences in body height had to be controlled for. Therefore the latency in milliseconds was divided by the height in metres (Jörg 1983). This quotient compensates for height but not for the scoliosis. Consequently an additional correction had to be introduced. Height in scoliotic individuals is diminished by scoliosis. With this circumstance in mind we measured the true length of the dorsal and lumbar spine between the seventh cervical and first sacral vertebrae following all the curvatures with a ribbon. Then, the direct distance between the two vertebrae was subtracted from the curved length. The result was added to the patient's height.

Table 3 shows these corrected latencies. It appears that the early components P_1 and N_2 may be slightly delayed in scoliosis, by up to 1–2 ms compared with normals, but this difference was not statistically significant. There was no significant difference between latencies under extension and under flexion of the spine in these scoliosis cases.

Discussion

Flexion of the cervical spine produces tension of the cord. Healthy individuals tolerate this tension without any discomfort. In some patients suffering from multiple sclerosis or myelopathy, however, the Lhermitte's sign can be elicited. Paraesthesiae are felt in the lower extremities or along the spinal column during maximum flexion of the neck. These acute neurological symptoms are possibly provoked by a build-up of pathological tension in the cord because of loss of normal elasticity. Breig (1984) showed that even longstanding neurological symptoms may be reversible, if flexion is completely avoided and constant extension is produced by the surgical technique of cervicolordodesis.

Provocation and alleviation of symptoms should be paralleled by changes in cord conductivity. We found no changes in latencies of early SEP components during full flexion and extension in symptomless, healthy individuals. In multiple sclerosis latencies appeared to be up to 2 ms longer during flexion, but this difference was not statistically significant. Even the pathological curvature of scoliosis did not produce significantly longer latencies as compared with normal individuals (Table 3). From these findings it can be concluded that there was no evidence of an acute decrease of cord conductivity by stretching of axons during flexion. Consequently, from the physiological point of view it does not appear that pathological conditions can be improved by extension. We can not, however, exclude the possibility that long-term changes of conductivity might be induced by regeneration of myelin sheaths during prolonged extension of the cervical spine.

At the present time we do not feel obliged to encourage our patients suffering from multiple sclerosis to undergo cervicolordodesis treatment.

Acknowledgement. We are indebted to Prof. Dr. H. Wagner, Head of the Orthopädische Klinik Rummelsberg, for making clinical data of scoliotic patients available to us.

References

Breig A (1984) Biomechanisches Verfahren zur Wiederherstellung neurologischer Ausfälle bei tetra- und paraplegischen Patienten mittels operativer Entspannung des traumatisierten Rückenmarkes. Fortschr Neurol Psychiat 52:2–5

Breig A, Renard M, Stefanko S, Renard C (1982) Healing of the severed spinal cord by biomechanical relaxation and surgical immobilization. Anat Clin 4:167–181

Breig A, Troup JDG (1984) Focal intramedullary tension in patients with cord lesion and its surgical relief by spinal cord relaxation. Lancet 1(8379):739–740

Jörg J (1983) Praktische SEP-Diagnostik. Enke, Stuttgart

Kurtzke JF (1965) Further notes on disability evaluation in multiple sclerosis, with scale modifications. Neurology (Minneap) 15:654–661

Assessment of Long Sensory Tract Conduction in Patients Undergoing Dorsal Root Entry Zone Coagulation for Pain Relief

S. J. Jones[1] and D. G. T. Thomas

Summary

Cortical somatosensory evoked potentials were recorded pre- and postoperatively, and conducted spinal cord potentials peroperatively, following posterior tibial nerve stimulation in patients undergoing dorsal root entry zone coagulation for pain relief. The majority were cases of brachial plexus avulsion. Cortical SEPs (initially normal in all patients except 2) showed significant deterioration postoperatively in 7 out of 10, correlating with the development of sensory and/or motor deficits in the lower limbs. Conducted spinal cord potentials, however, showed no such deterioration, indicating that the defect was unlikely to have been directly due to the surgical lesions but may have been a consequence of ischaemia or inflammation.

Introduction

Radio-frequency coagulation of the dorsal root entry zone (DREZ) has been found an effective treatment for otherwise intractable pain due to brachial plexus avulsion and other conditions involving peripheral nerves or the spinal cord (Nashold and Ostdahl 1979). In many cases, however, this also results in some degree of motor and/or sensory disturbance in the ipsilateral lower extremity (Nashold and Ostdahl 1979). In order to clarify the nature of the iatrogenic neurological impairment, and if possible to identify the patients at greatest risk, it was proposed to study cortical somatosensory evoked potentials (SEPs) to lower limb stimulation pre- and postoperatively. It was also proposed to examine conducted sensory potentials recorded from the dural or pial surface of the cord throughout the surgical procedure (Jones et al. 1982). The latter technique has been shown to be effective in detecting conduction defects arising during the surgical correction of scoliosis (Jones et al. 1983).

Material and Methods

Seventeen patients were studied, 12 male and 4 female, aged 21–78 years. In 12 cases DREZ coagulation was performed at C5–T1 levels, for the relief of pain following brachial plexus traction injury (myelography having indicated that the dorsal roots were avulsed from the spinal cord). Three patients received DREZ ablation for pain of other sorts, the source being post-herpetic neuralgia in 2 and an infiltrating tumour of the brachial plexus in 1. Two patients with a brachial plexus avulsion had the usual preoperative investigation but no surgery was performed.

[1] Medical Research Council and Gough Cooper Department of Neurological Surgery, The National Hospital for Nervous Diseases, Queen Square, London WC1N 3BG, Great Britain

Spinal Cord Monitoring
Ed. by J. Schramm and S. J. Jones

Cortical SEPs were recorded from the scalp preoperatively in 14 cases and postoperatively in 10. The stimulus was a square-wave constant voltage impulse of duration 0.2 ms, delivered at 3/s to the posterior tibial nerve at either ankle with an intensity sufficient to produce a slight plantar flexion. Silver/silver chloride disc recording electrodes were attached to the scalp with collodion, at locations on the midline 2 cm posterior to the vertex and 7 cm lateral to this site on either side. The reference electrode was located on the midline 12 cm above the nasion (Fz, International 10-20 system). The amplifier high-frequency response was less than 3 dB down at 5 kHz and the time constant was 1 s. Up to 1200 responses were averaged following stimulation of either leg, with 256 sampling points over a recording epoch of 64 ms starting 3–13 ms after the stimulus.

In 13 patients conducted spinal cord potentials were recorded peroperatively from the epidural (8) or subarachnoid (5) space, following exposure and opening of the dura mater at the appropriate levels. The electrical stimulus was similar to that described previously, but delivered at 10/s to the posterior tibial nerve in the popliteal fossa. The monopolar recording electrode consisted of a 3 mm length of stainless steel tubing 1 mm in diameter, or a stainless steel tab measuring 10 × 2.5 × 1.5 mm, at the end of an autoclavable lead. This was introduced at the rostral (in 8 cases also the caudal) end of the exposure at the midline and advanced by approximately 1 vertebral level. The reference electrode was a stainless steel needle inserted into adjacent muscle. The signals were amplified with a bandpass of 200 Hz to 2 kHz and averaged with 500 sample points over an epoch of 50 ms starting with the delivery of the stimulus.

Between 18 and 24 radio-frequency lesions were placed in the dorsal root entry zone at the appropriate levels at a depth of 2 mm, as described by Nashold and Ostdahl (1979). Conducted spinal cord potentials were recorded from the rostral site before, during and after the placement of lesions in 10 cases, with the stimulus delivered to the ipsilateral leg. In 8 cases responses to stimulation of the contralateral leg were also studied. Additional recordings were made at the caudal site following stimulation of the ipsilateral leg in 8 cases, and following contralateral stimulation in 4.

Results

1. Cortical SEPs

Cortical SEPs were recorded preoperatively in 14 cases, 11 with brachial plexus avulsion. Two had a mild sensory disturbance involving the lower limbs at the time of recording, one on the side contralateral to the brachial plexus lesion (due to cerebral trauma) and 1 on the ipsilateral side (attributable to posttubercular hyperaesthesia). In some avulsion cases direct questioning provoked recollection of mild sensory disturbance and/or weakness in the lower limbs at the time of the brachial plexus injury, but it was difficult to determine to what extent this may have been due to an intramedullary lesion.

The preoperative SEP waveform consisted of a sequence of positive and negative deflections, usually describing a W-shape. The most consistent feature was the first positivity, peaking at 33–48 ms. This was of largest amplitude at the midline elec-

Table 1. Clinical data plus pre- and postoperative cortical SEP findings in 17 patients, 15 undergoing DREZ coagulation

Patient	Side	Aetiology of pain	SEPs				LL symptoms postop.	
			Preop.		Postop. change			
			L	R	L	R	L	R
1. MF	R	Plexus avulsion	N	N	Nil	Nil	Nil	Nil
2. PM	R	Plexus avulsion	?N	N	Nil	+	Nil	+
3. MT	R	Plexus avulsion	N	N	Nil	+	Nil	+
4. MH	R	Plexus avulsion	N	N	+	++	Nil	++
5. DHo	R	Plexus avulsion	?N	N	+	++	Nil	++
6. MD	L	Plexus avulsion	N	N	Nil	Nil	Nil	Nil
7. DHa	R	Plexus avulsion	N	Abn	Nil	Nil	Nil	Nil
8. VW	R	Plexus avulsion	N/A	N/A	Abn	Abn	+	++
9. GB	R	Plexus avulsion	N	N	N/A	N/A	+	++
10. RD	R	Plexus avulsion	N/A	N/A	N/A	N/A	Nil	Nil
11. JMcC	L	Plexus avulsion	N/A	N/A	N/A	N/A	++	Nil
12. BL	L	Plexus avulsion	N	?N		Not operated		
13. SB	L	Plexus avulsion	N	N	N/A	N/A	Nil	Nil
14. AC	R	Plexus avulsion	N	Abn		Not operated		
15. AS	L	Post-herpetic	?N	N	Nil	+	+	+
16. HT	R	Post-herpetic	N	N	Nil	++	Nil	+
17. EF	L	Plexus tumour	N	N	N/A	N/A	Died	

trode 2 cm posterior to the vertex but was also present at the site 7 cm *ipsi*lateral to the stimulated limb (presumably due to the generator being located on the medial surface of the hemisphere). Preoperative SEPs were within normal limits of amplitude and latency in 8 cases, borderline on one side in 4 and unilaterally slightly delayed in 2 (Table 1). The last were the only patients in whom the SEP delay could have been due to a posterior column lesion sustained at the time of the plexus injury, since the 4 patients with borderline SEPs all had intact roots on the side of the possible electrophysiological abnormality.

SEPs recorded postoperatively in 10 cases (9 having had the preoperative investigation) were judged to be abnormal, either in absolute terms or relative to the earlier recording, in 7. SEP amplitude was abnormally low or had diminished by more than 50% in 4 cases, 3 with stimulation on the side ipsilateral to DREZ coagulation (see Fig. 1) and 1 bilaterally. Latencies were increased by more than 2.5 ms in 5 cases, 2 on the ipsilateral side, 2 contra- and 1 bilaterally (Fig. 1). The correlation between patients showing SEP deterioration and those exhibiting postoperative neurological deficits in the lower limbs was very good: the 7 patients with SEP abnormalities all had transient or persistent motor and/or sensory deficits, whereas in the remaining 3 no deficit was found. Sensory disturbance was less often noticeable than motor and the former, contrary to the findings of Nashold and Ostdahl (1979), did not always include demonstrable impairment of joint-position sense. The patients experienced

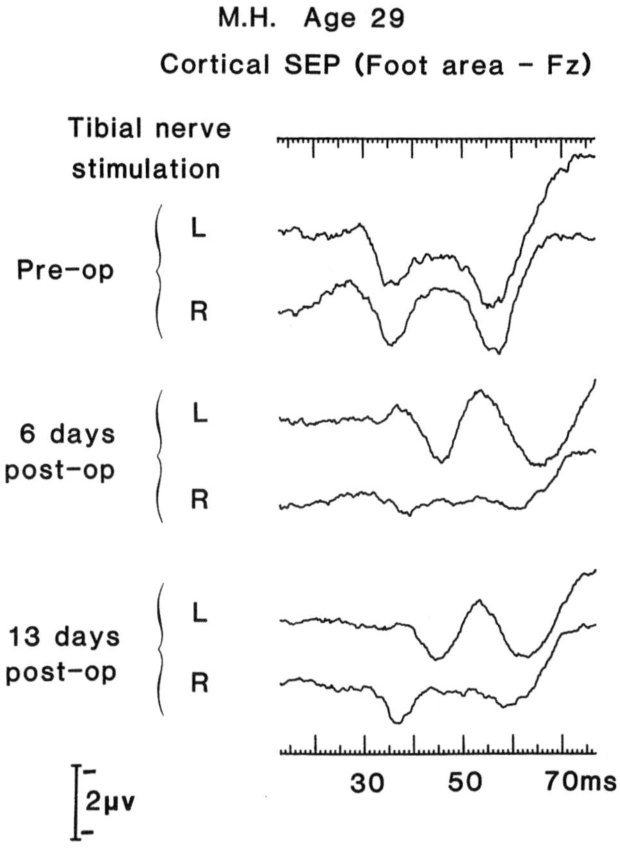

Fig. 1. Cortical somatosensory evoked potentials (negative polarity upwards in this and subsequent figures) recorded pre- and postoperatively in a patient underoging dorsal root entry zone coagulation for brachial plexus avulsion pain. The postoperative SEP deterioration was accompanied by weakness and sensory disturbance in both lower limbs (worse on the right) which improved considerably over the next 7 days. Note, however, the persistent marked SEP latency increase on the left

a variable degree of pain relief which was considered worthwhile in the majority. The postoperative outcome (with regard to neurological complications or pain relief) was not significantly predicted by abnormality or otherwise of preoperative SEPs, although pain alleviation was found to be particularly satisfactory in the 3 patients who experienced no neurological complications and had no postoperative SEP defects.

2. Conducted Spinal Cord Potentials

Evoked sensory spinal cord potentials were recorded extradurally in 8 cases and intradurally in 5, including 10 patients with brachial plexus avulsions. In neurologically normal scoliosis patients the waveform usually consists of 3 major components with latencies of 15–20 ms and a duration of 5–7 ms (Jones et al. 1982). In the older patients of the present study the initial component was sometimes of very low amplitude and the total duration sometimes as long as 15 ms. For 6 out of 7 cases of brachial plexus avulsion in whom the ipsilateral and contralateral legs were stimulated prior to the placement of DREZ lesions, the initial component and the largest subsequent peak recorded rostral to the avulsion ("first" and "late" components, see Table 2) were both smaller for the ipsilateral stimulus (Fig. 2). This, however, was

Table 2. Amplitude and latency (mean ± SD) of conducted spinal potentials (tibial nerve stimulation) recorded rostral to C5 before and after DREZ coagulation at C5–T1

	Before			After		
	First component		Late component	First component		Late component
	(μV)	(ms)	(μV)	(μV)	(ms)	(μV)
Ipsilateral stimulation ($n=10$)	0.9±1.3	17.7±2.2	1.2±0.9	1.0±1.2	17.5±2.2	1.2±0.9
Contralateral stimulation ($n=6$)	1.0±0.9	18.1±1.2	1.3±1.1	0.8±0.5	18.0±1.1	0.9±0.6

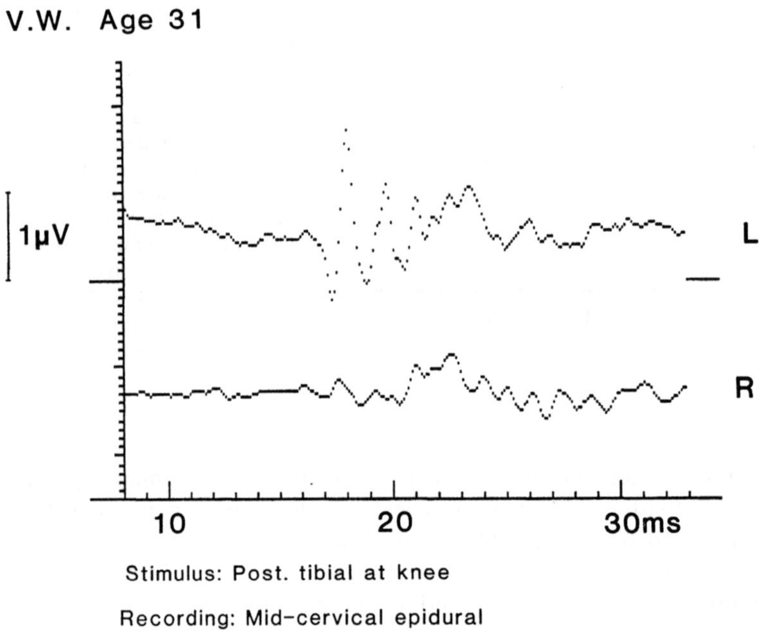

Fig. 2. Conducted spinal cord potentials recorded epidurally above C5 level, prior to dorsal root entry zone coagulation at C5–T1, in a patient with pain due to avulsion of the right brachial plexus. The early components were of much lower amplitude with stimulation of the posterior tibial nerve ipsilateral to the avulsion

not statistically significant on account of 1 patient who had much smaller amplitudes on the contralateral side. Ipsilateral response latencies were on average 0.2 ms longer, but this again was not statistically significant on account of 1 patient showing the opposite trend. The duration of the response was difficult to determine on account of the presence of small afterpotentials, but differed markedly between left and right in 2 patients (in one of the latter the longer duration response followed stimulation on the side *contralateral* to the avulsion).

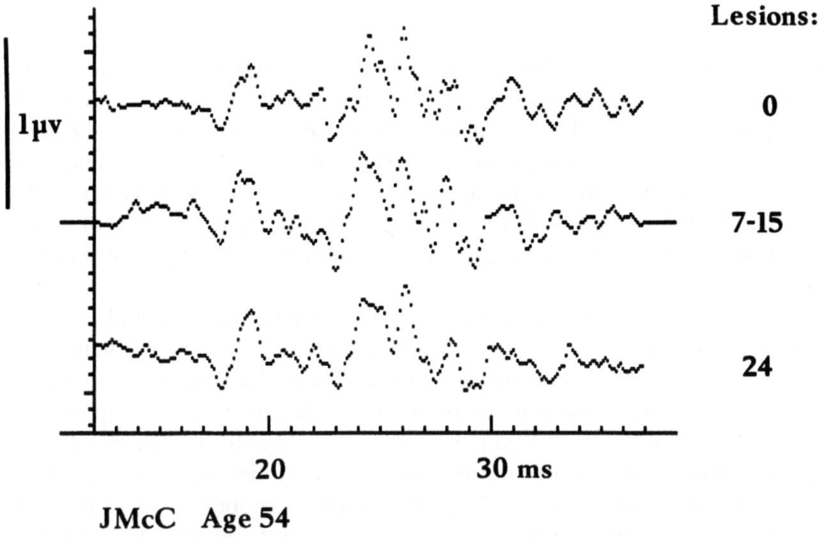

Fig. 3. Conducted spinal cord potentials recorded epidurally above C5 level before, during and after the placement of 24 lesions in the dorsal root entry zone from C5 to T1 (stimulation of the ipsilateral tibial nerve). The waveform was initially of low amplitude and very long duration, but showed no significant deterioration in spite of the development of marked weakness and numbness of the ipsilateral lower limb postoperatively

Responses were recorded at levels rostral and caudal to the exposure in 8 cases, 4 with stimulation on the side contralateral as well as ipsilateral to the pathology. Potentials recorded on the rostral side were on average about 50% smaller. This was not significantly different between 5 brachial plexus avulsion cases and 3 who had roots which were anatomically intact, nor between ipsilateral and contralateral responses in 4 avulsion cases. However, in the latter the first component was on average more than 50% smaller for the ipsilateral as compared with the contralateral stimulus, when recorded either rostral or caudal to the exposure (not statistically significant on account of the small number of cases). Rostral responses occurred on average approximately 1.8 ms later than caudal, but conduction time through the exposed segments bore no apparent relation to the nature or laterality of the pathology. A markedly prolonged conduction time of 5.9 ms was recorded for one patient (aged 78) in whom the roots and spinal cord were anatomically intact.

Spinal cord potentials recorded rostral to the exposure before, during and after DREZ coagulation showed no significant change, with the stimulus delivered to the ipsilateral leg in 10 cases and the contralateral leg in 6 (Table 2, Fig. 3). The magnitude of individual amplitude changes ranged from a reduction of 48% (affecting only the first component) to an increase of 78%. In one case responses which were unidentifiable before the placement of lesions were present thereafter, although with considerably longer latency than average. Changes observed peroperatively showed no apparent relation to the side or nature of the pathology, and were not correlated with the presence or absence of postoperative neurological symptoms.

Discussion

Severe pain is a frequent concomitant of brachial plexus traction lesions when the sensory roots are avulsed from the spinal cord. This is thought to be due to abnormal neuronal discharge patterns in the dorsal root entry zone. Treatment by thermo-coagulation of this structure at appropriate levels (usually C5 to T1) on the side ipsilateral to the avulsed roots is not yet widely practised, but has been reported to provide worthwhile and relatively long-lasting pain relief in a number of cases (Nashold and Ostdahl 1979). This is broadly in agreement with the experience of the present authors.

Relatively infrequently encountered in patients with brachial plexus avulsions are long tract signs, presumably reflecting intramedullary damage caused by violent traction on the rootlets. SEPs recorded in response to upper limb (median or ulnar nerve) stimulation have been used to assist with the diagnosis and prognosis of brachial plexus cases (Jones et al. 1981), but there have been no reports of SEP abnormalities following lower limb stimulation. The first finding of the present study, therefore, was that out of 11 avulsion cases studied preoperatively, 2 demonstrated lower limb SEP abnormalities which might be attributable to posterior column damage sustained at the time of the injury. Postoperatively, a much larger proportion demonstrated SEP abnormalities which were highly correlated with clinical sensory and/or motor impairment. Patients with abnormal SEPs were presumed to have involvement of the posterior columns (Halliday and Wakefield 1963), but it is noteworthy that not all such patients in the present study exhibited unequivocal impairment of joint-position sense.

By recording conducted spinal cord potentials peroperatively it was hoped, first to identify any patients with subclinical long tract involvement insufficient to give rise to abnormal cortical SEPs and second to see whether postoperative neurological defects were directly due to the placement of lesions in or around the dorsal root entry zone. It is likely that the components of the conducted response reflect, to a certain degree, activity in different tracts, and it was accordingly felt that changes in

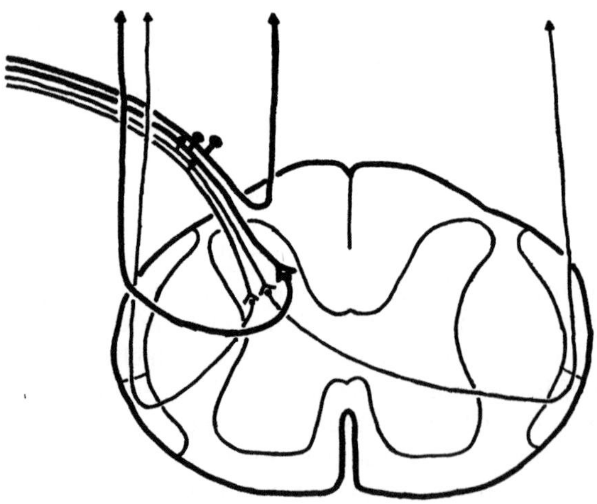

Fig. 4. Intramedullary projection of large diameter peripheral afferent fibres, either asynaptically in the posterior columns (medial to the dorsal root entry zone) or postsynaptically in more lateral tracts

the waveform seen after placement of 1 or 2 lesions might indicate whether the lesion site was too far lateral or too far medial (Fig. 4).

Prior to the placement of DREZ lesions there were features possibly suggestive of long tract involvement in several avulsion cases. These were broken up, long duration waveforms and amplitudes which were lower with stimulation on the ipsilateral (damaged) side. However, since potentials recorded just caudal to the exposure were also often of lower amplitude with the ipsilateral as compared with the contralateral stimulus, this may have been due to increased tissue resistance between the recording electrode and the active axonal elements, or to alteration of gross anatomy. In most cases the cord was observed to be hemiatrophic, fibrosed and dorsally rotated towards the side of the avulsion. However, long duration, delayed potentials were also recorded in some patients with anatomically intact roots and spinal cord, so it appears that identification of long tract damage in avulsion cases may be confounded by interindividual variability.

One fairly definite finding, however, was that conducted sensory spinal cord potentials showed little or no immediate deterioration as a result of DREZ coagulation. This suggests that long tract impairment may not be directly due to the surgical lesions but may reflect a more protracted ischaemic or inflammatory process initiated by the surgical procedure. Evidence which can be cited in support includes the observation that cortical SEPs following stimulation of the contralateral as well as the ipsilateral leg were sometimes affected. An inflammatory rather than an ischaemic mechanism could be suggested by the marked SEP latency increase sometimes observed: severely prolonged cortical SEP latencies are frequently encountered following the acute (inflammatory) phase of a spinal cord episode in multiple sclerosis (e.g. Trojaborg and Pedersen 1979), but compressive and ischaemic conditions such as cervical spondylotic myelopathy are generally associated with amplitude reduction and less marked slowing of conduction (Yu and Jones 1985).

References

Halliday AM, Wakefield GS (1963) Cerebral evoked potentials in patients with dissociated sensory loss. J Neurol Neurosurg Psychiat 26:211–219

Jones SJ, Edgar MA, Ransford AO (1982) Sensory nerve conduction in the human spinal cord: epidural recordings made during scoliosis surgery. J Neurol Neurosurg Psychiat 45:446–451

Jones SJ, Edgar MA, Ransford AO, Thomas NP (1983) A system for the electrophysiological monitoring of the spinal cord during operations for scoliosis. J Bone Joint Surg 65-B:134–139

Jones SJ, Wynn Parry CB, Landi A (1981) Diagnosis of brachial plexus traction lesions by sensory nerve action potentials and somatosensory evoked potentials. Injury 12:376–382

Nashold BS, Ostdahl RH (1979) Dorsal root entry zone lesions for pain relief. J Neurosurg 51:59–69

Trojaborg W, Pedersen E (1979) Visual and somatosensory evoked potentials in multiple sclerosis. J Neurol Neurosurg Psychiat 42:323–330

Yu YL, Jones SJ (1985) Somatosensory evoked potentials in cervical spondylosis: correlation of median, ulnar and posterior tibial nerve responses with clinical and radiological findings. Brain 108:273–300

Evaluation of Cord Function in Cervical Spondylosis by a Combined Method Using Segmental and Conductive Spinal Evoked Potentials (SEP)

H. Kotani[1], S. Hattori, F. Senzoku, S. Kawai, K. Saiki, H. Yamasaki, and K. Omote

Summary

In order to evaluate the cervical cord function and diagnose the level and severity of cervical cord lesions, both segmental and conductive spinal evoked potentials (SEP) were measured in 73 patients with cervical spondylotic myelopathy and/or radiculopathy. In normal subjects, segmental SEPs consisted of two waves (R and N waves). Ascending conductive SEPs consisted of two waves (first and second waves). Abnormal waves are discussed according to classification of myelopathy. Abnormal waves in cervical spondylotic radiculopathy are also discussed. The function of the cervical cord including roots, grey matter and white matter could be measured by the combined method with segmental and conductive SEPs.

Introduction

Recently spinal evoked potentials (SEP) have been applied clinically to evaluate spinal cord function. There are two main methods of measuring spinal evoked potentials from the human cervical cord. One is a segmental SEP recorded from the epidural space at the level of the enlargement of the cervical cord by stimulating the peripheral nerves. The other is a conductive SEP in which both the recording and stimulating electrodes are placed in the epidural space.

Since 1976 we had been using only cervical segmental SEPs to evaluate the cervical cord function (Hattori et al. 1979; Saiki et al. 1976, 1977; Saiki 1979). Since 1980 we have been using a combination of both methods (Kotani 1981; Kotani et al. 1981, 1983), that is, cervical segmental SEPs from the cervical epidural space by stimulating the median nerve at the elbow and ascending conductive SEP from the cervical epidural space by epidural stimulation of the thoracic cord (Fig. 1).

The purpose of this paper is to evaluate cervical cord function and diagnose the level and severity of cervical cord lesions using both segmental and conductive spinal evoked potentials.

[1]Department of Orthopedic Surgery, Tamatsukuri Kosei-Nenkin Hospital, Tamayu-cho, Yatsuka-gun, Shimane, 699-02, Japan

Spinal Cord Monitoring
Ed. by J. Schramm and S. J. Jones

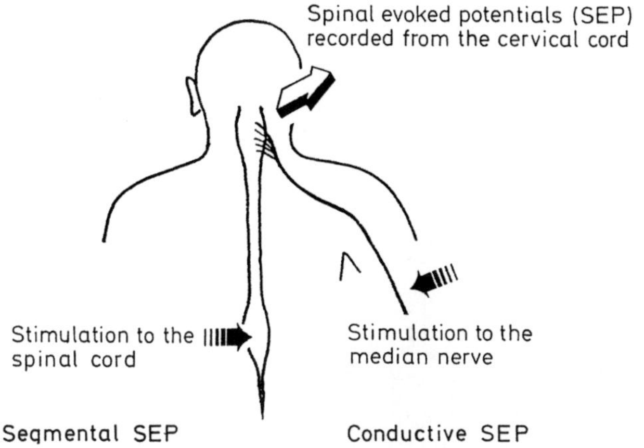

Fig. 1. Combined method with segmental and conductive SEPs

Materials and Methods

1) Subjects

From 1976 to 1983, SEPs were studied in about 300 cases with cervical spondylotic myelopathy and/or radiculopathy. Both segmental and conductive SEPs were recorded from the same levels of the cervical cord in 73 cases.

2) Electrodes (Fig. 2)

A monopolar electrode in a polyethylene tube (0.85 mm diameter) was used for recording or stimulating in the epidural space. An indifferent electrode was attached to the ear lobe for recording both the cervical segmental SEP and ascending conductive SEPs. Through a Tuohy needle, two electrodes were inserted into the cervical (C7–T1) and thoracic (T10-11 or T11-12) epidural space from the interspinous spaces according to the epidural anesthetic technique. After insertion the cervical electrode was progressively advanced as far rostral as possible, taking recordings at each intervertebral level, and the locations were confirmed by X-ray (Fig. 3).

3) Stimulation

A Nihon Kohden SEN-320 type stimulator and isolator were used. For recording the segmental SEP, stimuli of 0.5 Hz, 2 ms duration were delivered to the median nerve at the elbow percutaneously, and five to ten responses were averaged. For recording the ascending conductive SEP, stimuli of 3 Hz, 0.1 ms to 0.5 ms duration were delivered to the thoracic spinal cord from the epidural space and 15 to 20 responses were averaged.

Stimulus intensity at twice the threshold of the muscle twitch of the limb (100–200 volts) was used to produce the segmental SEP, and 1.5 times that of the paravertebral muscle twitch (50–100 volts) to produce the conductive SEP. Both segmental and conductive SEPs were recorded at each intervertebral level from C6-7 to C2-3.

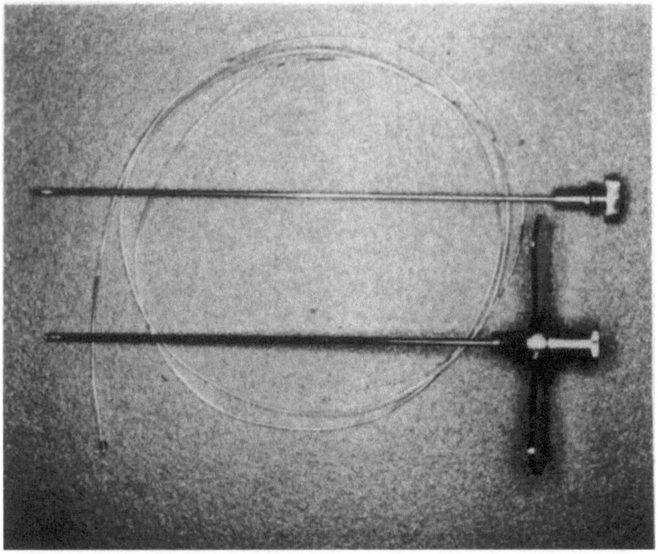

Fig. 2. A monopolar electrode and a Tuohy needle (E1)

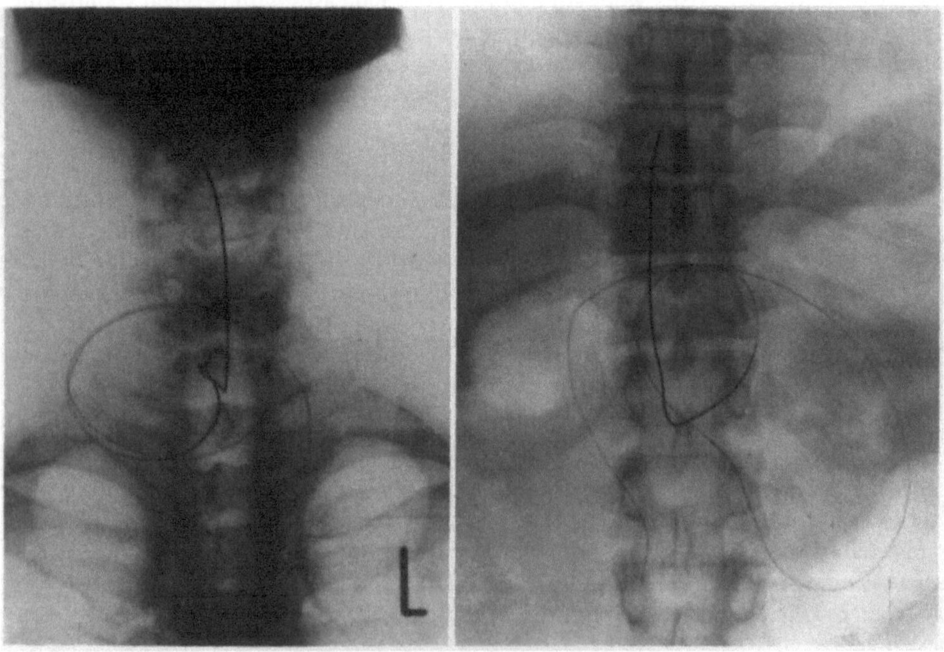

Fig. 3. By taking X-ray, the position of the electrodes was confirmed

4) Recording

Segmental and conductive SEPs were monitored on an oscilloscope, averaged by a computer (Nihon Kohden ATAC-350 type) and recorded on an X-Y plotter.

Results

1) Cervical Segmental SEP (Fig. 4)

In normal subjects, the segmental SEP obtained from the cervical epidural space by stimulation of the median nerve consisted of two waves. The latency of the first wave was 5 to 6 ms and the interval between two reactive waves was 1.9 ms. These two reactive components, R and N waves, were similar to those reported by Saiki (Saiki 1979; Saiki et al. 1977).

Stimulation of the median nerve produced clear R and N waves at the C4-5, C5-6 and C6-7 intervertebral levels. It was suspected that the main entry zone of the median nerve was at C4-5, C5-6 and C6-7 levels, though the median nerve contains fibres deriving from cervical nerves C4, C5, C6, C7 and C8. The R and N waves were obtained in 20 percent of subjects at C2-3 level, 56 percent at C3-4, 96 percent at C4-5, 100 percent at C5-6 and 97 percent at C6-7 (Table 1).

2) Ascending Conductive SEP (Fig. 4)

In normal subjects, two waves (first and second waves) were obtained from the cervical epidural space at each level by stimulating the thoracic cord. Only the first wave was obtained when the stimulus was weak. When the stimulus was intense enough the second wave also appeared.

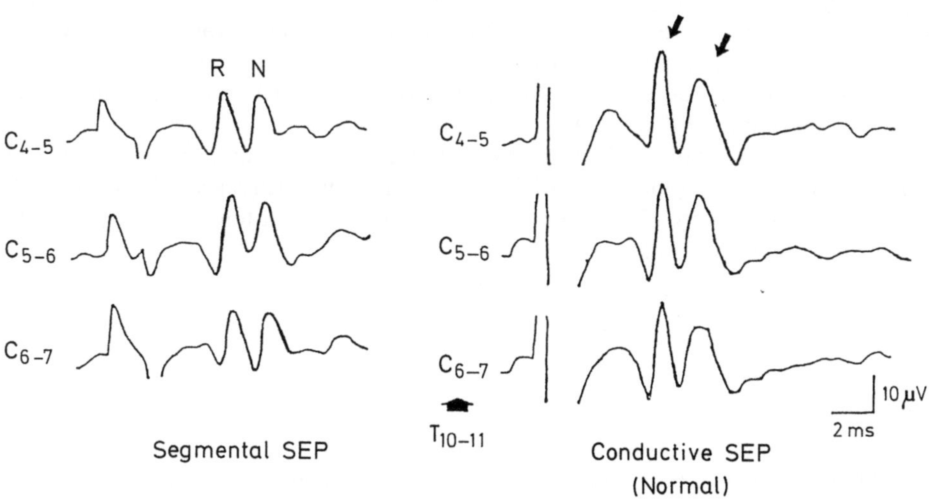

Fig. 4. Normal pattern of segmental and conductive spinal evoked potentials. Clear R and N waves are obtained at C4-5, C5-6 and C6-7 levels by stimulating the median nerve. Two waves *(arrows)* are obtained at each level by stimulation of the thoracic cord

Table 1. Distribution of segmental SEP in normal subjects by stimulating the median nerve

	No. of cases studied	SEP present in
C2-3	10	2 (20%)
3-4	24	14 (56%)
4-5	46	44 (96%)
5-6	55	55 (100%)
6-7	62	60 (97%)

The duration of the first wave was 1.2 to 1.8 ms and that of second 2.0 to 3.0 ms. The amplitude of the conductive SEP, 30–60 μV, was larger than that of the segmental SEP. As the distance from the stimulating electrode became longer, the latencies of the first and second waves were prolonged. This indicates that the first and second waves have different conduction velocities.

3) Abnormal Waves in Cervical Spondylotic Myelopathy or Radiculopathy

Abnormal components of the segmental and conductive SEPs in cervical spondylotic myelopathy are shown in Fig. 5.

Three intervertebral levels (C4-5, C5-6 and C6-7) were examined in 35 cases by the combined method with segmental and conductive SEPs. In type I (central lesion) of Hattori's classification (Hattori 1978) of cervical spondylotic myelopathy (Fig. 6), abnormal N waves were observed in four cases out of five, and the two waves of the ascending conductive SEP were normal in all cases. In type II (type I + postero-lateral column involvement) or type III (type II + antero-lateral column involvement), abnormal segmental SEPs were observed in six cases out of seven (86%) and 20 out of 23 (87%), respectively. Abnormal conductive SEPs were obtained in all cases of type II and III (Table 2).

In cervical spondylotic radiculopathy, polyphasic or low amplitude R waves were observed in 13 cases (46%) out of 28. In Keegan type (dissociated motor loss), abnormal R waves were observed in nine cases (90%) out of ten. Conductive SEPs, however, were normal in all cases (Keegan 1965).

Case Reports

Case 1 (Fig. 7)

A 50-year-old male complained of radiating pain in the right shoulder girdle and sensory disturbance in both hands. The function of the lower limbs was intact. Myelography revealed a filling defect at the level of C4-5 and C5-6, and of the right C5 nerve root. He was diagnosed as having a combination of cervical spondylotic myelopathy of type I and radiculopathy.

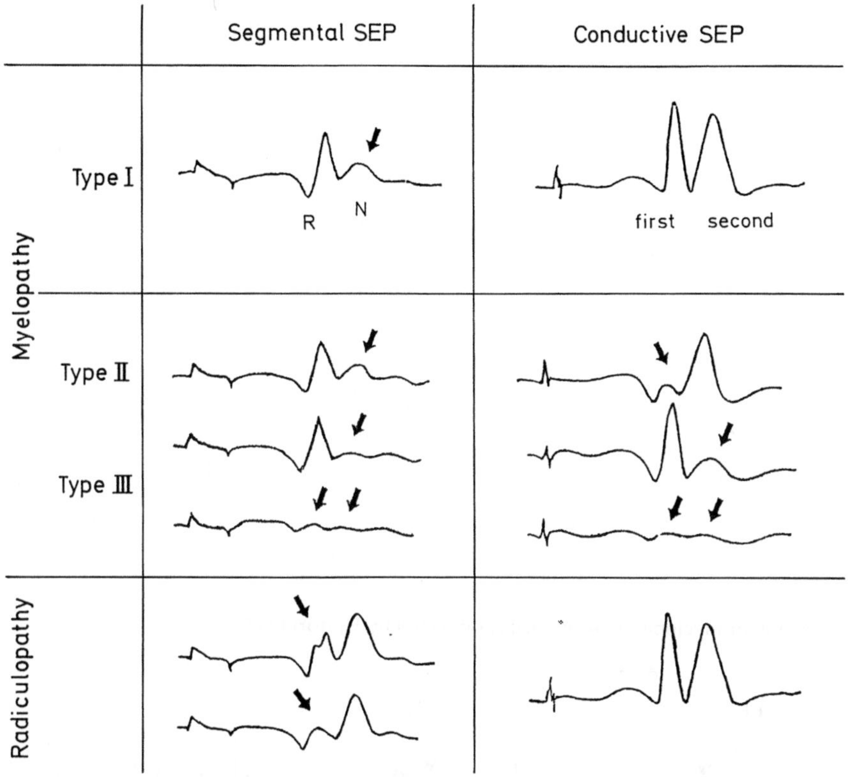

Fig. 5. Abnormal SEPs in cervical spondylotic myelopathy and radiculopathy. In type I of myelopathy, only abnormal N waves were observed. In type II or III, abnormal N waves and conductive SEPs were obtained. In radiculopathy, only abnormal R waves were observed

Table 2. Correlation between abnormal SEP findings and type of myelopathy (35 cases)

Type of myelopathy (classification of Hattori 1978)	Abnormal segmental SEP	Abnormal conductive SEP
Type I, 5 cases	4 cases (80%)	0 (0%)
Type II, 7 cases	6 cases (86%)	7 (100%)
Type III, 23 cases	20 cases (87%)	23 (100%)

In SEP examination, the N wave was reduced in amplitude at C4-5 level following stimulation of the left median nerve and the R wave was reduced in amplitude at C4-5 level following right median nerve stimulation. Two normal reactive components of ascending conductive SEPs were obtained at each level. The SEP examination indicated dysfunction of the grey matter of the spinal cord and dysfunction of the right nerve root at C4-5.

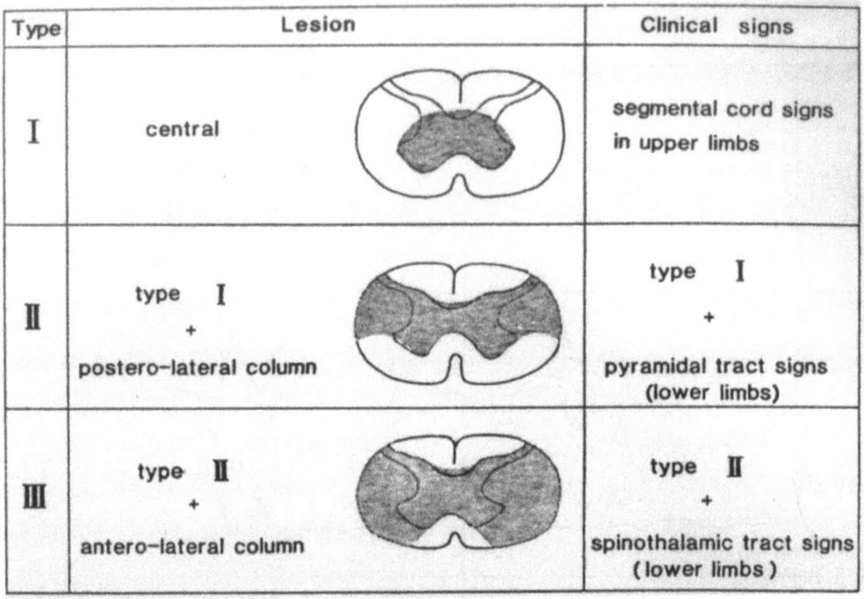

Fig. 6. Classification of myelopathy in cervical spondylosis by Hattori (1978)

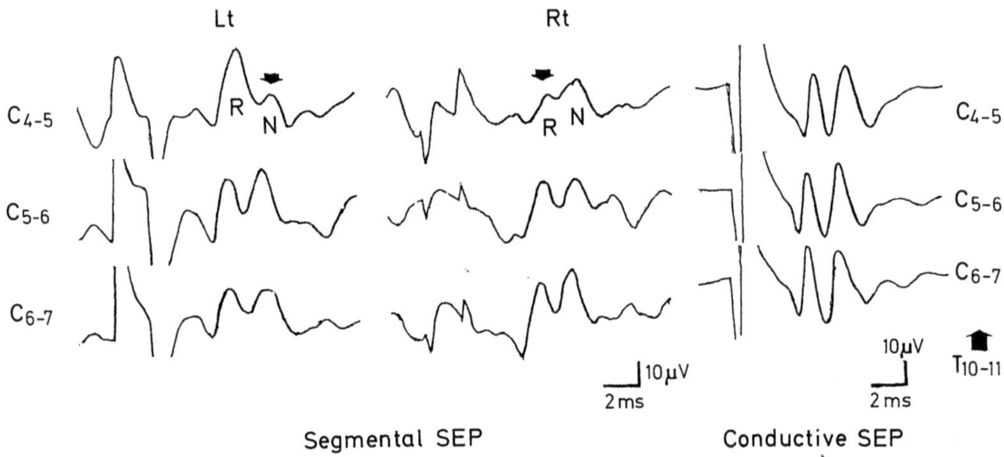

Fig. 7. Cervical spondylotic myelopathy (type 1) + radiculopathy (50-year-old male). N wave decreased in amplitude *(arrow)* at C4-5 level following stimulation of the left median nerve and R wave decreased in amplitude *(arrow)* following right median nerve stimulation. Normal ascending conductive SEPs were obtained at each level

Case 2 (Fig. 8)

A 45-year-old female complained of numbness and motor disturbance of both hands for two years. Exaggeration of deep tendon reflexes of both lower limbs was observed. X-ray showed ossification of the posterior longitudinal ligament at C2, C3,

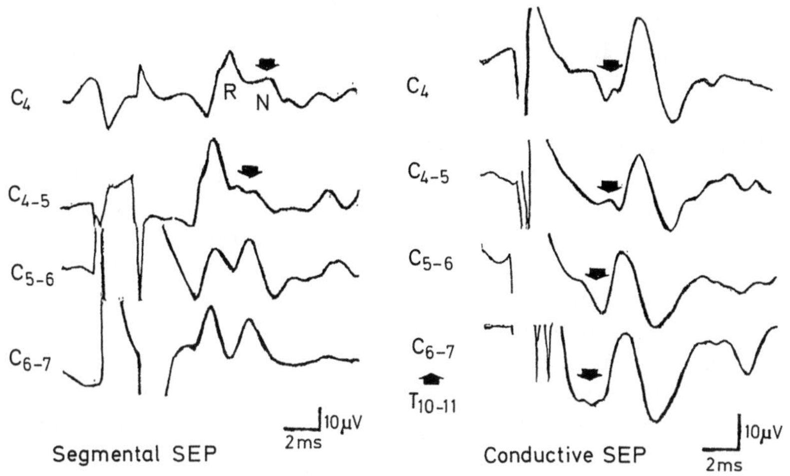

Fig. 8. Ossification of the posterior longitudinal ligament (cervical myelopathy type 2) (45-year-old female). Low amplitude N waves were obtained at C4 and C4-5 levels, and the first wave of the ascending conductive SEP was virtually absent at all levels

C4 and C5 levels, and myelogram revealed filling defects at C3-4 and C4-5 levels. She was diagnosed as having cervical myelopathy of type II.

SEPs in this case showed low amplitude N waves at C4 and C4-5 levels and disappearance of the first wave of the ascending conductive SEP at all levels. Dysfunction of both grey matter and the lateral funiculus of the white matter was indicated.

Discussion

Ever since Gasser and Graham (1933) first stimulated the lumbar spinal nerves of cats and obtained evoked potentials from the posterior roots, many researchers have recorded spinal evoked potentials in animals and man (Ertekin 1976; Magladery et al. 1951; Pool 1946; Sawa 1947; Shtark 1962). When Shimoji et al. (1971) showed that spinal evoked potentials (SEP) could be obtained safely from the human epidural space using averaging apparatus, clinical application of the SEP commenced.

During surgery, descending or ascending conductive SEPs have been recorded from the epidural space (Kurokawa 1972) or subarachnoidal space (Tamaki et al. 1972) by direct stimulation of the spinal cord with an epidural stimulating electrode. It was shown that SEPs were useful for spinal cord monitoring during spinal surgery. Cracco (1973) reported that SEPs could be obtained from the surface of the body by stimulating the peripheral nerve. There are several reports of SEPs recorded from the surface of the body (Satomi 1979; Shimoji et al. 1978; Delbeke et al. 1978), but the wave shapes are not very clear because the amplitude of the SEP is very small.

Saiki et al. (1976, 1977) recorded two negative potentials (R and N waves) from the cervical epidural space by stimulating the median nerve at the elbow, and reported that the R wave was related to both nerve root and posterior column potentials, whereas the N wave was related to the postsynaptic potentials of the grey matter of the spinal cord. Several investigators showed that it is possible to diagnose the level and severity of the cord lesion in cases of cervical spondylotic myelopathy by studying the cervical segmental SEP (Hattori et al. 1979; Kotani et al. 1981, 1983, 1984; Saiki 1979; Saiki et al. 1977; Shigematsu 1982; Yamasaki 1982).

Imai (1976) and Kurokawa (1979) recorded first and second waves in ascending conductive SEPs following epidural stimulation of the spinal cord. By performing a transection study they showed that the first wave was conducted in the superficial layer of the funiculus lateralis and the second in the funiculus posterior.

In recordings from the cervical epidural space we obtained both segmental SEPs following stimulation of the median nerve at the elbow, and ascending conductive SEPs following epidural stimulation of the thoracic cord.

By observing both segmental and conductive SEPs from the same level of the cervical cord it is possible to diagnose the status of the cervical cord accurately, since each component of the segmental and conductive SEPs indicates the function of a different part of the spinal cord.

In cervical spondylotic myelopathy type I of Hattori's classification, abnormal N waves of the segmental SEP were observed and ascending conductive SEPs were normal. This fact indicates localized lesions in the grey matter of the spinal cord. In type II or III, abnormal waves were observed in both segmental and conductive SEPs. This indicates extensive lesions involving the grey matter and white matter. Therefore, good correlation is gained between SEP findings and the type of myelopathy. As for the recovery of function after surgery, a case in which SEP findings were slightly abnormal made a good recovery but in a case for whom SEPs were not obtained the recovery was poor.

Only R waves were altered in approximately 50% of cases with cervical spondylotic radiculopathy. In Keegan's (1965) type abnormal R waves were observed in 90%. This fact suggests that the R wave is related not only to posterior roots but also to anterior roots. Since normal N waves and conductive SEPs were obtained in cases of radiculopathy, it is possible to discriminate between myelopathy and radiculopathy by our combined method using segmental and conductive SEPs.

In cases of extensive lesions, such as cervical spondylosis, ossification of the posterior longitudinal ligament, ossification of the yellow ligament or entire spinal canal stenosis, it is often difficult to diagnose the main source of neurological involvement. Consequently it is necessary to measure the function of the spinal cord itself to determine the main lesion. The function of the cervical cord, including roots, grey matter and white matter, can be assessed by our method, and it is possible to diagnose the level of the lesion by moving the electrode in the epidural space and observing both segmental and conductive SEPs.

References

Cracco RQ (1973) Spinal evoked response: Peripheral nerve stimulation in man. Electroenceph Clin Neurophysiol 35:379–386

Delbeke J, McComas AJ, Kopec J (1978) Analysis of evoked lumbosacral potentials in man. J Neurosurg 41:293–302

Ertekin C (1976) Studies on the human evoked electrospinogram. I. The origin of the segmental evoked potentials. Acta Neurol Scand 53:3–20

Gasser HS, Graham HT (1933) Potentials produced in the spinal cord by stimulation of dorsal roots. Am J Physiol 103:303–320

Hattori S (1978) Cervical spondylotic myelopathy. J Jpn Orthop Assoc 52:581–593

Hattori S, Saiki K, Kawai S (1979) Diagnosis of the level and severity of cord lesion in cervical spondylotic myelopathy (spinal evoked potentials). Spine 4:478–485

Imai T (1976) Human electrospinogram evoked by direct stimulation on the spinal cord through epidural space. J Jpn Orthop Assoc 50:1037–1056

Keegan JJ (1965) The cause of dissociated motor loss in the upper extremity with cervical spondylosis. J Neurosurg 23:528–536

Kojima Y, Yamamoto T, Ogino H, Okada K, Ono K (1979) Evoked spinal potentials as a monitor of spinal cord viability. Spine 4:471–477

Kotani H, Hattori S, Kawai S, Saiki K, Imagawa T, Miyamoto T (1981) Spinal evoked potentials for spinal cord lesion. Cent Jpn J Orthop Trauma Surg 24:679–680

Kotani H, Hattori S, Kawai S, Saiki K, Yamasaki H, Omote K (1983) Spinal evoked potentials—Technique and use in diagnosis for the spinal cord lesion. J Western Jpn Research Society for Spine 9:116–120

Kotani H (1981) Clinical studies on the segmental and conductive spinal evoked potentials. Arch Jpn Chr 50:426–444

Kotani H, Hattori S, Kawai S, Saiki K, Yamasaki H, Omote K (1984) Evaluation of spinal cord function—Using segmental and conductive spinal evoked potentials. In: Homma S, Tamaki T (eds) Fundamentals and clinical application of spinal cord monitoring. Saikon Publishing, Tokyo, pp 253–268

Kurokawa T (1972) Spinal cord action potentials evoked by epidural stimulation of cord—Report of human and animal records. Jpn J Electroencephalogr Electromyogr 1:64–66

Kurokawa T (1979) Clinical application of the evoked electrospinogram. Adv Neurol Sci 23: 409–420

Magladery JW, Porter WE, Park AM, Teasdall RD (1951) Electrophysiological studies of nerve and reflex activity in normal man. IV. The two neuron reflex and identification of certain action potentials from spinal roots and cord. Bull Johns Hopk Hosp 88:499–519

Pool JL (1946) Electrospinogram—spinal cord action potentials recorded from a paraplegic patient. J Neurosurg 3:192–198

Saiki K, Hattori S, Date Y (1976) Experimental studies on the extradural spinal cord potential changes at cervical cord compression in rabbits. Cent Jpn J Orthop Trauma Surg 19:160–163

Saiki K, Hattori S, Date Y, Miyamoto T (1977) Spinal electrogram at the cervical region. Cent Jpn J Orthop Trauma Surg 20:304–306

Saiki K (1979) Spinal evoked potential (SEP) obtained by stimulation on the median nerve—Experimental and clinical studies. J Jpn Orthop Assoc 53:1893–1913

Satomi K (1979) Spinal electrogram evoked by peripheral nerve stimulation in man. J Jpn Orthop Assoc 53:53–63

Sawa M (1947) Spontaneous electrical activities obtained from human spinal cord. Folia Psychiat Neurol Jap 2:165–176

Shigematsu A (1982) Clinical application of the segmental spinal evoked potentials of the cervical spinal cord.—An analysis about the patients with cervical osteochondrosis. Arch Jpn Chr 51: 276–288

Shimoji K, Higashi H, Kano T (1971) Epidural recording of spinal electrogram in man. Electroenceph Clin Neurophysiol 30:236–239

Shimoji K, Shimizu H, Maruyama Y (1978) Origin of somatosensory evoked responses recorded from the cervical skin surface. J Neurosurg 48:980–984

Shtark MB (1962) On the biopotentials of the human spinal cord in the normal state and under pathological conditions. Fiziol Zh 8:120–127

Tamaki T, Yamashita T, Kobayashi H, Hirayama H (1972) Spinal monitoring. Jpn J Electroencephalogr Electromyogr I:196

Tamaki T (1979) Basic analysis of evoked spinal cord potential elicited by direct spinal cord stimulation. In: Nash CL, Brown RH (eds) Spinal cord monitoring workshop data acquisition and analysis. Cleveland, pp 17–24

Yamasaki H (1982) Studies on spinal evoked potentials in cervical spondylotic myelopathy—Using both segmental and conductive SEP. Arch Jpn Chr 51:932–944

Somatosensory Evoked Potentials in Cervical Myelopathy

H. Baba[1], I. Shima, K. Tomita, S.-I. Umeda, Y. Sawada, S. Masuyama, K. Yonezawa, K. Kanekasu, S. Tsuji, S. Nomura, H. Yugami, and Y. Koshino

Summary

Somatosensory evoked potentials were recorded in 53 patients with cervical myelopathy due to spondylosis (32) or ossification of the posterior longitudinal ligament (21). The overall abnormality rate was 68 per cent, and slightly higher in OPLL. P14 and N18 components were the most often affected, even in less severe cases, but the P23 and N30 components correlated well with the severity of myelopathy.

Introduction

Various clinical and fundamental studies have been reported concerning somatosensory evoked potentials (SSEPs) in spine and spinal cord disease, since Dawson's pioneer work in 1947. Spielholz et al. (1979), Bennet (1984), Hall et al. (1978), and Nash et al. (1977) reported their clinical usefulness in spinal surgery.

These evoked responses, however, are strongly influenced by many events such as anaesthesia, recording techniques for elicitation and other recording circumstances. The amplitude of somatosensory evoked potentials is very low and excellent equipment is required to obtain clear responses. Thus, there seemingly still exists a limitation to their wide-spread clinical application. Furthermore, there are many evaluating methods in somatosensory evoked potentials. The authors, however, have utilized these potentials as one of the useful objective methods for the assessment of spinal cord dysfunction, since 1982.

It is our purpose here to study changes of somatosensory evoked potentials in cervical myelopathy of varying degrees, and to utilize these in the clinical assessment of spinal cord function.

Material and Methods

Fifty-three patients with cervical myelopathy due to various cervical spine diseases were studied: 32 cases with cervical spondylotic myelopathy (CSM), and 21 cases with ossification of the posterior longitudinal ligament (OPLL). In 9 cases with operated OPLL, serial monitoring of somatosensory evoked potentials was performed to investigate changes in peak latencies before and after surgery. There were 18 male and 3 female cases with OPLL, and ages ranged from 32 to 74 years with the average being 64 years. In cases with CSM, 24 males and 8 females were included with an average age of 55 years (38 to 77 years).

[1]Department of Orthopaedic Surgery, School of Medicine, Kanazawa University, 13-1 Takaramachi, Kanazawa, 920, Japan

Spinal Cord Monitoring
Ed. by J. Schramm and S. J. Jones

Monitoring of somatosensory evoked potentials was basically performed by the technique that Shagass and Schwarz described in 1964.

Rectangular wave impulses of 0.2 ms duration at 1 impulse per second were used for stimulation of the median nerve at the wrist, delivered by a Medelec NT6 stimulator. Stimulus intensity was set at just above the threshold at which the smallest visible muscle twich in the opponens muscle was produced.

A silver-silver chloride disc recording electrode was placed over the contralateral hand sensory area; 2 cm posterior and 7 cm contralateral in the line of the vertex to the external acoustic meatus. A reference electrode was placed on the ear contralateral to the side of the stimulation.

Two hundred and fifty-six sweeps synchronized with the stimulus were amplified with a bandpass of 10 Hz to 10 kHz and averaged with a Medelec AA6 amplifier and a Medelec DAV6 averaging computer, respectively. Components within 300 ms from the stimulus artifact were averaged and those within 60 ms were analyzed in the present study.

For the establishment of normative values of the SSEP peak latencies, 62 healthy adults were studied as the control group. As Yugami et al. reported in 1982, each peak in the SSEP was identified as P14, N18, P23, N30, P40, and N53, by statistical analysis.

Results

In cases with OPLL, the N18 and P23 components were frequently delayed or absent, but the P40 and N53 components were preserved. With regard to N18, 53 per cent were delayed and 23 per cent were absent, and to P23, 47 per cent were delayed and 23 per cent were absent. The N30 component, however, was normal in a higher rate of 45 per cent of cases with OPLL.

In cases with CSM the N18 component was most frequently delayed (51 per cent), and 48 per cent of P23 had normal latency. With regard to the N30 component, 67 per cent had normal latency. A lower incidence of abnormality was observed in CSM compared with OPLL.

Table 1. Changes in SSEP responses of the operated OPLL cases (values indicating percentage)

		P_{14}	N_{18}	P_{23}	N_{30}	P_{40}	N_{53}
Before surgery	A	17.2	16.7	17.1	15.8	25.0	24.7
	B	8.3	18.3	25.0	25.0	50.0	52.8
	C	74.4	75.0	57.9	60.2	25.0	21.5
After surgery (1–2 weeks)	A	25.0	37.4	46.6	67.6	58.3	24.0
	B	75.0	69.2	53.4	32.4	41.7	76.0
	C	0.0	0.0	0.0	0.0	0.0	0.0
After surgery (4–8 weeks)	A	64.6	50.0	77.7	100.0	81.7	50.0
	B	35.4	50.0	23.0	0.0	19.3	50.0
	C	0.0	0.0	0.0	0.0	0.0	0.0

A, normal; *B*, prolonged; *C*, disappeared

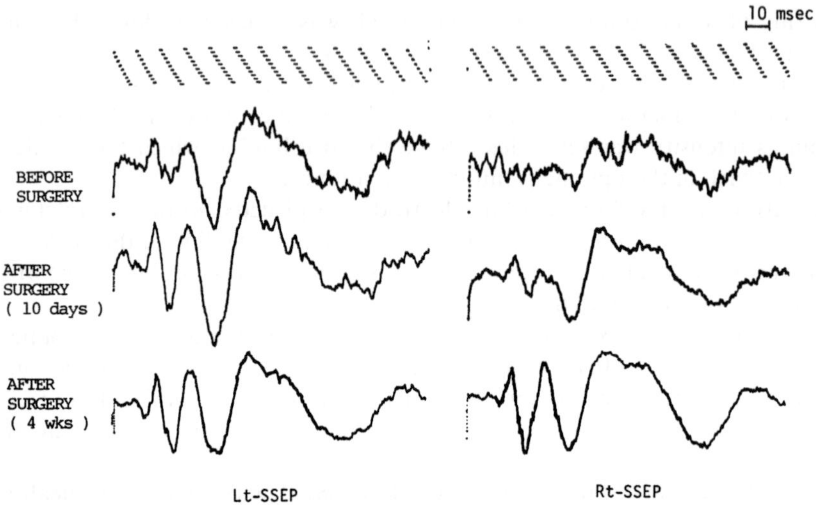

a

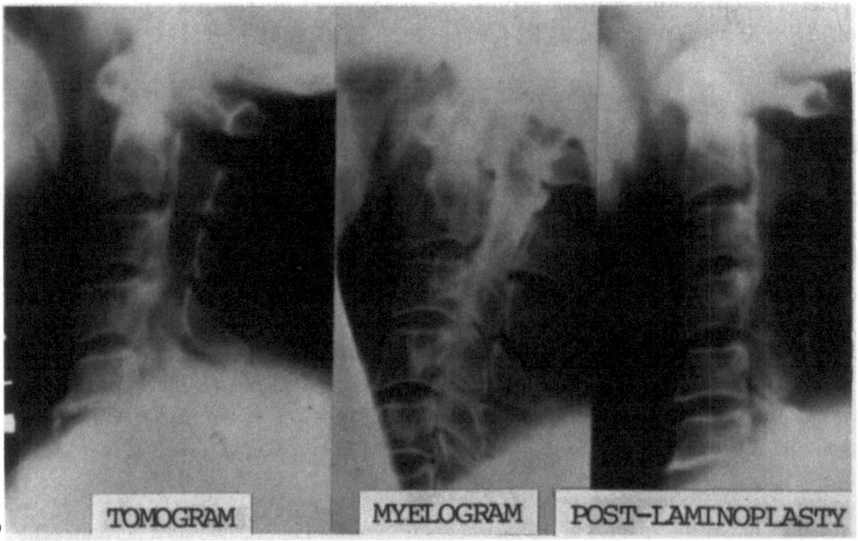

b

Fig. 1. a, b Pre- and postoperative SSEPs recorded in a patient with left-side predominant OPLL from C2 through C4, showing marked improvement in the right-sided response (stimulation on left)

In neurologically advanced cases, unlike those with moderate myelopathy, abnormalities in P23 and N30 were significant. In the present study, there were no significant abnormalities in side to side comparison of the SSEP amplitude.

Serial SSEP monitoring before and after surgery was performed in 9 operated cases with OPLL. Table 1 shows the abnormality rates in 18 responses recorded from those 9 operated cases. The majority of these cases showed marked neurological improvement after surgery. Before surgery, a high incidence of absent peaks was observed: 75 per cent for N18, 58 per cent for P23, and 60 per cent for N30. One to 2 weeks after surgery, however, none of the SSEP showed absent peaks and the

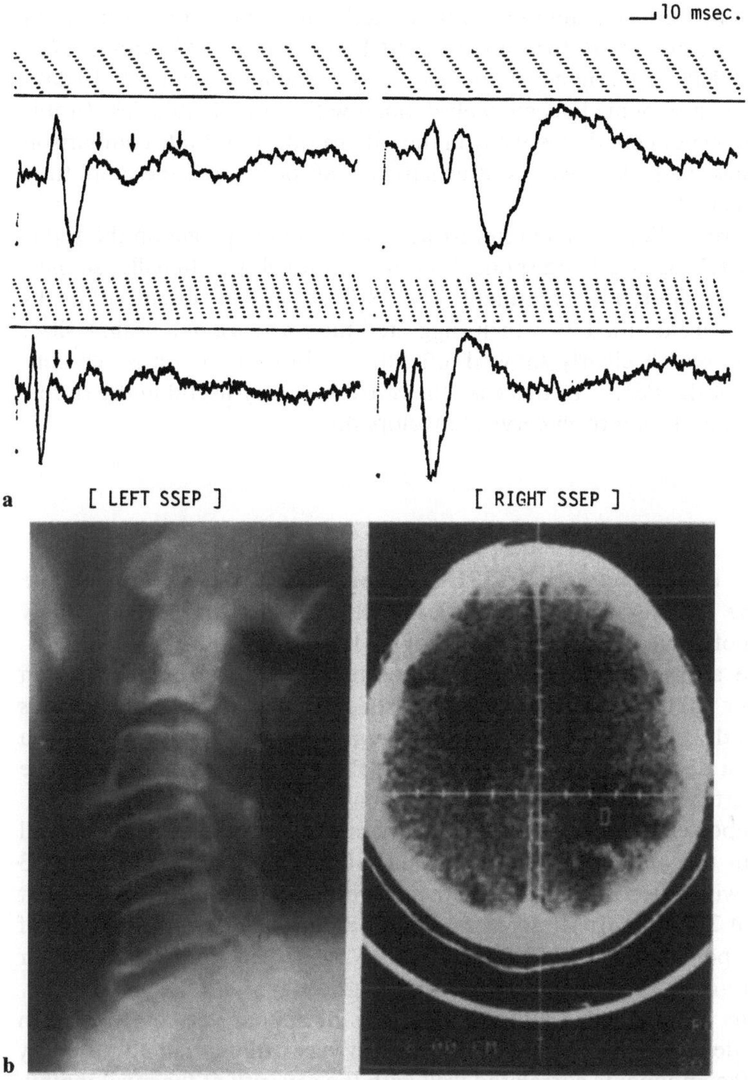

Fig. 2. a, b SSEPs recorded in a case of right hemiplegia with OPLL at C4 and spondylosis at C4 through C6. Abnormal late components in the left SSEP (stimulation on the right) suggested presence of a brain lesion, confirmed by CT

majority instead showed delayed latencies. A higher incidence of SSEP peaks with normal latency was observed 4 to 8 weeks after surgery. The adequacy of surgical decompression could be documented by this observation of a decrease in latency of conduction. Preservation or early return of SSEP peaks in the earlier postoperative period was also apparently associated with a generally good prognosis.

Traces in Fig. 1 were recorded from a male case aged 42 years with left-side predominant spastic quadriplegia due to ossification of the posterior longitudinal ligament from C2 through C4. Laminoplasty from C2 to C5 was performed to decompress the spinal cord posteriorly, and there was marked neurological improvement in the early postoperative period. Before surgery, SSEP peaks from P14 through N30 were absent in the SSEP on the right (left median nerve stimulated) whereas, 10 days after surgery, these components became well-defined with normal latencies. In this period, a more than 50 per cent discrepancy in the side to side amplitude comparison of the SSEP was observed. At 4 weeks after surgery, all peaks and the amplitude became normal and stable.

Traces in Fig. 2 were elicited in an interesting case with hemiplegia on the right. X-ray demonstrated OPLL at C4 and spondylosis at C4 through C6. Initially, surgical decompression was considered for treatment. The SSEP showed an absence of peaks with latencies over 30 ms on the left, which suggested the presence of a brain lesion. Further examination by CT clearly showed left-side cerebral infarction which confirmed the findings of the SSEP. In this case the SSEP was very useful in the differential diagnosis of brain lesion from cervical myelopathy.

Discussion

The somatosensory evoked potentials (SSEP) reflect mainly the integrity of the spinal cord and the leminiscus-thalamo-cortical pathways, and have been widely applied in the field of clinical neurology (Baba et al. 1984; Spielholz et al. 1979).

Since 1982, the authors have utilized somatosensory evoked potentials as an objective parameter in the assessment of cervical spine disease causing various degrees of myelopathy. The authors' preliminary report indicates that cases with a markedly delayed or absent second wave complex (P23–N30) in the median nerve evoked SSEP do better when surgically treated (Baba et al. 1984).

Giblin (1964) reported a 50 per cent abnormality rate in his series, and the overall abnormality rate in the authors' series was slightly higher (68 per cent; 36 of 53 cases). In 21 cases with extensive OPLL, however, a higher incidence of 95 per cent was observed (20 of 21 cases). In these cases, the frequencies of delay or absence of each peak were: 67 per cent for P14, 79 per cent for N18, 67 per cent for P23, 47 per cent for N30, 62 per cent for P40, and 78 per cent for N53.

The components with latency of 14 ms and 18 ms disappear very early even in cases with a lesser degree of myelopathy. However, it was striking that abnormality of the P23 and N30 components correlated well with the severity of the myelopathy. Similar findings were reported by Takagi and Kobayashi in 1975, the N30 component being most influenced by the cervical spinal cord disorder.

In conclusion, the P23 and N30 components are useful parameters in the clinical assessment of cervical myelopathy. However, how the physiological mechanism of these cortical components can be influenced by cervical spinal cord disease still requires much investigation.

References

Baba H, Shima I, Nomura S, et al. (1984) Somatosensory evoked potentials in cervical myelopathy (in Japanese). Cent Jap J Orthop Traumat 27(2): 772–774

Bennet MH (1984) SEP monitoring during spinal cord operation. In: Homma S, Tamaki T (eds) Fundamentals and clinical application of spinal cord monitoring. Saikon Publishing, Tokyo, pp 385–396

Dawson GD (1947) Cerebral responses to electrical stimulation of peripheral nerve in man. J Neurol Neurosurg Psychiat 10: 134–140

Dorfman LJ, Perkash I, Bosley TM, et al. (1980) Use of cerebral evoked potentials to evaluate spinal somatosensory function in patients with traumatic and surgical myelopathies. J Neurosurg 52: 654–660

Giblin DR (1964) Somatosensory evoked potentials in healthy subjects and in patients with lesions of the nervous system. Ann NY Acad Sci 112: 93–142

Hall JE, Levine CR, Sudhir KG (1978) Intraoperative awakening to monitor spinal cord function during Harrington instrumentation and spine fusion. J Bone Joint Surg 60-A: 533–536

Nash CL, Lorig RA, Schatzinger LA, et al. (1977) Spinal cord monitoring during operative treatment of the spine. Clin Orthop 126: 100–105

Rowed DW, McLean JAG, Tator CH (1978) Somatosensory evoked potentials in acute spinal cord injury: Prognostic value. Surg Neurol 9: 203–210

Shagass C, Schwarz M (1964) Recovery of function of somatosensory peripheral nerve and cerebral evoked response in man. Electroenceph Clin Neurophysiol 17: 126–135

Spielholz NI, Benjamin MV, Engler GL, et al. (1979) Somatosensory evoked potentials during decompression and stabilization of the spine: Methods and findings. Spine 4: 500–505

Takagi M, Kobayashi H (1975) Somatosensory evoked potentials in peripheral nerve and spinal cord disease (in Japanese). Clin Electroencephalography 17: 678–684

Yugami H, Koshino Y, Yamaguchi N, Matsubara F (1982) Somatosensory evoked potentials in normal young adults (in Japanese). Clin Electroencephalography 24: 330–335

Clinical Study of Cervical Spondylotic Myelopathy Using Evoked Spinal Cord Potentials

K.-I. Shinomiya[1], K. Furuya, I. Yamaura, H. Sato, A. Kaneda, and M. Yokoyama

Summary

1. Examination using both conductive and segmental SEPs is needed to diagnose correctly the function of the cervical spinal cord in cervical spondylotic myelopathy.
2. N1 originates from multiple roots caudal to the recording site.
3. Cases showing normal or slightly changed N1 and N2 recovered markedly after surgery, whereas cases showing conduction block had unsatisfactory results.
4. Cases showing definite changes of the conductive SEP at the site of the main lesion had satisfactory results with regard to function of the lower extremity.
5. Cases showing low amplitude, slow conduction velocity or a single wave caudal to the main lesion, or showing positive waves at the injured site in the conductive SEP, had low postoperative J.O.A. scores and an unsatisfactory recovery rate in the lower extremity.

Introduction

Diagnosis of cervical spondylotic myelopathy is made by careful neurologic tests and radiologic examinations such as myelography. Satisfactory results of surgical treatment can be expected once the main factor causing myelopathy is removed. However, there are some cases who show no improvement, contrary to preoperative expectations. It is presumed that the poor results are due to incorrect diagnosis of the site of spinal disorder or to irreversibility of the damage to the spinal cord.

Recently we have been using evoked spinal cord potentials (spinal SEP) for determining the site of spinal disorder, with the object of improving the results of surgery. At first we used conductive spinal SEP recorded from the cervical spinal cord following stimulation of the thoracic region to monitor the function of the spinal cord. Since 1982, we have used not only the conductive spinal SEP showing the function of the lower extremity, but also the segmental spinal SEP elicited by stimulating the median nerve to evaluate the function of the upper extremity (Shimoji et al. 1971; Shinomiya 1983).

Material and Method

On the day before operation, a recording electrode was inserted into the cervical epidural space through a Tuohy catheter. The recording electrode consisted of a

[1]Tokyo Medical and Dental University, Department of Orthopaedic Surgery, 5-45, Yushima, 1-Chome, Bunkyo-ku, Tokyo, Japan

Spinal Cord Monitoring
Ed. by J. Schramm and S. J. Jones

polyethylene tube with five platinum bands spaced 15 mm apart. Five SEPs can be recorded simultaneously using this five-pole electrode, and the conduction velocity of the SEP can be easily calculated. This electrode is inserted below the site which is presumed to be the main lesion and is pushed forward so that the five poles span the main lesion site. It is important that the recording electrode should be located at the center of the cervical epidural space, in order to record the segmental spinal SEP equally from both sides. The stimulating electrode for the conductive spinal SEP consists of a polyethylene tube with two platinum bands, and is inserted into the thoracic epidural space at T8 or T9 through a Tuohy catheter. For the segmental SEP, a pair of stimulating percutaneous needle electrodes is placed in the subcutaneous tissue at the level of the elbow to stimulate the median nerve (Brown and Nash 1979).

Pulses of 0.1 ms duration at supramaximal intensity are used for stimulation. The bandwidth of the recording system is from 20 Hz to 2 kHz. Fifty to 100 stimuli are applied and averaged by a Medelec MS 92 or DISA 1500 averaging system. Monopolar epidural recordings are acquired simultaneously from the five epidural electrodes and an indifferent electrode is placed in the subcutaneous tissue of the neck. In order to record the SEP clearly, it is important that the patient be placed in a supine posi-

Table 1. J.O.A. score (Japan Orthopaedic Association)

1) Motor function of upper extremity
   - 0: Patient cannot eat with chopsticks or spoon
   - 1: Patient can eat with a spoon, but not with chopsticks
   - 2: Patient can eat with chopsticks, though not easy
   - 3: Patient eats with chopsticks, but in an awkward manner
   - 4: Normal
2) Motor function of lower extremity
   - 0: Patient cannot walk
   - 1: Patient needs a support to walk on a flat area
   - 2: Patient needs no support on a flat area, but needs it when going up and down the stairs
   - 3: Patient can walk without a support, but in an awkward manner
   - 4: Normal
3) Sensation
   - A: Upper extremity
     - 0: Distinct sensory disturbance
     - 1: Slight sensory disturbance
     - 2: Normal
   - B: Lower extremity same evaluation as for upper extremity
   - C: Trunk same evaluation as for upper extremity
4) Bladder function
   - 0: Retention of urine
   - 1: Severe urinary disturbance
   - 2: Mild urinary disturbance
   - 3: Normal

Table 2. Recovery rate

$$\text{Recovery rate (all)} = \frac{(\text{postop. J.O.A. score}) - (\text{preop. J.O.A. score})}{17 - (\text{preop. J.O.A. score})} \times 100\%$$

$$\text{Recovery rate (upper extremity or lower extremity)} = \frac{(\text{postop. J.O.A. score}) - (\text{preop. J.O.A. score})}{6 - (\text{preop. J.O.A. score})} \times 100\%$$

tion, since EMG activity originating in muscles around the cervical spine decreases more in the supine than in other positions. Diazepam or pentazocine is administered to some patients who cannot otherwise tolerate the electrical stimulation.

The recording and stimulating electrodes are left in the epidural space and are used for monitoring the function of the spinal cord during surgery. At one week following the operation, the spinal SEP is recorded again to evaluate the correlation between the recovery of the spinal SEP and postoperative neurologic function.

Since 1982, 50 cases have been examined in this series under relatively fixed conditions. Of them, 33 cases were monitored during surgery and 25 cases were examined postoperatively.

In Japan, cervical spondylotic myelopathy is assessed using an evaluation score devised by the Japan Orthopaedic Association (J.O.A. score). Motor function of the upper extremity is graded from 0 to 4, 4 meaning normal and 0 that the patient cannot eat with chopsticks or spoon. In the same way, motor function of the lower extremity is graded from 0 to 4, 4 meaning normal and 0 that the patient cannot walk at all. Sensory condition is graded from 0 to 2 for the upper extremity, the lower extremity and the trunk, 2 meaning normal and 0 severe disturbance. Bowel and bladder function is evaluated by a score from 0 to 3, 0 indicating retention of urine and 3 meaning normal function. A normal examination scores 17 points in all, comprising six points each for the upper and lower extremity (Table 1).

The postoperative recovery rate is calculated by this formula, for the patient as a whole and the upper and lower extremity (Table 2).

Correlation between the SEP and the J.O.A. score or recovery rate was analyzed using a computer, with the object of improving the results of surgical treatment.

Results

The wave form of the normal spinal SEP is shown at the top of Fig. 1. The conductive spinal SEP consists of two waves and the segmental SEP consists of N1 and N2. As symptoms worsen the conductive SEP changes into a single wave with low amplitude, and finally turns positive or disappears. N1 and N2 of the segmental SEP also show a decrease in amplitude and finally disappear as symptoms get worse.

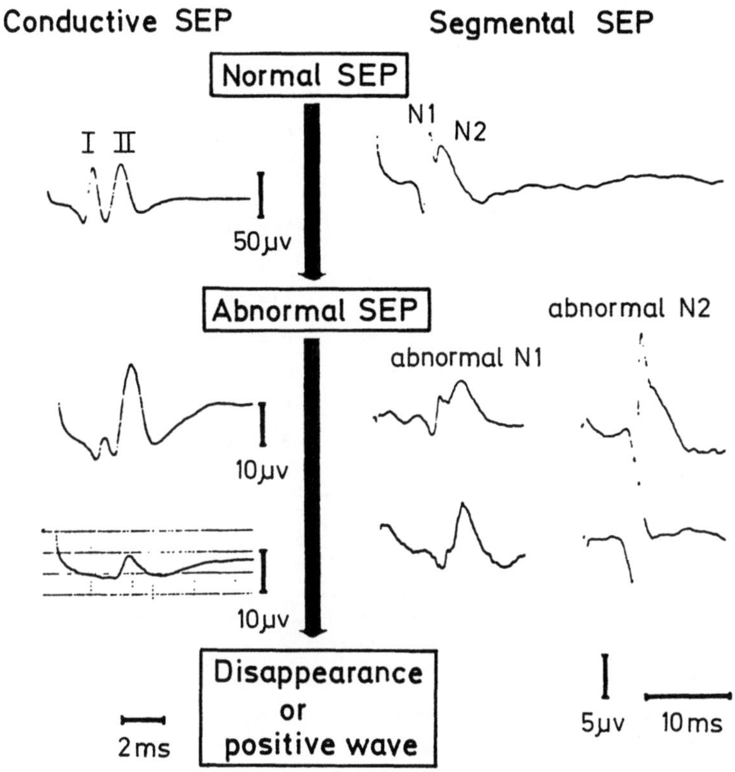

Fig. 1. The wave form patterns of the SEP

The wave form patterns of the segmental and conductive SEP were classified into several types. The segmental SEP was divided into four categories for N1 and N2 individually: 1. mild change, 2. moderate change, 3. severe change, 4. disappearance.

The conductive SEP was divided into seven types according to the amplitude decrement at the upper compared with the lower recording site: 1. normal (0–10%), 2. mild decrease (11–30%), 3. moderate decrease (31–50%), 4. severe decrease (51%–), 5. disappearance, 6. positive wave, 7. augmentation.

Segmental Spinal SEP

The segmental SEP was taken to represent the function of the upper extremity. In some cases N1, considered to originate in the primary afferent nerves, is not a single spike, but consists of multiple spikes. The latency of N1 is relatively constant from C7 to C5, but each spike is conducted at a certain fixed velocity above C5. N1 is likely to represent the potentials of primary afferent nerves in the posterior columns and the root at each level, but partly involves the potentials of primary afferent nerves from the lower roots.

Table 3. J.O.A. score (upper extremity, motor 4, sensory 2)

Type of segmental spinal SEP		(n)	Preop. J.O.A. score	Postop. J.O.A. score	Recovery rate
N_1	mild	(8)	3.75±1.47	5.25±0.66	60±30%
N_1	moderate	(18)	3.78±1.39	5.61±0.67	75±37%
N_1	severe	(9)	2.55±1.25	4.33±1.15	46±32%
N_1	disappeared	(8)	1.37±0.85	3.00±1.32	36±20%
N_2	mild	(6)	3.66±1.49	5.50±0.76	68±36%
N_2	moderate	(13)	3.46±1.59	5.30±1.06	64±41%
N_2	severe	(11)	3.09±0.99	4.90±0.89	65±24%
N_2	disappeared	(13)	2.38±1.82	3.84±1.60	44±32%
N_1, N_2 mild		(4)	3.75±1.78	5.25±0.82	52±35%
N_1, N_2 disappeared (block)		(7)	1.28±0.88	3.00±1.41	37±21%

Table 4. Recovery rate

Type of segmental SEP		0~30% (Poor) (n=12)	31~70% (Good) (n=13)	71~100% (Excellent) (n=18)
N1	Mild	8.3%	30.8%	16.7%
N1	Moderate	25.0%	23.1%	66.7%
N1	Severe	25.0%	30.7%	11.1%
N1	Disappeared	41.7%	15.4%	5.5%
N2	Mild	8.3%	15.4%	16.7%
N2	Moderate	33.3%	7.7%	44.4%
N2	Severe	8.3%	46.1%	22.2%
N2	Disappeared	50.0%	30.8%	16.7%

The segmental SEP changed when the stimulating frequency was increased from 5 to 50/s under anesthesia. The N2 wave showed a more marked decrement, supporting the general view that this component reflects postsynaptic potentials.

There was some correlation between the change of the segmental spinal SEP and the J.O.A. score of the upper extremity. The J.O.A. score decreased as the wave form of N1 and N2 gradually altered. Patients with conduction block, indicated by disappearance of the segmental SEP, showed the lowest J.O.A. score. There was also some correlation between the change of the segmental spinal SEP and the overall postoperative J.O.A. score. The "mild change" group showed a high postoperative J.O.A. score, whereas the score was low and the recovery rate unsatisfactory in the group for whom the SEP disappeared (Table 3).

Table 4 shows the frequency of various wave forms of the segmental spinal SEP within the operative region in three groups, divided according to the recovery rate (poor, good or excellent). In the "excellent" group the frequency of spinal SEPs with severe changes of N1 and N2 is low, while in the "poor" group the corresponding frequency is high.

Table 5 shows the correlation between the degree of change of the segmental SEP and the postoperative J.O.A. score. In the "poor" group disappearance and severe change of the spinal SEP are frequent. On the other hand, in the "excellent"

Table 5. Postoperative J.O.A. score

Type of segmental SEP		0~3 (Poor) ($n=7$)	4~5 (Good) ($n=19$)	6 (Excellent) ($n=17$)
N1	Mild	0	26.3%	17.6%
N1	Moderate	0	26.3%	76.4%
N1	Severe	28.6%	31.5%	5.9%
N1	Disappeared	71.4%	15.8%	
N2	Mild	0	10.5%	23.5%
N2	Moderate	14.3%	26.3%	41.2%
N2	Severe	14.3%	36.8%	17.6%
N2	Disappeared	71.4%	26.3%	17.6%

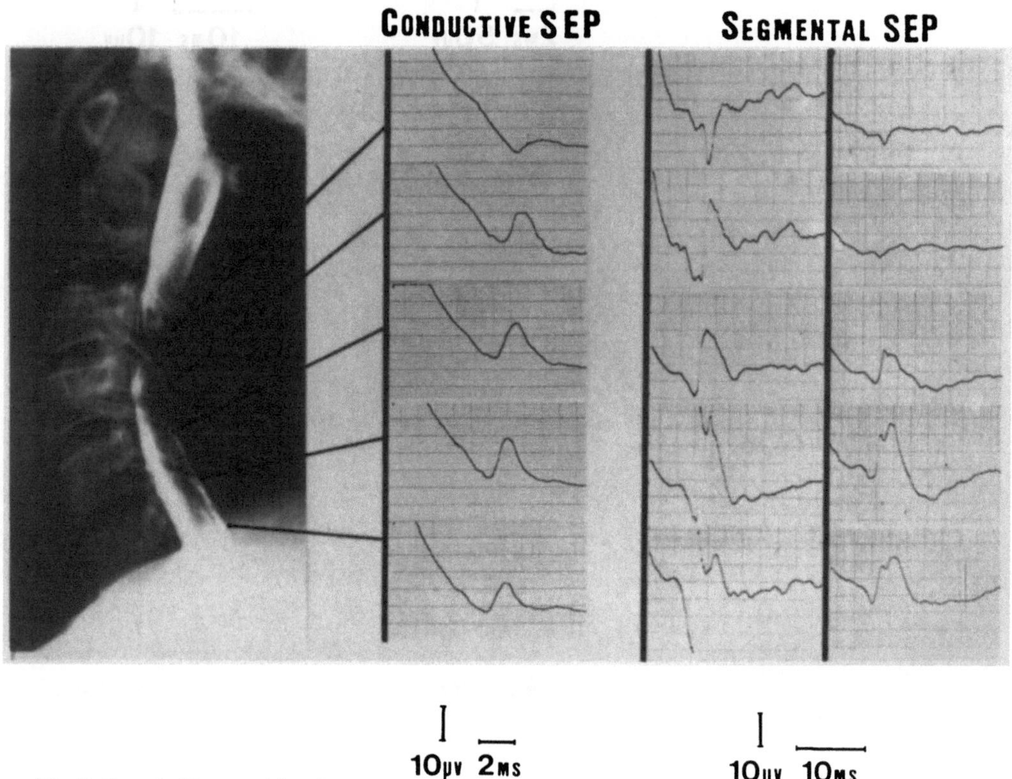

Fig. 2. Case 1, 50-year-old male

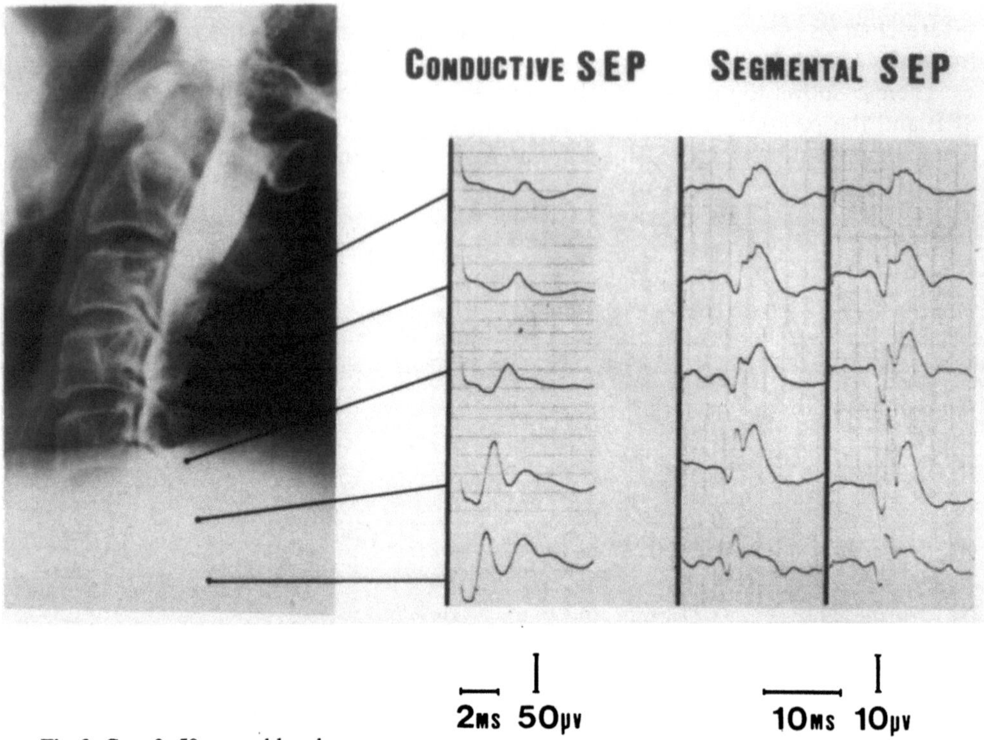

Fig. 3. Case 2, 50-year-old male

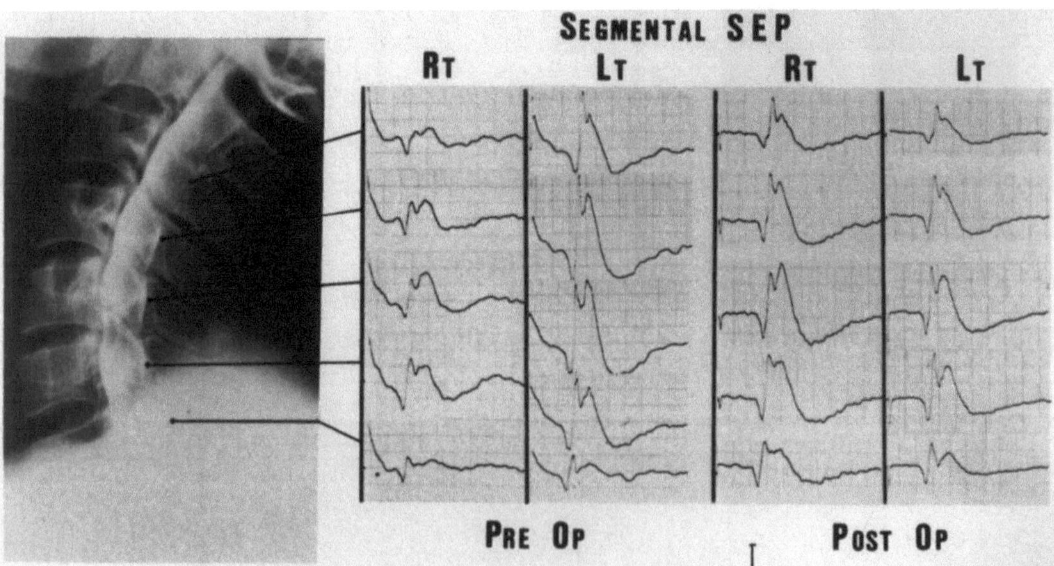

Fig. 4. Case 3, 39-year-old male

group the rate of mild and moderate changes is high. This tendency is prominent for N1.

Case 1, 50-year-old Male. The patient had been in a tetraplegic condition for the past two months. The left side segmental SEP showed conduction block at C4/5. The patient was operated by anterior decompression from C3 to C5, and immediately after the operation finger movement improved and sensation improved from anesthesia to hyperesthesia. Ten months after the operation, the J.O.A. score of the lower extremity was 2 and the recovery rate 33% (Fig. 2).

Case 2, 50-year-old Male. The J.O.A. score of the upper extremity was 3, and the segmental SEP showed only a slight change of the N1 wave. The patient recovered to normal after anterior decompression from C3 to C6 (Fig. 3).

Case 3, 39-year-old Male. Diagnosed as cervical disk herniation. The chief complaints were weakness of the right biceps muscle and hyperesthesia on the lateral aspect of the right forearm. In the preoperative segmental SEP the N1 wave on the right was slightly reduced. After anterior fusion at C5/6 the right segmental SEP recovered to almost normal, sensory disturbance disappeared and, three months after the operation, the power of the right biceps muscle recovered to normal (Fig. 4).

Conductive Spinal SEP

The conductive spinal SEP was taken to express the function of the lower extremity. There were some cases in whom the conductive SEP recorded caudal to the main lesion consisted of a single wave. In such cases the J.O.A. score of the lower extremity was lower than the cases who had two waves. Such a single-wave-type spinal SEP showed a lower preoperative conduction velocity (Table 6).

Table 7 shows the correlation between preoperative amplitude of the conductive SEP and the postoperative J.O.A. score. It is clear that the preoperative amplitude is high in cases who show a high postoperative J.O.A. score.

Table 8 shows the correlation between the preoperative conduction velocity and postoperative results. In the cases that show high postoperative J.O.A. scores, the preoperative conduction velocity is relatively fast. The postoperative conduction velocity (62 ± 20 m/s) tends to be faster than the preoperative conduction velocity (48 ± 18 m/s).

Table 6. Conductive spinal SEP (motor 4, sensory 2)

	Preop. J.O.A. score	Postop. J.O.A. score	Velocity
I Wave alone ($n = 13$)	2.1 ± 1.3	4.7 ± 1.4	36 ms $\pm$ 6
II Waves ($n = 20$)	4.2 ± 1.7	5.2 ± 0.9	55 ms $\pm$ 15

Table 7. Conductive spinal SEP

Postop. J.O.A. score	Amplitude	
1	12 ± 0 μV	(n = 1)
2		(n = 0)
3	26 ± 19 μV	(n = 3)
4	50 ± 51 μV	(n = 8)
5	49 ± 30 μV	(n = 12)
6	74 ± 73 μV	(n = 22)

Table 8. Conductive spinal SEP

Postop. J.O.A. score	Velocity	
1	36.3 ± 0 ms	(n = 1)
2		(n = 0)
3	22.7 ± 16.7 ms	(n = 3)
4	50.6 ± 16.4 ms	(n = 7)
5	48.7 ± 14.7 ms	(n = 12)
6	52.0 ± 15.0 ms	(n = 21)

Table 9. Conductive spinal SEP

Amplitude	Recovery rate	
	Poor (0∼30%)	Good (70∼100%)
Mild decrease (0∼30%)	78.5%	16.6%
Moderate decrease (31∼50%)	14.3%	25.0%
Severe decrease (51%∼)	7.2%	50.0%
Disappeared	0%	8.4%

Table 10. Positive wave (conductive spinal SEP)

	Postop. J.O.A. score	Recovery rate
Positive wave alone (n = 5)	3.8 ± 1.7	55 ± 27%
Positive wave + remnant wave (n = 5)	5.6 ± 0.8	68 ± 36%

Table 9 shows a comparison between the good recovery group and the poor recovery group, with regard to the degree of amplitude reduction of the conductive SEP at the site of operation. The "good" group tends to show a marked decrease of amplitude, whereas the "poor" group tends to show a slight decrease. Satisfactory results can be obtained by operation on the site where the conductive SEP shows a marked decrease.

There were 10 cases in whom the SEP was changed to a positive potential at the site of the main lesion. Five of these had residual SEP waves and the other five showed positive waves alone. It is clear that the cases who had positive waves alone showed markedly unsatisfactory results (Table 10).

Case 1 (continuation of history described above): This was a tetraplegic patient with a J.O.A. score of 0 in the lower extremity. The conductive SEP disappeared and changed to a positive potential at the site where myelography revealed severe obstruction. The patient was operated by anterior decompression from C3 to C5. His postoperative J.O.A. score of the lower extremity was 2 and the recovery rate was 33%. In such cases where the SEP shows a positive change, satisfactory results cannot be expected (Fig. 2).

Case 2 (continuation of history described above): The preoperative J.O.A. score of the lower extremity was 2. The conductive SEP had two waves caudal to C5/6, where myelography showed a partial block, but changed into a single wave with a marked decrease in amplitude at C5/6 level. After anterior decompression his J.O.A. score recovered to 5 and the recovery rate was 75% (Fig. 3).

Case 4, 49-year-old male, complaining of spastic gait. The conductive spinal SEP showed a marked decrease in amplitude and conduction velocity at C5/6. The chief complaint disappeared after surgery and the recovery rate was 100%. In such cases showing a clear decrease in amplitude and conduction velocity at the site of the main lesion, satisfactory results can be expected (Fig. 5).

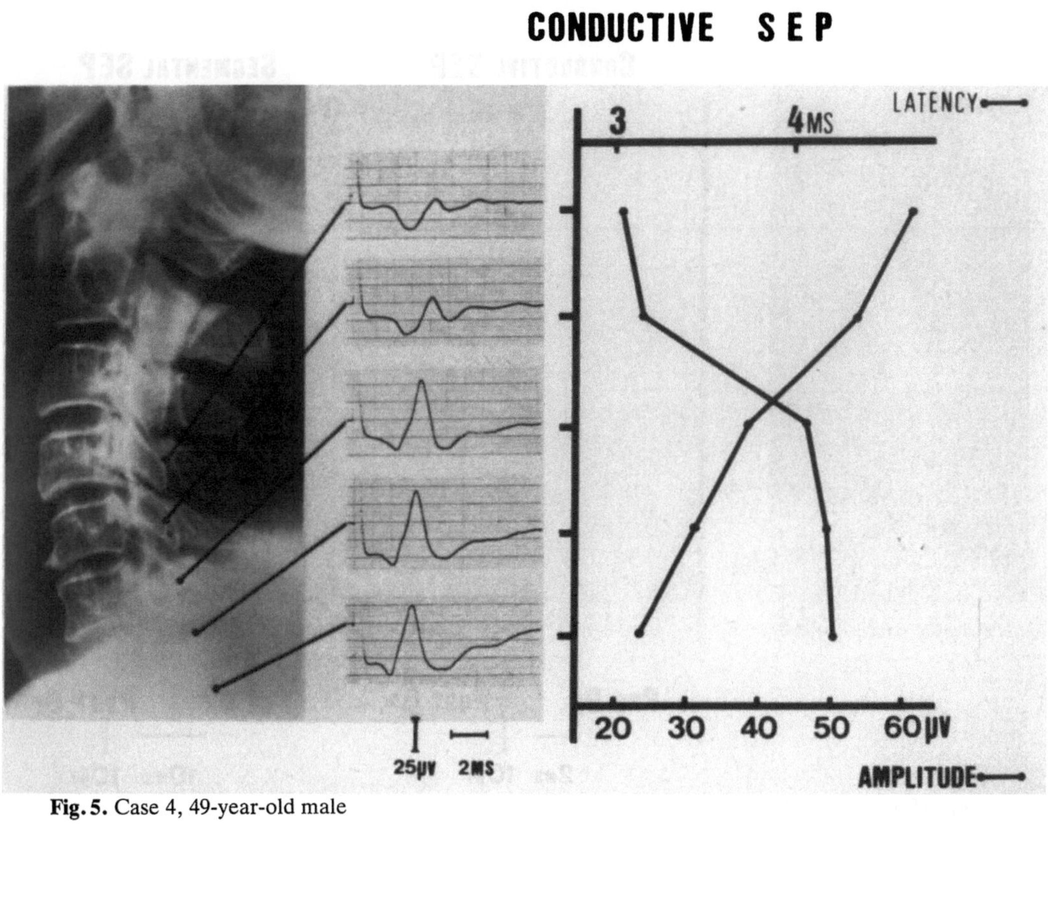

Fig. 5. Case 4, 49-year-old male

Discussion

In general a patient diagnosed as having cervical spondylotic myelopathy complains of symptoms in both the upper and the lower extremity. Therefore, examination using both segmental and conductive SEPs is considered essential to diagnose correctly the function of the cervical spinal cord.

Our system is invasive, but in cases who require surgery for myelopathy it is most important for the surgeons to know the function of the jeopardized cervical spinal cord and to determine correctly the site of the operation (Kurokawa 1978; Tamaki et al. 1981).

It was established by our fundamental study using cats that the N1 component of the segmental SEP originates in potentials of primary afferent nerves in the posterior columns at each level, or in multiple roots caudal to that level (Kaneda et al., this volume). In view of these findings it is difficult to determine the site of the main lesion by changes in N1 alone. On the other hand it is surmised that the site of damage to the cervical spinal cord can be determined more correctly by changes in the N2 component. This is because N2 is likely to originate in postsynaptic potentials of the posterior horns. In this study cases who showed minor changes of N1 and N2 recovered markedly after surgery, with respect to the function of the upper extremity. Conversely cases who showed disappearance of N1 and N2 had unsatisfactory postoperative results.

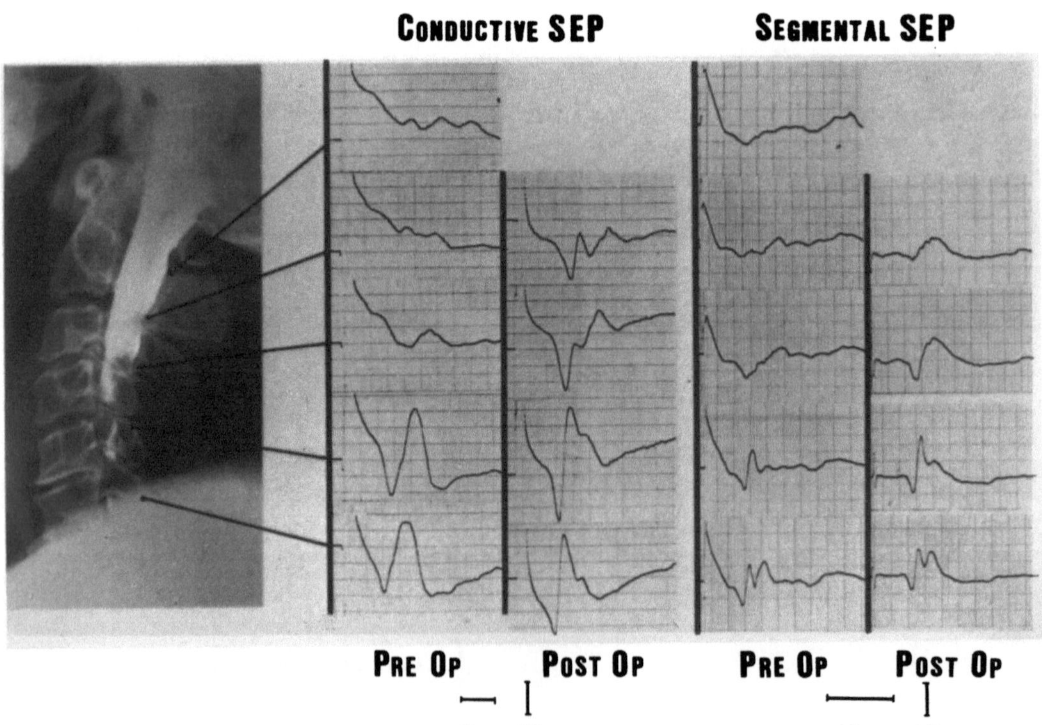

Fig. 6. Case 5, 65-year-old female

The conductive SEP is an aggregate of conducted potentials originating in both sensory and motor fibers of the white matter. Consequently, it can be suggested that the amplitude and conduction velocity of the conductive SEP is likely to decrease in proportion to the number of injured nerve fibers. According to our findings, in cases for whom the SEP showed low amplitude, slow conduction velocity or a single wave at the site caudal to the main lesion, or in whom the SEP was changed to a positive wave at the damaged region, the postoperative J.O.A. score of the lower extremity was low and the recovery rate was unsatisfactory.

Judging from this study, in order to obtain satisfactory results surgery must be performed before the function of the spinal cord has been severely damaged, namely while the SEP remains normal or shows only a slight change. Also in order to achieve good results, surgery must be performed on the site where there are definite changes of the SEP and where these findings are consistent with those of other examinations such as myelography.

As mentioned above, the conduction velocity had a tendency to become faster after surgery. By contrast, the wave form pattern of the SEP did not show any change for the better, excluding the cases of disk herniation that developed acutely such as case 3.

Case 5, 65-year-old female, diagnosed as cervical disk herniation. The total J.O.A. score was 7, improving to 16 postoperatively. The results were therefore thought to be satisfactory. The conductive SEP changed from a single wave to two waves and the block of the segmental SEP disappeared after surgery (Fig. 6).

References

Brown R, Nash C (1979) Current status of spinal cord monitoring. Spine 4:466–470

Kaneda A, Yamaura I, Kamikozura M, Shinomiya K, Sato H, Yokoyama M (1985) An analysis of spinal cord potentials evoked by median nerve stimulation. In: Schramm J, Jones SJ (eds) Spinal cord monitoring. Springer, Berlin Heidelberg NewYork Tokyo, pp 35–42

Kurokawa T (1978) Electrospinogram. Brain and Nerve 30:467–484

Shimoji K, Higashi H, Kano T (1971) Epidural recording of spinal electrogram in man. Electroenceph Clin Neurophysiol 30:236–239

Shinomiya K (1983) Monitoring of the spinal cord function using evoked spinal cord potentials during surgery of the cervical spine. Orthop Traum Surg 26:1961–1968

Tamaki T, Tsuji H, Inoue S, Kobayashi H (1981) The prevention of iatrogenic spinal cord injury utilizing the evoked spinal cord potentials. Int Orthop 4:313–317

SEP Following Tibial Nerve Stimulation in Spinal Cord Lesions

B. RIFFEL[1] and M. STÖHR

Summary

Somatosensory evoked potentials (SEP) following tibial nerve stimulation have been investigated in 30 normal subjects and in 118 patients with multiple sclerosis and various spinal cord diseases. Delayed sensory conduction indicated by prolonged absolute latencies and side-to-side differences of P40 were seen in 92% of MS patients. Additional recordings of spinal cord evoked responses helped to detect multifocal demyelination, to localize the demyelinating process, and to exclude peripherally delayed impulse conduction.

Spinal cord lesions due to tumors, cervical osteochondrosis or syringomyelia were mainly accompanied by a reduction in amplitude or even loss of the SEP, with or without only a slight or moderate delay of P40. In additional recordings of spinal evoked potentials the amplitude quotient of the cortical P40 and the lumbar N22 was the most important parameter to demonstrate a partial failure of impulse conduction in the spinal lemniscal pathway in these patients.

Introduction

Somatosensory evoked potentials (SEP) following median nerve stimulation have become widely used for diagnosis in various neurological disorders, whereas SEP following stimulation of mixed or sensory nerves of the lower limb (N. tibialis, peronaeus, suralis, saphenus) are less frequently investigated. For the examination of spinal cord diseases the latter stimulation site recommends itself because of the inclusion of the thoracic and lumbosacral sections of the spinal cord. We therefore investigated SEP following stimulation of the posterior tibial nerve in 49 subjects with definite MS, in 69 patients with space-occupying spinal cord diseases, and in a control group consisting of 30 healthy volunteers.

Subjects

The control group consisted of 30 subjects (mean age 34 years, mean height 1.68 m) without history, symptoms or signs of nervous system diseases (Riffel and Stöhr 1982).

[1] Abteilung für Neurologie und Klinische Neurophysiologie, Zentralklinikum, D-8900 Augsburg

Spinal Cord Monitoring
Ed. by J. Schramm and S. J. Jones

The MS group included 49 patients (mean age 35 years, mean height 1.68 m) with definite MS according to the criteria of McAlpine et al. (1972). The group of patients with different space-occupying lesions of the spinal cord ("tumor group") (mean age 49 years, height 1.67 m) comprised 36 subjects with spinal cord tumors, 13 patients with syringomyelia causing neurological signs in the legs, and 20 subjects with herniated cervical disc leading to medullary compression.

Methods

For stimulation of the posterior tibial nerve at the ankle, rectangular pulses of 0.1 ms duration were regularly applied via a bipolar surface electrode (cathode proximally) with a frequency of 5 Hz. The stimulus strength was set to 4 mA above the motor threshold of the flexor hallucis brevis muscle. Evoked potentials were recorded with platinum needle electrodes located 3 cm behind Cz (= Cz′). Additional spinal recordings above the spinous processes at C2, L1 and L5 were obtained in 30 normals, in 11 MS-patients, and in 8 patients of the "tumor group". The reference electrode for Cz′ and C2 was placed at Fz and for L1 and L5 over the iliac crest. The bandpass of the amplifier was set to 10–1000 Hz (scalp recording) and to 50–1000 Hz (spinal recording). On each side 1000 to 4000 sweeps were averaged. Latency and amplitude (baseline-to-peak) measurements were done from the screen with the appropriate cursor and waveforms were plotted by a system-integrated X-Y plotter.

Results

Control Group

The normative data have been described previously (Riffel and Stöhr 1982) and are summarized in Table 1a–c.

MS-Patients

Table 2 summarizes the mean latencies and amplitudes of cortical SEPs in this group. The mean latency of P40 was 55.0 ms with a range from 32 to 87 ms, the mean side-to-side difference of the P40 latency was 5.5 ms (range 0.1–34 ms). Seventy-three percent of the patients showed a prolonged latency of P40, 68% an increased side-to-side difference. Combining information from both measures, 92% of the patients exhibited abnormal results.

The latencies of the spinal evoked potentials at L5 and L1 were normal in all patients. Well-defined cervical potentials could only be obtained in 2 of the 11 patients and both showed a marked latency increase. The interpeak-latency (N22 at L1 to P40) was 32.9 ms on average and was significantly increased in 81% of measurements.

The mean amplitude of P40 was slightly reduced in comparison with the normal value (1.3 ± 1.0 μV versus 1.8 ± 1.3 μV). Using the amplitude quotient P40/N22 as an additional parameter (abnormal in 28% of patients), all patients with spinal recordings showed a pathologic SEP.

Table 1a. Latencies, amplitudes and side-to-side differences in 30 healthy volunteers (mean age 34 years, mean height 1.68 m)

Recording site	Compo-nents	Mean latency ± SD (ms)	Mean side-to-side difference of latency ± SD (ms)	Mean amplitude ± SD (μV)	Mean side-to-side difference of amplitude ± SD (μV)
L5	N18	18.4 ± 1.2	0.5 ± 0.4	0.3 ± 0.2	0.14 ± 0.15
		23.0 ± 1.5	0.6 ± 0.4		
L1	N22	21.7 ± 1.6	0.4 ± 0.3	0.6 ± 0.3	0.15 ± 0.14
C2	N30	29.5 ± 1.9	0.6 ± 0.5	0.6 ± 0.3	0.17 ± 0.15
Cz′	P40	38.8 ± 2.0	0.6 ± 0.6	1.8 ± 1.3	0.78 ± 0.69

Table 1b. Interpeak latencies and upper limits in 30 healthy volunteers

Recording sites	Components	Interpeak latencies ± SD (ms)	Upper limit (mean +2.5 SD) (ms)
L5, L1	N18–N22	3.8 ± 0.9	6.0
L1, C2	N22–N30	7.9 ± 1.0	10.4
C2, Cz′	N30–P40	8.8 ± 1.6	12.9
L1, Cz′	N22–P40	17.0 ± 1.7	21.3

Table 1c. Amplitude ratio and lower limits in 30 healthy volunteers

Recording sites	Components	Amplitude ratio (mean ± SD)	Lowest control value of amplitude ratio	Lower limit of amplitude ratio (mean −2.5 SD)
Cz′, L1	P40, N22	4.9 ± 1.77	0.85	0.47

Patients with Various Space-Occupying Lesions

Table 3 summarizes the mean latencies and amplitudes of cortical SEPs in this group of patients. The mean latency of P40 was 42.3 ms as compared with 38.8 ms in the control group, with a range from 35.5 to 53.2 ms; the mean side-to-side difference of P40 was 2.2 ms (range 0–8.6 ms). In 60% of all recordings (total $n = 136$) the latency of P40 was within the normal limits, and in only 15.4% slightly to moderately prolonged. In 24%, however, the cortical evoked response was missing.

The mean amplitude of P40 was markedly reduced in comparison with the normal value (0.8 ± 0.94 μV versus 1.8 ± 1.3 μV). When spinal evoked potentials were recorded additionally, in 65% the quotient of the cortical P40 and the lumbar N22 amplitudes was below the lowest value encountered in control subjects (0.85). In 36% of patients P40/N22 was below 0.47 (control mean −2.5 SD) (Fig. 1).

Table 2. Latencies, amplitudes and side-to-side differences in 69 patients with definite MS

	P40-Latency (ms)	Side-to-side difference P40-latency (ms)	P40-Amplitude (μV)	Side-to-side difference P40-amplitude (μV)
Mean	55.0	5.5	1.3	0.5
SD	13.7	5.9	1.0	0.5

Table 3. Latencies, amplitudes and side-to-side differences in 69 patients with space-occupying lesions of the spinal cord

	P40-Latency (ms)	Side-to-side difference P40-latency (ms)	P40-Amplitude (μV)	Side-to-side difference P40-amplitude (μV)
Mean	42.3	2.2	0.8	0.4
SD	3.8	1.7	0.9	0.4

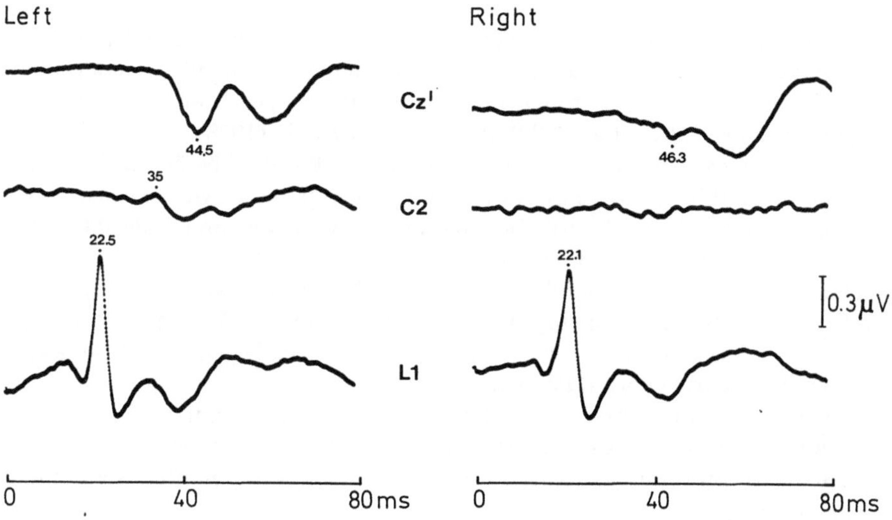

Fig. 1. 20-year-old male with spinal cord compression due to an extradural tumor (Th9–11). Tibial SEP following left and right side stimulation recorded at L1, C2 and Cz'. The lumbar N22 is normal on both sides. The interpeak latency N22 to N30 on the left hand side and the absolute latencies of P40 on both sides are slightly delayed. The amplitude quotient P40/N22 is below the lowest value of the control group on the left, and below the control mean −2.5 standard deviation on the right. This indicates a partial failure and a slight delay of impulse conduction in the spinal lemniscal pathway (Riffel and Stöhr 1982)

Discussion

MS-Group

Focal or multifocal demyelination produces some delay in sensory impulse conduction (McDonald and Sears 1970; Waxmann and Brill 1978). Using latency parameters, the total frequency of abnormal tibial SEPs was 92%, similar to the results of Trojaborg and Petersen (1979) who used median and peroneal nerve stimulation and found an abnormal SEP in 86% of their patients with definite MS. In other series of cases of definite multiple sclerosis (Eisen and Odusote 1980; Mastaglia et al. 1977; Namerow 1968) there were abnormal evoked responses following median nerve stimulation in 50% to 83%. The statistical probability of detecting pathological demyelination in multiple sclerosis seems to be higher after stimulation of the lower limb as compared with arm nerve stimulation, because the longer funiculus gracilis is more often affected by demyelination than the funiculus cuneatus.

The additional consideration of the interpeak latency N22 to P40 enables us to localize the site of the lesion within the central nervous system and to exclude the effects of peripheral neuropathy or a low skin temperature. If the cervical evoked response at C2 can be recorded, which is difficult even in normals, the demyelinating process can be located to the spinal or supraspinal part of the lemniscal sensory pathway. A prolongation of both interpeak latencies N22–N30 and N30–P40 allows the suggestion of multifocal demyelination.

In some cases of acute MS, conduction block may be marked and can be detected by using amplitude parameters (side-to-side difference or P40/N22-quotient).

Tumor Group

Latency parameters seem to be not very helpful in detecting a partial failure of impulse conduction due to cord compression. P40 latency was slightly to moderately prolonged in only 15.4% of recordings in the tumor group. This is in agreement with our previous results (Riffel et al. 1982) and those of Ertekin (1978) using intrathecal recording techniques. The findings are in contrast to those of Noël and Desmedt (1980), who invariably found delayed SEP-latencies in patients with extrinsic compression of the cervical spinal cord. In some cases a very low amplitude P40-wave may be overlooked and P58 may erroneously be taken for P40.

In patients with cervical spondylosis and spinal stenosis mainly amplitude abnormalities (of responses following median nerve stimulation) were observed by Caccia et al. (1976) and Siivola et al. (1981), whereas El-Negamy and Sedgwick (1979) described clearly delayed latencies.

In patients with spinal cord lesions, clinically judged to be complete, leads rostral to the lesion may yield no response (Cracco 1973; Riffel et al. 1984; Stöhr et al. 1982). If standard procedures are used, the loss of an evoked potential is a definite pathologic result.

The absolute amplitudes of the cortical evoked responses are not very helpful because of their large person-to-person variability. A more useful parameter seems to be the amplitude quotient P40/N22. Quotients below normal limits as defined by the absolute range of control values (0.85) or the mean -2.5 SD (0.47) seem to be important in demonstrating a partial failure of impulse conduction, for example by compression.

References

Caccia MR, Ubiali E, Andreussi L (1976) Spinal evoked responses recorded from the epidural space in normal and diseased humans. J Neurol Neurosurg Psychiatr 39:962–972

Cracco RQ (1973) Spinal evoked response: peripheral nerve stimulation in man. Electroenceph Clin Neurophysiol 35:379–386

Eisen A, Odusote K (1980) Central and peripheral conduction times in multiple sclerosis. Electroenceph Clin Neurophysiol 48:331–340

El-Negamy E, Sedgwick EM (1979) Delayed cervical somatosensory potentials in cervical spondylosis. J Neurol Neurosurg Psychiat 42:238–241

Ertekin C (1978) Evoked electrospinogram in spinal cord and peripheral nerve disorders. Acta Neurol Scand 57:329–344

Mastaglia FL, Black JL, Cala LA, Collins DWK (1977) Evoked potentials, saccadic velocities, and computerized tomography in diagnosis of multiple sclerosis. Br Med J 1:1315–1317

McAlpine D, Lumsden CE, Acheson ED (eds) (1972) Multiple sclerosis: A reappraisal, 2nd edn. Churchill Livingstone, Edinburgh

McDonald WI, Sears TA (1970) The effects of experimental demyelination on conduction in the central nervous system. Brain 93:583–598

Namerow NS (1968) Somatosensory evoked responses in multiple sclerosis patients with varying sensory loss. Neurology (Minneap) 18:1197–1204

Noël P, Desmedt JE (1980) Cerebral and far-field somatosensory evoked potentials in neurological disorders involving the cervical spinal cord, brainstem, thalamus and cortex. In: Desmedt JE (ed) Clinical uses of cerebral, brainstem and spinal somatosensory evoked potentials. Karger, Basel NewYork, pp 205–230

Riffel B, Stöhr M (1982) Spinale und subcorticale somatosensorisch evozierte Potentiale nach Stimulation des N. tibialis. Arch Psychiatr Nervenkr 232:251–263

Riffel B, Stöhr M, Petruch F, Ebensperger H, Schleglmann K (1982) Somatosensory evoked potentials following tibial nerve stimulation in multiple sclerosis and space-occupying spinal cord diseases. In: Courjon J, Mauguière F, Revol M (eds) Advances in neurology, vol 32: Clinical applications of evoked potentials in neurology. Raven Press, NewYork, pp 493–500

Riffel B, Stöhr M, Körner S (1984) Spinal and cortical evoked potentials following stimulation of the posterior tibial nerve in the diagnosis and localization of spinal cord diseases. Electroenceph Clin Neurophysiol 58:400–407

Siivola J, Sulg I, Heiskari M (1981) Somatosensory evoked potentials in diagnostics of cervical spondylosis and herniated disc. Electroenceph Clin Neurophysiol 52:276–282

Stöhr M, Dichgans J, Diener HC, Buettner UW (1982) Evozierte Potentiale SEP – VEP – AEP. Springer, Berlin Heidelberg NewYork

Trojaborg W, Petersen E (1979) Visual and somatosensory evoked cortical potentials in multiple sclerosis. J Neurol Neurosurg Psychiatry 42:323–330

Waxmann SG, Brill MH (1978) Conduction through demyelinated plaques in multiple sclerosis: computer simulations of facilitation by short internodes. J Neurol Neurosurg Psychiatry 41: 408–416

Single and Double Stimuli SEP Examinations in Spinal Tumors and in Multiple Sclerosis

H. Gerhard[1], K. Wurzer, G. Demmer, K. H. Jordan, and J. Jörg

Summary

In order to improve the prospect of differential diagnosis between multiple sclerosis and spinal tumor cases using tibial nerve SEP, responses to single and double stimuli were studied in up to 117 patients. No significant differences between MS and tumor groups were found for the single stimulus responses. However, with the double stimuli definite MS cases were found to have a significantly increased relative refractory time, whereas in 6 out of 8 tumor cases the absolute refractory time was abnormal.

Introduction

If there is a compression of the spinal cord, spinal tumors lead to a more or less fully developed sensory disorder in addition to motor deficiency symptoms, bladder- and rectum complaints. Phlogistic diseases of the spinal cord, the majority of which occur within the course of multiple sclerosis, also show transverse lesion symptoms with more or less developed sensory disorder. In the differential diagnosis of spinal tumors and multiple sclerosis electrophysiological examinations are increasingly applied (Cracco 1980; Dimitrijevic et al. 1980; Ertekin et al. 1980; Ganes 1980; Riffel and Stöhr 1982; Stöhr et al. 1982). The afferent system can be examined with the help of somatosensory evoked potentials. The tibial nerve SEP is of especial value, as through it the whole afferent system can be examined. Another method of examination of the afferent system, requiring more time, is to study the effect of double stimuli on the tibial nerve SEP and to determine the relative and absolute refractory times.

The present study was performed with the object of determining whether a differential diagnosis between multiple sclerosis with spinal involvement and spinal tumor is possible through scalp SEP recording, as some authors (Riffel and Stöhr 1982; Stöhr et al. 1982) suggest.

Another objective was to discover whether the differential diagnosis between multiple sclerosis and spinal tumor can be improved through the application of double stimuli SEP examinations.

Material and Methods

The single stimulus examination of the scalp SEP was performed in 33 control subjects of the age of 15 to 60 years. In 25 control subjects the refractory time of the tibial nerve SEP was examined.

[1]Neurologische Universitätsklinik, Hufelandstr. 55, D-4300 Essen

Spinal Cord Monitoring
Ed. by J. Schramm and S. J. Jones

The case material consisted of 96 patients with definite, probable or possible multiple sclerosis and 23 patients with a spinal tumor. In all these patients the single stimulus SEP was examined. In 15 MS-patients and 8 patients with a spinal tumor double stimuli SEP were examined. Stimulation of the nerve trunk at the ankle was performed at an intensity defined by the sum of the motor and sensory thresholds leading to a strong, though not painful, contraction of the muscle. Rectangular impulses were delivered with an intensity of 4 to 5 mA, a duration of 0.2 ms and a frequency of 3 Hz. Cortical SEPs were recorded from a site midway between C_z and P_z (10-20 system), with the reference at F_z. Band pass was 2–2000 Hz, the AD sampling rate 10 kHz. The SEP of the single stimulus was the referential potential for the SEP of the double stimuli. The double stimuli were applied with varying interstimulus intervals (60, 50, 40, 30, 25, 20 to 10 ms in 1 ms intervals). For determining the absolute refractory time the interstimulus intervals were varied in steps of 0.5 ms. The absolute refractory time corresponds to the greatest interstimulus interval which does not show a response to the second stimulus. The relative refractory time was determined for the initial positive potential (P_1), corresponding to the interstimulus interval at which the latency of P_1 in the double stimuli SEP differed from that in the single stimulus SEP by more than 2 ms. P_1 latencies and amplitudes measured from P_1 to the following negativity (N_2) were compared in patients with MS and patients with a spinal tumor with the help of the medians, since by this method missing SEPs could also be included in the comparison. The two groups of patients were compared by means of the Mann-Whitney U-test.

Results

Control Subjects

The cortical SEP of the 33 control subjects showed 4–6 peaks during the first 100 ms analysis time in most cases. The value of the latency of P_1 can be seen in Table 1. The correlation coefficient for body height and the peak latency of P_1 was $r^2 = 0.53$ ($y = 0.13x + 18.5$).

The SEP refractory period was determined through the change of the latency of P_1 (Table 1). At shortening interstimulus interval a lengthening of latency occurred. The relative refractory period was defined as the interstimulus interval at which the latency increased by at least 2 ms. The absolute refractory time in normal subjects was 25.3 ± 3.9 ms and the relative refractory time was 33.2 ± 6.5 ms (Table 1, Fig. 1).

Table 1. Mean values and standard deviations of P_1 latency, N_2 latency and P_1/N_2 amplitude and absolute and relative refractory times of the P_1 peak in the tibial nerve SEP in the control group

P_1 latency (ms) $\bar{x} \pm s$	N_2 latency (ms) $\bar{x} \pm s$	P_1/N_2 amplitude (μV) $\bar{x} \pm s$	Absolute ref. time (ms) $\bar{x} \pm s$	Relative ref. time (ms) $\bar{x} \pm s$
41.4 ± 2.8 ($n = 33$)	49.5 ± 3.2 ($n = 32$)	2.5 ± 1.0 ($n = 32$)	25.3 ± 3.9 ($n = 25$)	33.2 ± 6.5 ($n = 25$)

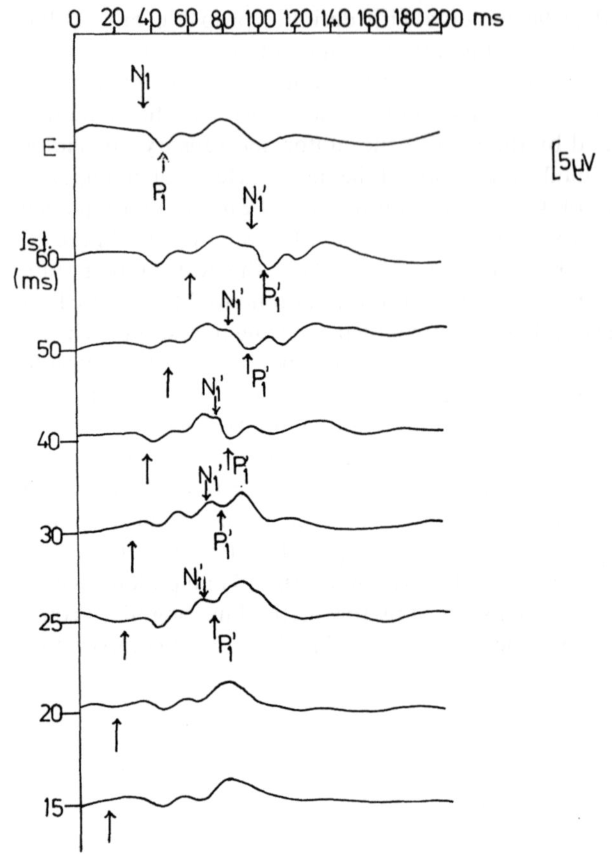

Fig. 1 Cortical SEP following single and double stimulation of the tibial nerve in a 31-year-old control subject. The arrows indicate delivery of the second stimulus and the N_1' and P_1' components. *E*, Single stimulus SEP; *Ist*, interstimulus interval

Table 2. Mean values and standard deviations of P_1 and N_2 latencies and P_1/N_2 amplitude of the tibial nerve SEP, plus number of missing SEPs in patients with definite, probable and possible MS and patients with spinal tumors

	Definite MS ($n=32$)	Probable MS ($n=32$)	Possible MS ($n=32$)	Spinal tumors ($n=23$)
Latencies (ms)				
$\bar{x} \pm s$				
P_1	54.0 ± 15.3	50.1 ± 8.7	43.3 ± 5.5	48.1 ± 5.8
	($n=26$)	($n=27$)	($n=30$)	($n=20$)
N_2	63.7 ± 17.7	60.5 ± 13.8	51.6 ± 6.4	56.3 ± 3.4
	($n=27$)	($n=27$)	($n=31$)	($n=19$)
Amplitude (μV)				
$\bar{x} \pm s$				
P_1/N_2	2.3 ± 1.4	1.7 ± 1.1	2.4 ± 1.8	1.6 ± 1.3
	($n=22$)	($n=23$)	($n=30$)	($n=19$)
Number of missing scalp SEP	5	5	1	3

Table 3. Median values of the P_1 latencies and P_1/N_2 amplitudes of the tibial nerve SEP in patients with definite, probable and possible MS and with a spinal tumor

	Positive MS		Probable MS		Possible MS		Total MS		Spinal tumors	
	ri.	le.	ri.	le.	ri.	le.	ri.	le.	ri.	le.
Latency (ms)										
P_1	50	48	50	48.5	43	42.5	48	47	46	47
Amplitude (μV)										
P_1/N_2	1.5	1.5	1	1	2	2.5	1.5	1.5	1	1

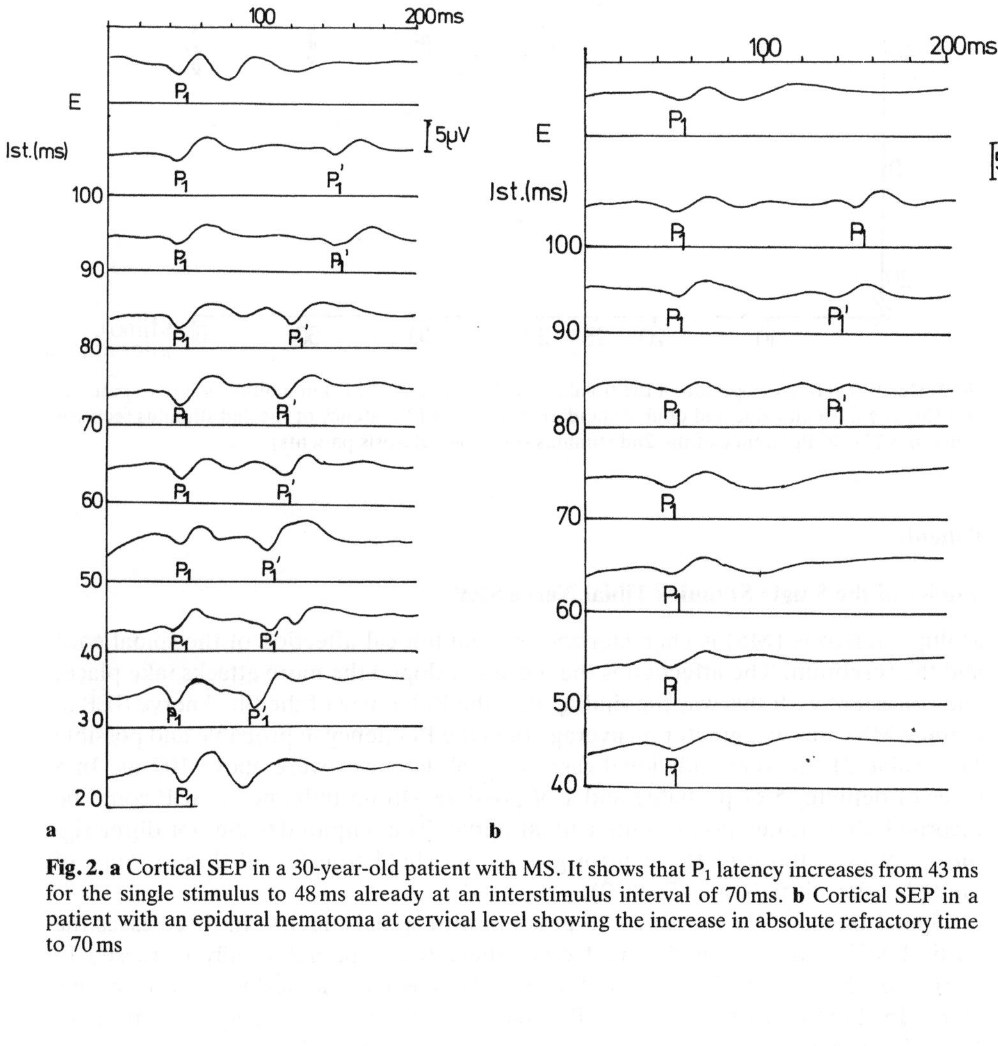

a b

Fig. 2. a Cortical SEP in a 30-year-old patient with MS. It shows that P_1 latency increases from 43 ms for the single stimulus to 48 ms already at an interstimulus interval of 70 ms. **b** Cortical SEP in a patient with an epidural hematoma at cervical level showing the increase in absolute refractory time to 70 ms

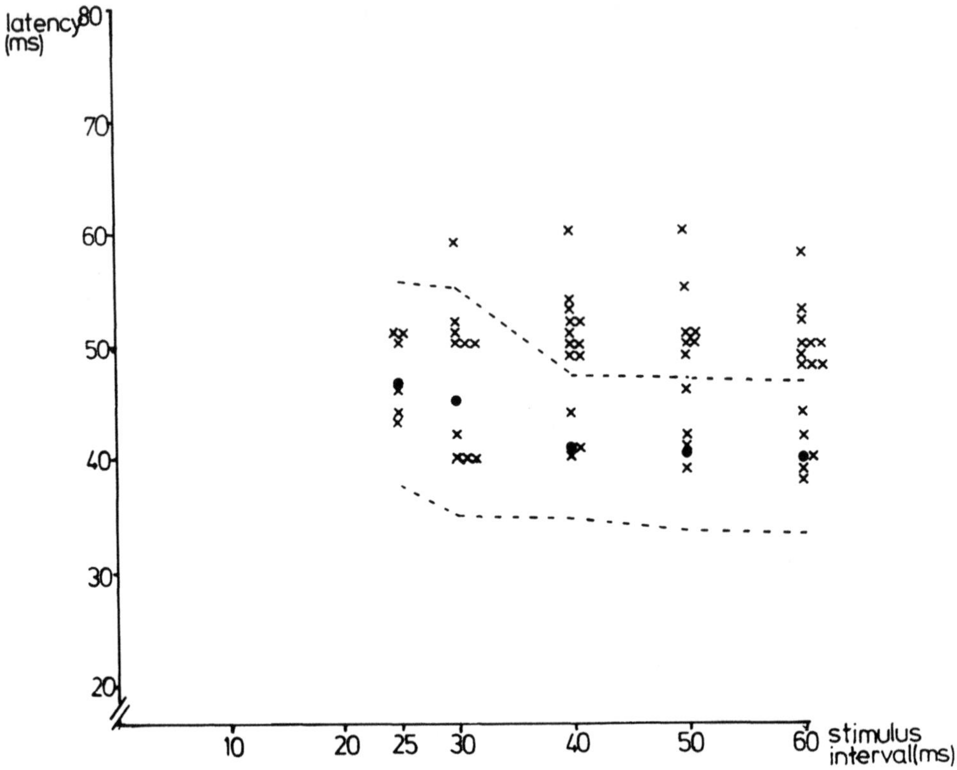

Fig. 3. Relative refractory period of the tibial nerve SEP (latency of P_1) in multiple sclerosis patients ($n = 15$). –•– Mean value and double standard deviation of P_1 latency of the 2nd stimulus (control group; $n = 25$). × P_1 latency of the 2nd stimulus (multiple sclerosis patients)

Patients

Studies of the Single Stimulus Tibial Nerve SEP

Multiple sclerosis (MS) is characterized by a multifocal affection of the spinal cord and the cerebrum. The affection is the more developed the more attacks take place, and consistent with this was the finding that the P_1 latency of the tibial nerve SEP in definite MS cases was greater on average than the P_1 latency in probable and possible MS (Table 2). In some individual cases P_1 peak latencies were above 100 ms. In 5 cases of definite, 5 of probable and 1 of possible MS no tibial nerve SEP could be recorded after stimulation of either tibial nerve. The amplitudes did not differ significantly from those of the control group. This held true for all three groups of multiple sclerosis patients (Table 2).

Spinal tumors: We examined 23 patients with spinal tumors. In three cases the cortical SEP was missing. In eight cases there was a pathologically increased P_1 latency of the scalp SEP. The amplitude P_1/N_2 was pathologically reduced in nine cases. In three cases the spinal SEP made localization of a spinal tumor possible through recording above and below the tumor. In one case it became evident that, in

Table 4. Relation of SEP latency to interstimulus interval, plus absolute and relative refractory times of the tibial nerve SEP in control subjects, patients with definite multiple sclerosis and with a spinal tumor

	Normal subjects	Multiple sclerosis patients	Patients with spinal tumors
Single stimulus	40.0 ± 3.4 ($n = 25$)	45.7 ± 5.9 ($n = 15$)	45.9 ± 3.5 ($n = 8$)
Double stimuli			
Interstimulus interval (ms)			
100	—	—	46.5 ± 4.2 ($n = 8$)
80	—	—	46.3 ± 4.2 ($n = 8$)
60	40.1 ± 3.4 ($n = 25$)	47.2 ± 5.7 ($n = 15$)	48.2 ± 5.8 ($n = 7$)
50	40.5 ± 3.4 ($n = 25$)	48.5 ± 5.7 ($n = 15$)	48.1 ± 5.9 ($n = 3$)
40	41.2 ± 3.3 ($n = 25$)	49.7 ± 5.4 ($n = 13$)	—
30	45.1 ± 5.1 ($n = 21$)	50.0 ± 6.7 ($n = 10$)	—
25	46.1 ± 4.6 ($n = 14$)	50.5 ± 3.6 ($n = 6$)	—
Absolute refractory time	25.3 ± 3.9 ($n = 25$)	27.3 ± 7.0 ($n = 10$)	50.5 ± 4.8 ($n = 8$)
Relative refractory time	33.2 ± 6.5 ($n = 25$)	45.3 ± 11.7 ($n = 10$)	—

spite of a missing spinal SEP, a scalp SEP was present with reduced amplitude and prolonged latency (Table 3).

When the two groups were compared (spinal tumors and the 3 MS groups combined) with respect to latency of P_1 and P_1/N_2 amplitude with the aid of the U-test, there was found to be no significant difference.

Studies of Refractory Time

Examining the relative refractory time of the SEP following double stimulation of the tibial nerve in patients with definite MS ($n = 15$), it was demonstrated that the increase of latency defining the relative refractory time took place earlier than in control subjects, whereas the absolute refractory time was not significantly different (Figs. 2, 3; Table 4).

In contrast, a pathological absolute refractory time was found in six out of eight cases of spinal tumor (Fig. 2). The MS patient with a pathological absolute refractory time in combination with a normal single stimulus SEP and pathological relative refractory time is illustrated in Fig. 2.

Discussion

Through use of electrophysiological techniques, amplitude-reduced spinal and cortical SEP recorded above the site of compression can be found in most cases of spinal tumors, and in some cases latency-increased spinal and cortical SEP also. However, the value of determining the amplitude of the cortical SEP is limited on account of the great interindividual variability. In MS, on the other hand, more latency-increased cortical SEP are found than in spinal tumor cases.

Our results correspond to those of other authors (Noël and Desmedt 1980; Stöhr et al. 1982; Riffel and Stöhr 1982; Riffel et al. 1984). These results show, however, that a differential diagnosis of MS with spinal tumor using the tibial nerve SEP has to be made with care in individual cases. There was no significant difference in the single stimulus SEP between the two groups of patients with MS and spinal tumor as regards P_1 latency or P_1/N_2 amplitude. An improvement of the technique with regard to the possibility of differential diagnosis has been attempted through the recording of the spinal conduction time (Dorfman and Bosley 1979; Eisen and Odusote 1980; Ganes 1980; Riffel et al. 1984). The value of this method is reduced through the fact that the spinal SEP cannot always be recorded in a satisfactory way.

A further possibility for determining the differential diagnosis of multiple sclerosis with spinal tumor is to measure the refractory time of the scalp-recorded tibial nerve SEP. Up to now such refractory times have mainly been studied and used diagnostically with respect to the median nerve SEP (Shagass and Schwartz 1964; Jörg et al. 1980a, b; Allison 1962). Further observations of the refractory time of the cortical SEP following stimulation of the tibial nerve have been made by Tsumoto et al. (1972), Sauer and Schenck (1977) and Gerhard et al. (1982, 1983, 1984).

In the initial phase of MS the thicker myelinated fibres are preferentially affected. Thus the faster conducting fibres are partly non-functioning (Namerow 1970). This is evident through the increase of the relative refractory time.

In acute spinal compression a more or less developed injury through compression across the whole spinal cord appears. This becomes evident through the reduced amplitude of the scalp SEP.

It has been suggested by Noël and Desmedt (1980) that an amplifying effect at the cortex might increase the amplitude of the SEP in patients with spinal tumors and thus make the defect less marked. Through determining the absolute refractory time this effacing effect of the cortex can be cancelled again. It was demonstrated in the present study that the absolute refractory time of the scalp SEP was pathological in all cases of spinal tumors. Therefore, based on the study of the refractory time, an improvement of the differential diagnosis might be attained.

References

Allison T (1962) Recovery functions of somatosensory evoked responses in man. Electroencephalogr Clin Neurophysiol 14:331–343

Cracco RQ (1980) Spinal evoked potentials. In: Desmedt JE (ed) Clinical uses of cerebral, brainstem and spinal somatosensory evoked potentials. Karger, Basel, pp 1–14

Dimitrijevic MR, Prevec TS, Sherwood A, McKay WB, Butinar D, Brecelji J (1980) Somatosensory perception and cortical evoked potential in established paraplegia. Paper presented at: Symposium International. Applications Cliniques des Potentials Evoqués en Neurologie, Lyon

Dorfman LJ, Bosley TM (1979) Age-related changes in peripheral and central nerve conduction in man. Neurology 29:38–44

Eisen A, Odusote K (1980) Central and peripheral conduction times in multiple sclerosis. Electroencephalogr Clin Neurophysiol 48:253–265

Ertekin C, Mutlu R, Sarica Y, Uckardesler L (1980) Electrophysiological evaluation of the afferent spinal roots and nerves in patients with conus medullaris and cauda equina lesions. J Neurol Sci 48:419–433

Ganes T (1980) Somatosensory evoked responses and central afferent conduction times in patients with multiple sclerosis. J Neurol Neurosurg Psychiatry 43:948–953

Gerhard H, Jörg J, Selter I, Jansen H (1982) Spinale und cerebrale SEP des N. tibialis. In: Struppler A (Hrsg) Elektrophysiologische Diagnostik in der Neurologie. Thieme, Stuttgart New York, S 162–163

Gerhard H, Jörg J, Jansen H, Selter I (1983) Das spinale und cerebrale SEP bei Einzel- und Doppelreiz des N. tibialis. Arch Psychiatr Nervenkr 233:297–306

Gerhard H, Jörg J, Hartjes H, Jansen H (1984) Vergleichende Untersuchung von somatosensorisch evozierten Potentialen, Ganzkörpercomputertomographie und Myelographie bei spinalen Raumforderungen. In: Hohmann D, Kügelgen B, Liebig K, Schirmer M (Hrsg) Neuroorthopädie 2. Springer, Berlin Heidelberg New York Tokyo, S 82–88

Jörg J, Gerhard H, Lehmann HJ (1980a) Cerebral refractory period of the somatosensory system in diseases of the central nervous system. EEG-EMG 4:218–222

Jörg J, Gerhard H, Lehmann HJ (1980b) Somatosensorische Reizantwortpotentiale bei Normalpersonen nach Einzel- und Doppelstimulation des N. medianus. EEG-EMG 11:211–217

Namerow NS (1970) Somatosensory recovery functions in multiple sclerosis patients. Neurology 20:813–817

Noël P, Desmedt JE (1980) Cerebral and far-field somatosensory evoked potentials in neurological disorders involving the cervical spinal cord, brainstem, thalamus and cortex. In: Desmedt JE (ed) Clinical uses of cerebral, brainstem and spinal somatosensory evoked potentials. Karger, Basel, pp 205–230

Riffel B, Stöhr M (1982) Spinale und subkortikale somatosensorisch evozierte Potentiale nach Stimulation des N. tibialis. Arch Psychiatr Nervenkr 232:251–263

Riffel B, Stöhr M, Körner S (1984) Spinal and cortical evoked potentials following stimulation of the posterior tibial nerve in the diagnosis and localization of spinal cord diseases. Electroencephalogr Clin Neurophysiol 58:400–407

Sauer M, Schenck M (1977) Vergleichende Untersuchung somatosensibler spinaler und kortikaler Evoked Potentials bei Kindern. Arch Psychiatr Nervenkr 223:295–308

Shagass C, Schwartz M (1964) Recovery functions of somatosensory peripheral nerve and cerebral evoked responses in man. Electroencephalogr Clin Neurophysiol 17:128–133

Stöhr M, Dichgans J, Diener HC, Buettner UW (1982) Evozierte Potentiale. Springer, Berlin Heidelberg New York

Tsumoto T, Hirose N, Nonaka S, Takahashi M (1972) Analysis of somatosensory evoked potentials to lateral popliteal nerve stimulation in man. Electroencephalogr Clin Neurophysiol 33:379–388

Peripheral Autonomic Surface Potential — Applications in Clinical Neurology

S. BAJADA[1] and W. KNEZEVIC

Introduction

Electrophysiological tests of spinal cord function have developed rapidly over the past few years, initially with sensory evoked potentials and latterly with methods of testing motor pathways. In seeking a simple reproducible test of sympathetic function we have examined a previously little-studied skin potential which we have named the peripheral autonomic surface potential (PASP). It was initially described by Tarchanoff in 1890. We believe this potential may provide a basis for examining lateral column spinal cord function.

Method

The PASP was recorded in relaxed subjects and patients, with lights dimmed, ambient temperature controlled at 20–23°C, and recorded limb temperature at 34°C. Recording electrodes were tin surface electrodes attached to the palmar surface of the hand (cathode), palmar surface of the middle finger (anode) (Fig. 1), and to the foot on the sole and plantar surface of the second toe (Fig. 2). A ground plate was placed at the opposite wrist and randomly timed, 15–20 mA, 0.2 ms duration electrical

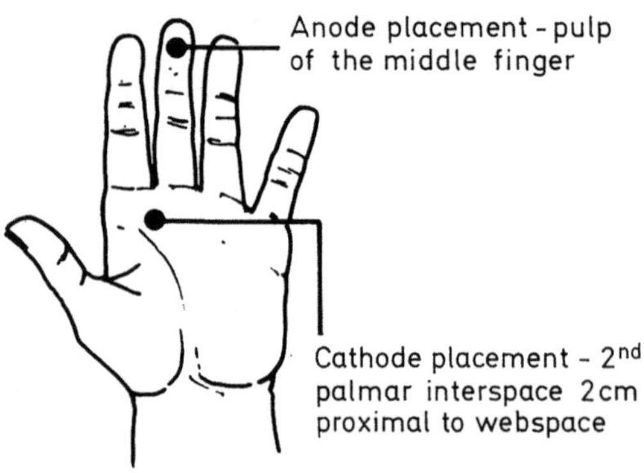

Fig. 1. Recording electrode placement for the hand

[1] Department of Neurology and Clinical Neurophysiology, Fremantle Hospital, Fremantle, Western Australia

Spinal Cord Monitoring
Ed. by J. Schramm and S. J. Jones

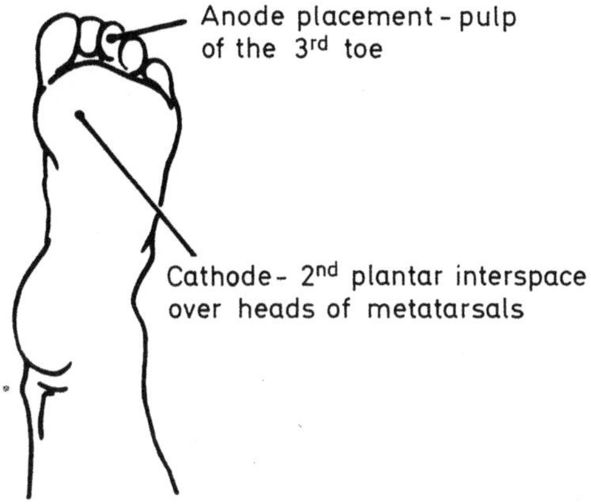

Fig. 2. Recording electrode placement for the foot

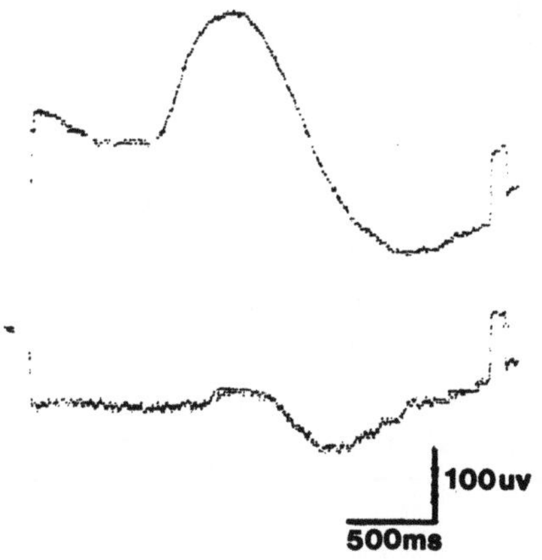

Fig. 3. Normal palmar *(upper trace)* and plantar *(lower trace)* PASPs (averaged × 4)

stimuli were given to the median nerve at the wrist contralateral to the recording side. Responses were recorded on a DISA 1500 Digital EMG system, band pass 2 Hz to 5 kHz, amplification sensitivity 50 to 500 μV/cm and sweep speed 500 ms/cm. The distance from the C1 vertebra to the hand (dH) and the C1 vertebra to the foot (dF) was measured in each subject.

The PASP was recorded in 30 healthy volunteers, 10 diabetic patients, 5 patients with unilateral cervical or lumbar sympathectomy, and 5 patients with spinal cord lesions. In 5 normal volunteers the PASP was recorded before and after iontophoresis of atropine into the skin under the recording electrode.

Results

The normal PASP consists usually of a biphasic potential with an initial negative peak (Fig. 3). Table 1 shows the normal PASP parameters. Atropine iontophoresed through the skin in 5 subjects abolished the potential in each for at least 20 min.

Table 1. Mean onset latencies and peak-to-peak amplitudes of palmar and plantar PASPs in 30 normal subjects

	Latency (s)	Amplitude (μV)
Palmar PASP	1.52 ± 0.13	479 ± 105
Plantar PASP	2.07 ± 0.16	101 ± 40

All values expressed as mean ± standard deviation

Table 2. Patient data. Spinal cord lesions

Patient no., sex, age (years)	Diagnosis	Duration of illness (years)		PASP			
				Palmar		Plantar	
				Lat. (s)	Ampl. (μV)	Lat. (s)	Ampl. (μV)
1, F, 19	Traumatic cervical myelopathy	4.5	R	1.58	180	N.R.	
			L	1.26	1000	N.R.	
2, M, 28	Traumatic cervical myelopathy	2.0	R	1.09	300	2.91	2
			L	1.34	360	N.R.	
3, M, 25	Traumatic cervical myelopathy	3.0	R	1.89	180	N.R.	
			L	1.73	400	1.28	50
4, F, 47	Thoraco-cervical transverse myelitis	0.3	R	3.60	50	N.R.	
			L	N.R.		N.R.	
5, M, 54	Lumbar cord ependymoma	10.0	R	1.82	110	N.R.	
			L	1.59	220	2.25	100

N.R., No response

Abnormality of PASP parameters was found in all diabetic patients. Mean palmar and plantar PASP amplitudes were $179 \pm 158\,\mu$V and $17 \pm 16\,\mu$V respectively and were significantly different from controls ($P < 0.001$).

In 3 patients with cervical sympathectomy (6 weeks to 4 years prior to testing) the PASP was present in the normal upper limb and absent on the sympathectomized side. In 2 patients with lumbar sympathectomy (2 weeks prior to testing in each) the PASP amplitude was markedly decreased or absent on the sympathectomized side.

Five patients with spinal cord lesions have been studied. The patient data and PASP results are shown in Table 2. Plantar PASPs are either absent or markedly reduced in amplitude in 4 patients. In patient 5 with the lumbar ependymoma, absent right ankle jerk and absent right 'H' reflex, the plantar PASP is present on the left and absent on the right.

In 30 normal subjects the following formula was used to calculate a central autonomic velocity (CAV):

$$\mathrm{CAV} = \frac{(\mathrm{dF} - \mathrm{dH})\ \mathrm{m}}{(\mathrm{Lpf} - \mathrm{LpH})\ \mathrm{s}}$$

where dF and dH are the respective distances in metres from the C1 vertebra to the foot and hand recording electrodes, and Lpf and Lph are the foot and hand PASP onset latencies in seconds.

The mean CAV of the normal subjects was 1.28 ± 0.27 m/s.

Discussion

We have studied a poorly recognized skin potential first described in 1890 (Tarchanoff). The data support previous evidence that the potential is of sweat gland origin and evoked by presumed efferent electrical signals travelling via the spinal cord and sympathetic pathways. The results in diabetic patients suggest that the PASP may be a good indication of sympathetic dysfunction in polyneuropathy and this view is supported by a recent publication (Shahani et al. 1984).

The results have shown an approximate 0.5 s difference in the onset latency for the palmar and plantar PASP. If, as the data in the spinal patients suggest, the evoking signal passes to the thoraco-lumbar sympathetic outflow via the sympathetic tracts of the lateral columns, then part of the 0.5 s delay to the plantar PASP is a result of central conduction. It is postulated that the central autonomic velocity, as described, may prove a valuable measure of lateral spinal cord function. It should be stated that there is some experimental animal work which suggests that the PASP may be depressed in deep anaesthesia and this may limit some potential applications (Delerm et al. 1982). We are confident, however, that the PASP will find application in many areas of diagnostic electrophysiology.

References

Delerm B, Delsaut M, Roy JC (1982) Mesencephalic and bulbar reticular control of skin potential responses in kittens. Exp Brain Res 46:209–214

Shahani BT, Halperin JJ, Boulu P, Cohen J (1984) Sympathetic skin response—A method of assessing unmyelinated axon dysfunction in peripheral neuropathies. J Neurol Neurosurg Psychiat 47:536–542

Tarchanoff J (1890) Uber die galvanischen Erscheinungen an der Haut des Menschen bei Reizung der Sinnesorgane und bei verschiedenen Formen der physischen Thätigkeit. Pflüger's Arch Ges Physiol 46:46

Subject Index

Page numbers in *italics* refer to Figures

MIX
Papier aus verantwortungsvollen Quellen
Paper from responsible sources
FSC® C105338

If you have any concerns about our products,
you can contact us on
ProductSafety@springernature.com

In case Publisher is established outside the EU,
the EU authorized representative is:
Springer Nature Customer Service Center GmbH
Europaplatz 3, 69115 Heidelberg, Germany

Printed by Libri Plureos GmbH
in Hamburg, Germany